Thomas Sproston

Biophilosophy

Biophilosophy

BERNHARD RENSCH

TRANSLATED BY C. A. M. SYM

 COLUMBIA UNIVERSITY PRESS

NEW YORK AND LONDON 1971

BERNHARD RENSCH is Professor emeritus
of the University of Münster.

This book is a translation of the
German edition, *Biophilosophie*,
published in 1968.

Foreword

Several biologists have dabbled in philosophy. They were rewarded at best with tolerant smiles of professional philosophers, who could hardly overlook lacunae in the erudition of the amateurs. Philosophers dabbling in biology have fared scarcely better. The philosophy of science is a respected field of specialization. It happens however that, at least in English-speaking countries, philosophers of science have their scientific training mostly in physical sciences and in mathematics. Biology has some problems to which physical sciences offer no good parallels or adequate preparation. Professor Bernhard Rensch is that *rara avis* who is equally at home in biology and philosophy. For many years the eminent professor and director of the Institute of Zoology at the University of Münster, he is the author of much original research in evolutionary biology, comparative anatomy and neurology, and behavior studies, as well as in philosophy.

The present book is not easy reading, but it will amply reward a serious student of either biology or philosophy. It is a product of lifelong research, study, and contemplation. It is truly synthetic, making biology and philosophy integral parts of a harmonious world view, a "Weltanschauung." The need for such a book has long been felt—and now it is available to those who must, or prefer, to read it in English.

The Rockefeller University

Theodosius Dobzhansky

vii

Contents

Biophilosophy

*Om. May Brahma protect us both! May Brahma
bestow upon us both the fruit of knowledge! May
we both obtain the energy to acquire knowledge!
May what we both study reveal the Truth! May we
cherish no ill feeling towards each other! Om.*

*Invocation from the opening of the
Katha-Upanishads, c.600* B.C.

Chapter One

Introduction

The past few decades have witnessed enormous progress in the analysis of the phenomena of life. The invention of the electron microscope has opened up whole new areas of cytological research, extending to the macromolecular field. Molecular genetics, based on the study of viruses, has led to astonishing new views. Electrophysiology, radioactive tagging of compounds, and many other new techniques in physiology and chemistry have promoted our understanding of life processes and have already led the investigations into the realm of chemistry and physics. There have been significant advances in developmental physiology, research into sexuality, population genetics, evolution, and various other branches of biology. All this has made it possible to arrive at a causal explanation of many life processes which up to the beginning of the present century seemed to be insoluble and must have been regarded as "marvels."

Many of these findings have a significance that extends far beyond the limits of biology. The problem of what life is—its specific laws, its emergence and evolution in a largely inanimate universe—is one of the basic questions of natural philosophy. Besides, biology is already linked to philosophy, and particularly to epistemology, by the fact that conscious phenomena are correlated with certain processes in the nervous system (and possibly also in sense organs) or, rather, that there is complete correspondence between them. So it is no exaggeration to say that a philosophy which fails to take sufficient account of the state of relevant biological knowledge cannot be adequate to reality.

Yet any philosophical assessment of biological findings faces great difficulties. The peculiarities of life are so multifarious that biology has been obliged to split up into more and more separate branches. Each of these has developed its own methods of study, established its own specialized institutes, and produced a literature which seems almost limitless. The biologist finds it increasingly difficult to keep pace with advances even in his own field of research, and he cannot hope to acquire more than an outline of neighboring disciplines. The specialist in developmental physiology can no longer survey sufficiently the findings in metabolic physiology or electrophysiology, nor can the cytologist keep abreast of the literature on genetics or evolution. Specialization is bound to increase with the advances in knowledge; and it has led many research workers not only to concentrate on their own limited sphere but to forego any attempt to acquire a more comprehensive biological ideology—still less, one based on natural philosophy. But is this not regrettable "modesty," to limit oneself in this way, to renounce any idea of spending the life one has in inquiring into the most profound problems and draining the cup of knowledge? It is possible at least to trace the main lines of scientific progress and to become acquainted with the findings of neighboring disciplines essential to a total view of the world.

However, the biologist who aspires to integrate his special knowledge into some larger view of life must possess at least some philosophical and, in particular, some epistemological knowledge. He will in any case have to examine what limits man's ability to perceive and to think impose upon his knowledge, how far he can claim that his results represent the truth, and what part is played by psychic and logical laws beyond his own largely causal research. He will also need to consider whether or in what measure his findings accord in a general way with other opinions and convictions.

It is regrettable that most present-day biologists—and in fact most scientists in general—are almost totally without any philosophical training. The specialization we have mentioned, as well as internal and external delimitations, and in particular the separation of the scientific faculties from those to which philosophy belongs, have all contributed to a cleavage between the arts and the sciences. Yet, as recently as fifty years ago this cleavage would have seemed incredible. We must make some attempt to bridge the gap again, if we wish to acquire a soundly based idea of the world and the meaning of life.

Yet the prospects that such efforts may have quick success are far from bright. Present-day biological research is intent upon uncovering a greater and greater number of verifiable facts and relationships, and theoretical discussions tend to be confined to questions closely connected with concrete results. There is general distrust of more far-reaching theoretical conclusions and speculations, even though they may be perfectly correct. Theoretical inferences reached in the course of scientific discussion are often shrugged off as "mere philosophizing." (The widely current and often totally unscientific Anglo-American use of the word "philosophy" has contributed to this state of affairs.) Many biologists have therefore still to be convinced that epistemology, the basic science of philosophy, is absolutely precise, because it is built upon the most reliable factual material known to us: the reality of experienced phenomena.

On the other hand, philosophy itself has been in a state of crisis for the past decades. There has been relatively little research into its most fundamental field, epistemology. Instead, the philosopher's attention has been focused on historical, logistical, and ethical questions and especially on those concerned with cultural and religious philosophy. After a final flowering of philosophy at the end of the nineteenth century and the beginning of the twentieth, when a number of well-based and comprehensive systems were worked out, there followed a phase of restriction. Philosophy in our time hardly gives the impression of an all-embracing discipline able to combine the most general findings in *all* branches of knowledge and to work them into a consistent theory. The scientist in particular often finds that little if any regard is paid to the philosophically significant implications of his field; and he sees that the great achievements of the Age of Enlightenment are often depreciated or even largely ignored. Yet if a philosopher no longer aims at unbiased "enlightenment," in order to get nearer to the truth, he cannot count upon any general response. In the past decades, too many philosophers have unfortunately sacrificed exactitude and clarity in favor of brilliant phrases. Brilliancy, however, is a surface phenomenon. One cannot help suspecting that much of the terminology is needlessly involved. Indeed, many biologists, chemists, and physicists might agree with Chamfort's comment: "When it comes to the metaphysicians, I feel like quoting Scaliger's remark about the Basques: 'They are said to understand one another, but I can hardly credit it.'"

On the other hand, we must not fail to recognize that the philosopher trained in the faculty of arts is scarcely in a position to acquire, in addition to his own studies, enough knowledge of the most essential results of science and to make a critical assessment of them. The spheres of physics, chemistry, and biology are too varied; the methods, terminology, and symbols are too complicated; and the rate of progress is too rapid. The scientist is in a rather better position in this respect. As already mentioned, all he can command is a limited field of study or a few specialized areas; but his initial scientific training in neighboring fields makes it easier for him at least to follow the more important advances in science as a whole, or at least in biology and biochemistry. And if the problems dealt with lead him to discuss findings on a wider basis of natural philosophy, he is able to acquire the fundamental essentials of epistemology; for in this sphere the main general problems have not altered very much for centuries, and the methods of research have remained the same in principle. Unfortunately, additional study of this type is often somewhat imperfect, if it is undertaken at all. As a result, despite some quotations from the great philosophers, expositions of biological philosophy often reveal that the author has hardly progressed beyond a certain naive realism.

The present work seeks to avoid this kind of onesidedness. Its author, a biologist who from his student days has also been devoted to the study of philosophy, has endeavored to work up biological, psychological, and philosophical findings into the most consistent ideology possible. Of course, it cannot give a definitive picture, but it offers currently appropriate solutions to certain basic philosophical problems. His starting point has been the fundamental and successful principle of Descartes: *de omnibus est dubitandum*—that is to say, he has not based his conclusions on maxims, unsupported philosophical claims, or "eternal truths." On the contrary, he starts from what we may describe as indubitable reality—in other words, psychic phenomena, their interrelations, and the direct inferences to be drawn from them.

As most biologists, however, in common with many other scientists and "educated people" of different professions (to whom this book is also addressed) are unaccustomed to thinking along epistemological lines, the work opens with a discussion of entirely "factual" statements about life processes. At this stage it is irrelevant whether the reader interprets these findings in the sense of "naive realism," "functional materialism," or "idealism" ("spiritualism"). It is only when the problems of psychic phenomena are treated that the epistemological bases and limits of our knowledge will be discussed, so that a comprehensive biophilosophical picture can be built up. The author then offers a panpsychistic, identistic, and polynomistic interpretation in the hope that it does justice both to the realism of the scientist and to the legitimate epistemological demands of the philosopher.

Numerous scientists and philosophers have already treated many of the questions under discussion here; and they have reached very different conclusions. The present writer has made no attempt to compile and critically discuss all these views. His aim has rather been to find new solutions on the basis of the factual material now at our disposal. As the work is addressed not only to biologists but to others less well versed in this field, it has seemed appropriate to recapitulate some facts briefly when they are treated in a different context in later chapters. This will avoid the necessity for cross-reference while reading. Besides, most chapters are more or less self-contained and can be read independently. Thus, at a first reading one can omit the chapter on research methods; or one can begin with the sections on epistemology. The biologist will not pay so much attention to the discussions concerning his field, while the philosopher may dispense with the introductory section on epistemology.

The bibliography relates exclusively to works cited in the text. Any comprehensive reference to the relevant papers would fill a whole volume.

Chapters 7 and 8 on the practical application of the philosophical conception put forward, in relation to ethics and religious philosophy, are not intended to offer more than an outline.

Attempts to solve the ultimate problems of existence can be based on different fields of knowledge. As only the biologist can be a competent judge of biological problems, one should concede that he is entitled to seek for solutions which also need philosophical interpretations. Moreover, this task is one of particular urgency at the present era of great cultural upheavals and general reassessment.

The most striking achievement of our own planet, and indeed of the whole solar system during the milliards of years of its gradual development, has been the production of living substance, from which there finally emerged that strange species *Homo sapiens*. Man has not only succeeded in finding his way in the world with the help of sense organs and brain, as other organisms have done; he has also attained the ability to understand his own nature, his phylogenetic development, and the development of the earth; and he has gained an insight into laws that underlie and govern all things. Among the aims that a present-day specimen of *Homo sapiens* may set himself, there can hardly be a worthier one than to share as far as he can in this momentous knowledge, and possibly even to make some contribution to it himself. May this book arouse or deepen the urge to possess such knowledge in all who realize that the uniqueness of our short existence is a marvel that allows us to aspire toward this kind of goal. May their eager search for knowledge in no way be damped by the limitations in man's capacities and the brief span allotted to him, which led Faust's attendant, Wagner, to exclaim:

> Wie schwer sind nicht die Mittel zu erwerben,
> Durch die man zu den Quellen steigt!
> Und eh' man noch den halben Weg erreicht,
> Muss wohl ein armer Teufel sterben!

> How hard it is to gain the means
> To reach the fountain-head;
> Before his journey's half-way done
> Poor devil!—he is dead.

Chapter Two

Methods of Biological Research and the Establishment of Theories

A. The Framing of Concepts, Induction, and Deduction

Very early in life we become aware that conscious processes are not totally chaotic. In perception and mental imagery something the same or at least similar is appearing or recurring. As our mental images develop, these identical elements come together, reducing our varied experiences to some degree of order. The outside world becomes increasingly intelligible. Gradually, more or less stable concepts are framed, and the child then learns to connect these with certain words, at first with their sound and sound image, and later by pronouncing the words himself.

This existence of identity, similarity, and difference, or "objective basis of all phenomena," as Kant describes it in his *Kritik der reinen Vernunft* (1787), is something which we must accept as a "given" feature of experienced reality, a basic fact of our phenomena.

We can trace various stages in the *framing of concepts*, which are based on the comprehension of identity, similarity, and difference. The first is a primary recollection of something perceived, but usually involving only essential features or those which have caught the attention. Repetition of identical or similar elements of perception strengthens these particular and distinctive features in the mental image (accentuation) and promotes the dwindling of unessential features (repression). In most cases it is the qualitative and intensive properties (like shape or color and its intensity) that hold the attention. On repetition, however, we come to disregard the varying temporal and spatial features, and to recognize the constant characters, wherever and whenever they are presented, as qualitatively the same. The result is a primary mental image of an individual "thing" (see Ch. 6 F II). To take an example: we may suppose this "thing" to be a certain individual mountain zebra which we have often seen. By observing similar mountain zebras differing somewhat in size and coloring, we can arrive at a completely generalized concept of this species of zebra, a concept limited to the distinctive characters divorced from any individual example and associated with the expression "mountain zebra."

In much rarer cases the stable temporal and spatial attributes attract attention, and the properties are disregarded. This occurs, for instance, when we frame such biological concepts as "area of distribution" or "embryonic stage." Or again, attention may be concentrated on purely affective identities, rendered by such words as "anxiety," "joy," and "mood." All concepts formed in this way have a *suprapersonal significance*.

The process of abstraction, carried a stage further, leads on to a superior order of *generalized concepts*. To take the same example: by concentrating on the characteristics common to all species of zebra, one can progress from "mountain zebra" to the concept "zebra" and further, by comparison with donkeys, horses, antelopes, cows, etc., to "ungulate," "mammal," "vertebrate," and finally "animal." In a corresponding manner we can generalize the spatial and temporal concepts still more and speak of "zoogeographical region," "life span," and finally "space" and "time." Or, again, if we concentrate on processes, we arrive at activity concepts such as "running," "flying," "swimming," or, on a higher level of abstraction, "moving."

There is no psychological explanation for the capacity to discern identical or similar elements or to analyze them. We have therefore to accept these *functions of differentiation*, i.e., the functions of comparison, analysis, and synthesis, as something "given," as basic phenomenal processes which we can only describe in terms of our own experience. We are therefore not capable of defining identity, similarity, and difference. Most definitions hitherto attempted have implicitly presupposed the concept of identity (see Ziehen 1934 §§ 12 and 22). It has also not yet proved possible to relate these functions to specific physiological processes in the brain, though partly discussible hypotheses have been put forward (Ch. 6 C IV).

Experiments have shown that higher animals also can comprehend equality and similarity; they can pick out certain details, a particular object within their field of vision for instance, and they can act as though capable of more or less abstract "averbal" thinking (Ch. 6 H III). So we may assume that selection has been operating during the long course of animal phylogeny to adapt the functions of differentiation to the inanimate world, in the same way as it has adapted physiological functions.

In almost every system of philosophy, the framing of concepts is regarded as a function of logical thought. So it is worth pointing out that *not only basic concepts but also most abstract concepts can be framed without any cooperation of logical processes.* As long as a concept is not yet connected with a word and remains averbal, as it does with higher animals and in man's early childhood, it is solely a matter of repeated identical or similar experience. Adult humans, too, occasionally frame averbal concepts—of an unfamiliar plant which we have frequently seen but whose name is unknown to us, or of a tune, or an unfamiliar scent, or the like. *It is experience alone that determines the meaning and reliability of a concept.*

As they grow older, children learn to attach words to most concepts, and this makes thought much easier; for example, many similar mental images, of different kinds of butterflies are replaced by the sound image of the word "butterfly," occasionally accompanied by some vague visual memory. Quite apart from its meaning for the individual, association with a certain word also "fixes" the concept for many other people as well. None of this necessarily involves logical processes.

Abstraction and conceptualization, which bring some measure of *order* to our perception and mental imagery of "things" and the changes they undergo, already take place *at a prescientific stage.* Thus every language distinguishes different species of animals; different categories (e.g., monkeys, birds, fishes, and butterflies) ; and different activities (e.g., flying, swimming, and feeding).

These concepts only acquire *scientific meaning* when they are expressed as *judgments* and are more or less definitely *defined.* Only then is it possible to discover if they are true or false, too particular or too general—in other words, if further experience confirms them sufficiently. This means that they can be said to be adequately adapted to external factors. *Reliable concepts have an empirically based validity*

*which we may describe as the primary stage of generalization, i.e.,
generalization at the level of the initial arrangement of the diverse
elements of experience.* It was this general validity that led Plato to
his doctrine of "ideas" in the sense of supraindividual archetypes which
are variously reflected as the perceived "things."

Insofar as they are expressed as judgments, however, the concepts
are also subject to the *laws of logic* employed in all branches of thought.
Yet these are not only rules governing human thinking, but universal
laws to which man's thought has adapted itself during the course of
his phylogeny (Ch. 6 K VI). A *judgment* involves an association between two or more mental images or concepts and establishes a coincidence of temporal or spatial properties (or, more rarely, some
other stable relationships). For instance, when we say "This blackbird
is black," the two terms "blackbird" and "black" are coincident in
time and space. The same is true of generalized judgments: in "adult
male blackbirds are black," the two elements are coincident. In a corresponding manner, negative judgments express the absence of such
coincidence, as in "songthrushes are not black" (see Ziehen 1913 §
87-88, who critically discusses Kant's derivation of categories from
the different types of judgment).

Two kinds of judgment in particular are important in the sphere of
scientific knowledge: the *problematic* and the *sejunctive*. In the first
of these, the associative components or the connection between them
is considered uncertain or doubtful: "Blackbirds may perhaps feel
anxiety." Ziehen makes a sharp distinction between this type and the
second, which is intentionally somewhat vague: "Male blackbirds may
be black," i.e., they may occasionally be white or fawn. No doubt is
being expressed here, as in the problematic judgment; it is simply a
positive judgment not expressed in an apodictic form. As by far the
majority of biological findings involve exceptions, they ought to be
expressed more or less in this way.

Many judgments are derived from previous ones, or several may be
combined to form a *conclusion.* Most of these consist of coincident
concepts expressed in a series of judgments: "Blackbirds have feathers,"
"Sparrows have feathers," "Ducks have feathers." Conclusion: "Birds
have feathers." This method of conclusion is typical of *induction,* and
it forms the basis of our biological—indeed all our scientific—knowl-

edge. A number of isolated cases lead on to a general statement. Almost all our "biological facts"—that is to say, everything that implies more than a single observation—are the outcome of this type of conclusion. And this in turn is based on judgments, the framing of concepts, and a recognition of identities, similarities, and differences in a number of situations either perceived or imagined.

It is by induction that science at first usually proceeds to the *ordering* of statements, in other words to *classification*. That is to say, scientific knowledge is arrived at empirically, through *experience*. And yet all we are dealing with here is *probability judgments*, and conclusions drawn from these. Often the degree of probability is so great that we treat and express the judgments in an apodictic manner as facts, though strictly speaking we are not entitled to do so. When classifying, we always look for some more general concept, a generic term, and then distinguish the characteristics of the more particular concepts subsumed under it. "Blackbirds are thrushes (generic term) whose males are black"; "Thrushes are song-birds (superior generic term) which have a certain structure of palatine bones and certain proportions of pinion"; "Song-birds are birds characterized by certain song muscles on the syrinx"; "Birds are vertebrates which are covered with feathers"; and so on. In this manner classification leads to a greater and greater degree of universal validity and represents the second stage of generalization.

At each stage of classification, however, it is possible to proceed in the opposite direction, by *deduction*. In that case we draw conclusions from the general to the particular, testing the validity of a classification in a number of instances. For example, after establishing several families of birds—the Sylviidae, Timeliidae, Muscicapidae and Turdidae—other species were then examined to see if they fitted into this classification. The fact that they did not always do so led many authorities to modify or discontinue this type of classification.

Almost all generalization in the sphere of biology makes use of this combination of deduction and induction. Every order, every classification, begins as a hypothesis and has to be proved by testing it in additional instances. It follows that every classification embodies a potential *prediction* concerning further instances. But this prediction is no more than a logical possibility. When it had been established than many insects—beetles, butterflies, locusts, bees, etc.—have six legs, it was reasonable to conclude that all insects have six. But that does not rule out the possibility of discovering some with more than six; for instance, very primitive insects or fossil species possessing functional legs instead of vestigial abdominal stumps. *Not only does induction come first; it is also more decisive than deduction* (see John Stuart Mill 1843).

Deduction does, however, represent an important *heuristic principle*. After it had been established by inductive methods, for example, that whales are mammals by virtue of their skeletal structure, lacteal glands, and the fact that they are viviparous and warm-blooded, it could be concluded by deduction that their forebrain probably has a cortex consisting of several layers of neurons, their red corpuscles are anucleate, and more urea than uric acid is present in their excreta. Examination based on these deductions led to confirmation along inductive lines, and the classification was shown to be correct. Thus far we may agree with Riehl (1908) when he writes of the "unity of induction." But it would be wrong to conclude, for instance, that because all known carnivores, ungulates, rodents, monkeys, bats, etc., have four limbs, therefore all mammals have four; whales, in fact, have only two.

M. Hartmann (1948) described the normal method of concluding from the particular to the general as "generalizing induction"; and he distinguished it from "exact induction" in which a generalization arrived at by induction is followed by a deductive forecast and this is then verified inductively by one single example. Following up what Riehl had written in 1908, Hartmann declared (p. 131): "Erfüllt dieser streng deduktiv erschlossene synthetisch konstruierte Einzelfall im Experiment die gefolgerten Voraussagen, so wird dadurch das zunächst hypothetisch angenommene Allgemeine als allgemeines Gesetz bewiesen und so im Prinzip *durch die Analyse eines einzigen Falles die gesetzliche Konstitution aller besonderen Fälle und Zusammenhänge der gleichen Art erbracht.*" ("If this rigorously deduced individual case based on analysis and synthesis proves on experiment to be in line with what had been predicted, then the general statement which had hitherto been hypothetical may be accepted as a universal law. Thus analysis of a single case may establish the laws governing all particular cases and relationships of the same type.")

It is true that this traditional method of deduction applied to further instances and then checked by examination may do much to strengthen a generalization already assumed. But I do not see how we can assert that a single case can furnish evidence for *all* others of the same kind. There is always the possibility of exceptions; in other words, some hitherto unknown variation may invalidate the generalization. In my view, therefore, the expression "exact induction" adds more confusion than clarity, and the contrast with "generalizing" induction tells us no more of the interaction of induction and deduction than we have already learned from John Stuart Mill.

B. Hypotheses, Theories, Laws

Classification largely consists in a systematic arrangement of relevant and more or less strictly defined subjects; for instance, when we group adult animal species sharing the same or similar characters in a genus. But we learn a great deal more when *processes* are arranged in this way. When we become aware of a succession of phenomena with identical features, especially if the spatial relations remain constant, we conclude that we are perceiving the "same" object all the time. In so doing we reach out beyond the pointlike individual act of perception to one of the widest of all concepts, that of "being." Parmenides had already drawn attention to this, and Aristotle gave it concise expression in his *Metaphysics*, Book IV: "Change presupposes a being in so far as something becomes something else." (Ch. 6 I, L)

The method in both kinds of classification is the same: Induction is interspersed with deductive processes, and general principles are inferred from particular ones. The conclusions are at first no more than probabilities, so they are mostly expressed as hypotheses. These hypotheses, reinforced or combined with others, may develop into theories. Finally, the theories in turn can be verified to such a degree that they can be called laws. It becomes evident, then, that all statements in the spheres of biology, chemistry, physics, and astronomy indicate that universal laws exist which govern all processes. In particular, the *law of causality* is of the widest significance in all scientific research. As cause and effect are seen to be operative throughout the material world, we deduce that causality determines every kind of process. Apart from some disputed exceptions (Ch. 6 K III), this heuristic principle has proved its value in every case. This means that *all scientific research is predominantly research into causes.* Causal factors are absent only in the case of pure description, simple classification, and logical inference. Statistical laws are in part logically determined (the laws of probability).

Knowledge has widened and the development of comprehensive theories has become possible mainly because instruments and methods have been devised which reveal structures and processes beyond what our sense organs can perceive. In the realm of biological research the invisible has become visible through light and electronic microscopy, phase contrast and elective staining methods, marking of processes by radioactive tracers, cathode ray ocillography, slow motion photography illustrating rapid processes, ultracentrifuges, and a host of other biophysical and biochemical methods.

Hypotheses. Repeated observations of a regular sequence of phenomena in identical subjects often lead to the assumption of a connection in the sense of cause and effect. We deduce a propter hoc from a post hoc in this way because the causal relationship is one that always seems to fit the conclusions. And conversely, we tend to regard causal applicability as constituting at least a provisional "explanation."

As long as a generalization is of doubtful validity, we normally speak of it as a *working hypothesis.* This term is used to indicate that the generalization may be somewhat premature but is of value because it determines the lines of future investigation.

Further valid instances are normally needed to strengthen a hypothesis. Especially *experiments* based on deduction have great conclusive force. The result, based as it is on causal relationships, can be forecast. An accurate forecast, if it proves right, carries a good deal of weight because every experiment may serve as an example for corresponding instances. Experiments are particularly valuable in the case of biological processes, which normally have a complex background, for the premises can be restricted to decisive elements and the causal connections are usually unambiguous.

Sometimes the empirical evidence underlying a hypothesis can also be supplemented *paradigmatically,* that is, by mental verification alone. This is done by replacing judgments arrived at through experience by others based on probability (problematic or sejunctive judgments discussed above). This type of deductive verification purely by devised examples can be carried out most easily when the causal relationships behind the essential processes leading up to the hypothesis can be assessed, as in many biochemical and biophysical reactions. Thus, if we have established that in a few cases vertebrates die when given certain quantities of hydrocyanic acid (because this inhibits their respiratory enzymes), we do not need to test this on many species belonging to different families and orders. A few individual experiments are sufficient to set up a train of deduction which will convince us that hydrocyanic acid in general has a lethal effect on vertebrates. By following through the biological processes in the mind in this way, we often arrive at a conviction that the hypothesis must be correct because our experience tells us a contradictory result is out of the question.

Some biological hypotheses have been based largely on *speculation.* Many, however, could not be verified; for instance, the hypothesis that certain lower animals originated by spontaneous generation from mud or decayed matter, or the view held by the "animalculists" that hereditary factors are transmitted through the spermatozoon alone. A certain amount of observation had preceded these hypotheses, but they lacked any sufficient empirical basis. On occasions, however, speculative hypotheses have turned out to be substantially correct. At the end of the eighteenth century, Erasmus Darwin built up a theory of evolution on quite insufficient evidence and on a largely speculative basis; but it proved remarkably accurate.

We must also bear in mind that many if not most important discoveries and findings in biology have first been based on fairly bold speculation. Speculation is in fact of considerable heuristic value, for it can lead to planned investigation and then to inductive verification of the hypothesis put forward. Indeed, this kind of "scientific imagination" has often proved more useful than an accumulation of empirical data. The imposing edifice of modern science was not built up stone by stone; it owes its main structural features to projected schemes which have dictated specific lines of research.

A hypothesis may be the result of *intuition*; that is to say, it suddenly springs to life in the mind of the scientist without his going through the intermediate stages of imagery. Many confused and almost mystical explanations of an intuitive flash of this kind have been attempted. But it is probably determined by the so-called narrowness of consciousness. Reflection is the outcome of cerebral excitations, and these are only in part accompanied by conscious mental imagery, especially when the attention is turned to it. It is possible, however, that unconscious excitation may lead over into the complex of excitations which accompanies the stream of consciousness; a pattern of imagery is then suddenly set up, and a solution to the problem may spring into the mind (Ch. 6 F V).

Theories. In the first book of his *Metaphysics* Aristotle wrote that "a theory is produced when from many notions based on experience, a general view is formed with regard to like objects." This is an apposite characterization of theory, and contrasts in this respect with many of the definitions which the past century has offered.

The "many notions based on experience" may lead to the framing of a theory in several ways.

1. A broadly applicable hypothesis may be so reinforced by further evidence, mainly inductive, that it amounts to a theory. The theory of the linear arrangement of genes along the chromosome may serve as an example. When he tried to design a chromosome map of *Drosophila melanogaster*, Morgan was at first only putting forward a hypothesis presuming that parts of two chromosomes were exchanged by crossing-over during meiosis. It was only later, when giant chromosomes were examined and the structure of the chromosome was studied with the aid of special staining, that the linear arrangement of the genes, or rather of portions of the chromosomes that correspond to them, could be regarded as a valid and well-authenticated theory.

2. A theory may also be more largely deductive in origin, when a more general statement is based on several particular hypotheses of an empirical nature. Wegener arrived at his theory of continental drift when he realized that hypotheses and findings from the spheres of geophysics, geology, palaeontology, and palaeoclimatology, as well as biogeography, all pointed toward the same conclusion.

3. Finally, some theories are purely deductive and derive from established laws. We must assume that causal and logical laws preceded the evolution of any organism possessing the power of thought (Ch. 6 K). So it has been possible to propound the theory that in the course of

phylogenetic development, the psychological processes, or rather the processes in the brain which underlie them, have become adapted to these universal laws. This would explain how humans became able to think logically and in accordance with causal relationships.

The value of a theory is based on the fact that it makes possible *a more general and more certain forecast* than a hypothesis would allow. Moreover, a theory can be used to "explain" relationships, in other words to show how the corresponding statements are in accordance with more universal interpretations. Theories can therefore claim to be more in line with "truth" than hypotheses, so they are often taken to express scientific "facts." Yet we must remember that even apparently well-founded theories have sometimes proved erroneous.

Laws. On analyzing our experience, if we become aware of certain generally valid and apparently invariable relationships, and if we are convinced that no exceptions exist, we then refer to these as laws. Laws, like hypotheses and theories, are largely built up empirically by induction and by deduction, which is then confirmed by induction. They derive from experience and therefore, strictly speaking, are only theories which have been found to be extremely probable in all known cases. They are usually expressed in apodictic form, because it appears cogent to apply them, by analogy, to future instances and because, with few exceptions (Ch. 6 K II), all of the special natural laws can be subordinated to a more general law of causality. Ultimately, all laws express constant relationships between the phenomena experienced, or rather between the components of an ultimate "something" which constitutes the transsubjective world underlying these phenomena (Ch. 6 K). This means that in formulating laws we are attempting to comprehend the nature of transsubjective "being." That is why so-called eternal laws were so often mixed up with religious imagery. Following Heraclitus, they were called the logos which permeates and directs all things, a divine world-reason; or else they were interpreted panentheistically as a godhead immanent and active in all being (Ch. 8).

Biological research is concerned with a multitude of very different laws. Analyses have shown that all sorts of processes in living organisms —metabolism, neural and sensory excitations, the circulation of the blood, growth, as well as hereditary and other processes—are controlled by chemical and physical laws. If it is a matter of simpler reactions, these laws appear to hold good without exception. But this is not true of the majority of more complex processes. We may establish all kinds of more or less generally applicable *rules*; but it seems doubtful that we can call them biological laws. For instance, mutation is constantly producing exceptions even to something as generally considered to be valid for plants, animals, and humans as Mendel's first two "laws."

This distinction between typically biological processes and the more elementary chemical and physical processes has led many scientists to hold the view that the essential phenomena of life are controlled by certain "vital" factors which science is unable to analyze and which H. Driesch writes of as entelechies. It was also suggested that microphysical processes, which so far cannot be explained by causal laws, could produce "freedom" in biological processes. The presumed effect of psychological factors on physiological processes, too, especially in the expression of the will in man and the higher animals, seems to many scholars to be incompatible with the causal law. As we shall see in Chapter 6 K IV, however, it is highly doubtful if these vital factors actually exist, and it is at least widely contested whether acausality is in fact operative in the microphysical realm. The results of a host of biological investigations show it to be increasingly likely that every life process is causally conditioned. The only "exceptions" occur when the context is so complex that the interplay of different laws makes it impossible to establish anything but rules.

The causal nature of biological processes is often made more difficult to understand because we are dealing with *system laws*, in the sense of laws relating to mutual relations in complicated systems. Living organisms are individuals, in other words composite entities, in whom every reaction is partly conditioned by its relationship to the whole. Every chemical or structural alteration in the individual results in new relationships and often new properties. This is true even of the simplest chemical processes. When a certain number of protons, deuterons, and electrons combine to form an atom, new elements with new properties come into being. The same applies to the fusion of atoms into molecules; for instance, when the light metal sodium combines with the gas chlorine to form salt (sodium chloride). But system laws such as this do not constitute a special category contradicting the causal law. They simply express the innumerable complex relationships, each one causally determined, within, for example, a single living cell. These relationships are almost unimaginably varied.

A number of examples do not constitute a sufficient basis on which to establish biological rules. Every relevant case must be examined and the percentage of exceptions calculated. This means that a statistical method is necessary. In the biological sphere, results can often be expressed only as *statistical laws*, but they are of the same nature as any other laws. Only the methods are statistical; that is to say, they utilize the laws of probability which belong to the sphere of the logical laws and apply to all knowledge acquired by induction. But the underlying causal laws are absolutely valid (Ch. 6 K III, IV, VI).

To sum up: everything goes to show that biological rules or laws, like all chemical and almost all physical and astronomical ones, fall within the scope of *a universal causal law acting without any interruption*. We cannot "explain" this law. We can only accept it as something which is deduced from the whole of our experience, something which exists and is a characteristic of all being. The philosopher O. Liebmann (1876) was in error when he spoke of the "logic of facts."

Logical laws also are of great importance in biological research, in questions of judgment, conclusion, and proof. They apply in even the simplest act of classification. Not only do we humans find them indispensable; the higher animals too are capable of averbal judgments and conclusions. In the course of their phylogeny, animals have apparently been adapted to logical laws in the same way as to causal ones, in particular in the matter of cerebral reactions and accompanying psychological processes (Ch. 6, H III).

Logical laws, unlike almost all causal ones, are concerned with simultaneous factors and not with successive ones. As already mentioned, they are not to be identified with the laws of human thought alone. The statement "Two quantities equal to a third are also equal to each other" must hold good of objects not yet perceived by creatures capable of thought. Otherwise no physical, chemical, or cosmological research would be possible. Mathematical laws form a branch of these logical laws.

Biological research is further concerned, it seems, with a third system of basic laws which determine the invariable correlation of psychic processes, sensations, and mental images, with certain physiological processes in the brain and sense organs. For instance, the excitations produced in

the retina when exposed to light waves of 670 nm,* or rather the corresponding excitations of the visual center in the forebrain, are correlated
with the sensation "red," just as that of waves of 520 nm are with
"green." This correlation, which applies not only to sensory excitations
and sensations but to many other cerebral excitations and to the imagery
and thought processes that run parallel to them, has been called *parallel
laws* by Ziehen. As we shall see, however, it is not necessary to presume
the existence of such laws if we adhere to a realistic, identistic interpretation (Ch. 6 K V).

When analyzing human brain processes, we have to do with psychical
phenomena directly. What sort of perceptions and imagery animals have,
we cannot know. In higher animals at least, judgments by analogy with
humans are cogent; we may safely say that a dog "sees," "hears," "feels
pain," "is pleased," can "remember," and even that a chimpanzee can
"understand"—in other words, can comprehend a causal connection
and can "plan" its actions accordingly (i.e., has mental images referring to the future and acts with "foresight"). Research into behavior is
thus in part "animal psychology," and for that reason it must be concerned with the problem of whether or not parallel laws exist.

All laws, and above all the basic laws we have mentioned, represent
the highest stage in our levels of generalizations. Their universal validity
points to their being features of some extra-subjective being, and suggests that we may regard them as "truths." In Chapter 6 D IV we shall
discuss how far in fact we are justified in doing so.

C. The Significance of Mathematical Methods

As language developed, identity and similarity in many objects and
processes produced *concepts of number.* Presumably numbers were first
used simply as units (cardinals) and only later as ordinals, when the
need arose to indicate series differences. The establishment of a graded
quantitative scale then led to *measuring* and *weighing.* It became clear
that many objects and processes in nature could be quantitatively assessed. Further scientific observation showed that since many relationships between material objects and processes are constant and regular,
they could be expressed in a simplified and abstract way by symbols

* nm is an abbreviation for nano-meter, 10^{-9}m ($= 10$ Å, angstrom units).

and partially by *mathematical formulas,* which could be used for many series of observations.

As this method of representation was precise and the formulas, when logically applied, led to the discovery of further relationships, the mathematical methods proved invaluable for most scientific statements. Pythagoras and his followers had already regarded numbers and the relationships between them as the principle behind all things. This led to a certain amount of untenable speculation, but it marked the beginning of the realization that quantitative differences form one of the most essential features of the material world. In the Renaissance, when the world of nature was once more the object of intensive study, Leonardo da Vinci noted that there could be no certainty in any field where the mathematical sciences were neither applicable nor relevant. Descartes tried to apply mathematical methods to the solution of problems not only in physics but in physiology as well, and Spinoza developed his philosophical system "more geometrico." Kant, too, held that exact empirical knowledge is arrived at solely by the mathematical method.

As already mentioned (Ch. 2 A), scientific laws and rules are mainly inductive, and based on a large number of identical cases. But empirical findings, especially those concerned with living organisms, are seldom identical. The great merit of Laplace was in shrewdly pointing out that *the inductive method is not sufficient, because its results are probabilities, not definite certainties.* In the introduction to his *Essai philosophique sur les probabilités* (1814) he insists: "On peut même dire, à parler en rigueur, que presque toutes nos connaissances ne sont que probable; et dans le petit nombre des choses que nous pouvons savoir avec certitude, dans les sciences mathématiques elles-mêmes, les principaux moyens de parvenir à la vérité, l'induction et l'analogie se fondent sur les probabilités." ("Strictly speaking, one can even say that almost all our knowledge is merely probable; in regard to the few things we can know with any certainty, in the mathematical sciences themselves, our principal means of arriving at the truth, that is to say induction and analogy, are based on probabilities.") It is worth noting that many of Laplace's own mathematical results are based on induction. Although he does not enlarge on it, he apparently held the opinion that mathematical principles and even axioms are not independent of experience, not purely aprioristic, but are confirmed by imagined experiments and paradigmatic conclusions. This view has more recently been expounded by Ziehen (1934, pp. 142-151), in contrast to what Kant had expressed in his *Kritik der reinen Vernunft* (1787).

The task before the scientist is to devise methods or formulas which will produce the most reliable results from findings which at best are no more than highly probable and are subject to individual variation. This led Laplace to formulate his principle of a general *probability calculus* for arriving at reliable and definite results from inductively acquired data. He based his ideas on such earlier seventeenth-century thinkers as de Fermat, Pascal, Bernoulli, and others. As scientific knowledge increased, it became more and more important to state the *significance of all statistically acquired knowledge*. This was particularly necessary in biology because of the many complex interrelating factors involved.

Biostatistics (biometry) as a largely independent discipline grew out of the special requirements of zoology, anthropology, and botany. Biologists and physicians like F. Galton, K. Pearson, F. Martius, R. A. Fisher, S. Wright, J. B. S. Haldane, S. Koller, K. Mather and many others have helped to develop the mathematical treatment of findings in such various fields as morphology, physiology, studies in development and growth, genetics, evolution, and ecology.

Mathematics is intrinsically a completely theoretical and formal science. It is founded on certain axioms, and it works with "pure" numbers and abstract structures to build up formulas and computations with the help of logical rules alone. Mathematical concepts and operations do not refer in the first instance to anything in the world of reality, but they can be applied to relationships within that world. Being abstract, mathematical operations can cover a very wide range of subjects including physics, chemistry, astronomy, the various branches of biology, and political economy. In all these fields, numbers are meaningful only insofar as they correspond to concrete objects or their properties or to concrete processes. Much the same is true of mathematical structures. Insofar as geometry is concerned, they are often thought of as concrete, but their universal significance rests on the fact that they are completely abstract. This also explains how it was that metageometrics could be elaborated.

With regard to microphysical investigations, the mathematical treatment has a special significance. As we shall see (Ch. 6 I), matter is ultimately only comprehensive as a spatial and temporal system of relations. The description of these relations in the form of mathematical symbols and their logical connection therefore allows us to comprehend the nature of matter quite immediately. Moreover, we must bear in mind that the applied logical laws are not only laws of human thinking but laws of the universe to which our thinking has been adapted in the course of phylogeny (Ch. 6 K VI).

In biostatistics, both "pure" and "applied" mathematics are employed. Most statements are based on data obtained from specific objects, the reliability or the relationships of which are to be examined. Biological concepts are therefore translated into mathematical symbols and treated in a purely mathematical, i.e., logical, manner. Only when results are obtained are they related back to the objects concerned.

This treatment, which provides exact results beyond the scope of inductive findings, has led to statements being expressed quantitatively whenever possible. If repetition of observations or testing of many cases yields sufficient data which can be expressed in numbers, then biostatistical methods can be applied. Even when structures, processes of development, or animal behavior are described, it is necessary to indicate the percentage of exceptions to the norm. This explains why writings on biological subjects now contain a far greater number of tables, formulas, and graphs than they did in the first two decades of the century.

Biostatistics can be applied in a wide variety of fields. In regard to our biophysical considerations it may be sufficient to point to a few significant examples. We have already noted that it is always very important to indicate whether variable data acquired by observation are sufficiently reliable—in other words, *statistically significant*. These calculations are based on the statement that in a normal case by far the greatest number of variable data are grouped around a mean, whereas extreme values are very rare. If normally distributed data are correlated on a graph with the values on the abscissa and the frequencies of cases on the ordinate, where a large number of values are considered, the result usually takes the form of a bell-shaped binomial curve. This Gaussian curve of normal probability is obtained whether variations occur naturally or are subjective, the result of technical inadequacy or failures in observation.

It is a comparatively simple matter to calculate whether a series of results corresponds to this curve of probability and whether two such series show any significant difference. For example, a group testing the relevance of some factor might be compared with a control group excluding this factor. For statistical validity, a 99.7 percent probability is usually required. This percentage is dependent on the fact that in a normal distribution 99.7 percent of the results lie within three times the standard deviation. However, in many cases it is sufficient to state that two sets of results may be significantly different whether the probability is only by 99, 98, or even 95 percent. (For more detailed treatment, see R. A. Fisher 1925; G. Just 1928; H. Hosemann 1949, and others.) It is also important to consider what is required if only relatively small groups of data are available (see, for instance, S. Koller 1939).

The degree of *interdependence between two values*—for example, the size of an organ in relation to body size—can be calculated by other methods. Here too, data obtained inductively can only yield results of a certain degree of probability. If the coefficients of correlation are calculated, the result can also be related to other correlated coefficients (for example, comparison of the coefficients of correlation of the length of femur, shank, and foot each with reference to that of the body).

When studying biological processes it is most important to measure to what extent a given character alters when a correlated character increases or decreases. In such cases, calculation of regression coefficients can give us a precise assessment of correlated processes within the wholeness of the body. Much-studied examples include those of the *allometric growth* of organs in relation to that of the body, or of histological structures to the organs or parts of organs (for instance, cortical regions in relation to the whole cortex). This can be expressed by a relatively simple function formula: organ $= b \cdot \text{body}^\alpha$. The exponent α indicates the degree of correlation. Only calculations of this kind have furnished evidence that in many cases the relative rate of growth remains constant for some considerable period during embryonic or juvenile development. They have also revealed that these allometric tendencies are often found to an equal or similar degree in many related species and genera.

These methods also help to explain phylogenetic trends and parallel phylogenetic processes in related forms. When an allometric exponent, or at least the allometric tendency, remains constant (allomorphosis), positive allometric growth of an organ must lead to its relative enlargement in larger species. This has often been established or seen to be probable. It also explains why phylogenetic growth in body size may be accompanied by the development of "luxuriant" and excessive organs, such as the overgrown antlers of the ice-age Giant Elk, the giraffe's neck, and the excessive length of jaws, antennae, and legs in some of the larger species in several families of beetles. Without consideration of such growth gradients this would not be intelligible.

All the above examples also illustrate the great heuristic value of biostatistical formulas. They serve as abstract models, and other cases can be studied in regard to whether or how far they may correspond. They therefore often supply a basis for new paths of research, especially when formulas are developed which embrace several variables to which it is difficult to assign concrete values and whose main use is to clarify the principles underlying complex processes. For instance, if it is important to estimate how long it will take for an advantageous variant produced by mutation to become predominant in a given population, one can advance complex formulas whose symbols represent dominance or recessivity, the mutation rate, the percental selective advantage, and the size of the population. Almost never do these formulas correspond to any single example in nature, because among these factors only recessivity or dominance and in rare cases the mutation rate of the character in question can be calculated with any certainty—not, however, the exact selective advantage and the changes of population size. Besides, there are other influential factors outside the scope of the formulas: the effect of a different genetic milieu brought about by introducing the mutated gene in different gene combinations; loss of alleles through migration, changes, and fluctuations in population size and of selective factors in the sequence of generations. Yet in spite of these limitations, these formulas are of great value, for they can exemplify how even a very minor selective advantage, as little as 1 percent, may lead to a mutant gradually becoming established in a population, and how very greatly the prospects for recessive and dominant mutants differ. They may also show what a widespread influence population size exerts on racial change, and to what extent hypothetical calculations can indicate the length of time it takes for races and species to develop.

To sum up: *Mathematical treatment of biological data can yield more than purely biological methods, for it can establish statistical signifi-*

cance, clarify functional relationships, lead to more exact and abstract measurement of correlations, to more precise formulation and improved interpretation of biological laws, and to production of models of thought which may have considerable heuristic value.

Sometimes, however, biological research has been dazzled by the very precision of mathematical methods. The formal correctness of a calculation, it must be remembered, is only of value if it rests on adequate *material* correctness; that is to say, if the premises for applying particular formulas and mathematical proofs are appropriate. Sometimes the formula has been a wrong one, or observation has been based on insufficient numbers of data, or inadequate series have been compared, or data were selected in an inadmissible manner. These and similar defects creep in especially when the biological and mathematical methods of thinking are very heterogeneous.

The methods are much more likely to be reliable, however, when biological and mathematical studies not only have the same basis but also follow a parallel course of development. This is especially the case in *cybernetics*, a science mainly developed by N. Wiener (1948), which is likely to be widely applied in the future. H. Frank (1964) has defined it as follows: Cybernetics is a theory which examines the range and limitations of systems of information detached from their physical, physiological, and psychological features. It is then applied to certain systems embodying these features, and it attempts to translate such systems into reality in pursuance of circumscribed aims. The definition shows that this is a field of research concerned with problems which are similar in the fields of physiology, psychology, physics, and, we may add, genetics, ecology, evolutionary studies, and sociology. It deals with the *processing of information and with control techniques.* That it covers such a wide range of very different fields is clear from terms like "electronic brain," "thinking machine," "hereditary code," "information transference of hereditary characters."

Some of the methods used in cybernetics had already been foreshadowed by the attempts of biologists to find mathematical formulas for relationships within the wholeness of an individual, which owing to its metabolism is an "open system." Allometric growth is just such a case (Ch. 4 B). Von Bertalanffy's "general theory of systems" was the fruitful result of such efforts (see L. von Bertalanffy 1945, 1950, and contributors to *Human Biology* 23, 302-361, 1951).

The processes to be considered, mainly in the fields of physiology and genetics, correspond so closely to *technical* ones that when research in biological cybernetics developed, the technical terminology could be taken over unaltered. This was of great advantage in avoiding such errors as easily follow from the use of different terms. The processes are not of course totally identical, for biological systems are normally much more complex than technical ones, and such factors as metabolism, growth, and reproductive and regenerative ability make them more variable. But the basic principles of the two branches of knowledge present the same problems.

Processes of information are involved in a great number of biological problems, such as the effect of hereditary factors and cerebral activity, and in a simpler way in every automatic adjustment made by the organism. *Closed-loop systems* have frequently been developed; that is, feedback systems in which determinant factors intervene to maintain certain conditions. In other cases some external *steering factor* acts as a regulator.

It may be sufficient to take a few examples. A vertebrate is continually controlling its stance by reflex adjustments in the midbrain and cerebellum. These brain parts receive sensory information (regulative factors) on deviations from the normal position and initiate corresponding activity of certain groups of muscles (adjustment features), and the normal posture is restored. The reflex of the pupil is another instance. When the eye is exposed to a stronger light (the regulative factor), the visual cells are stimulated to a corresponding degree. This excitation (the control signal), passing through the nucleus of the nervus oculomotorius in the midbrain (the regulator), causes the iris muscles (the adjustment feature) to contract to the required extent. The result of this feedback process is a reduced incidence of light (see B. Hassenstein 1964, 1966). Blood pressure and temperature, respiration, kidney function, the release of pancreatic enzymes, and other functions are also kept normal by regulative processes, some of them effected by hormonal and other humoral signals. In this way the organs and the whole body continue to function properly (fatigue acting as a signal), or the limbs are made to move in certain ways. An ecological example is offered by the steering effect exerted by the percentual relationship between enemies and prey. The enemies increase to a point where excessive reduction in their victims, acting as a regulatory factor, reduces the number of predators. An important aspect when considering problems of evolution is that selection for harmful variants keeps animal and plant species vigorous.

In several cases it was possible to show not only that biological processes of this kind can be treated and more precisely characterized by methods of cybernetics but also that these made it possible to bring about a causal explanation. Concepts such as proportional and integral control and the frequency pattern of a closed-loop system, when applied to biological research, have led to new means of solution and new formulations of problems.

Another advantage of such methods is that many processes which had seemed to support a vitalistic or holistic interpretation could now be explained in terms of cause and effect. In this connection it is enough to refer to the concise and admirable exposition of biological cybernetics by B. Hassenstein (1966) with its impressive examples, especially in relation to sensory physiology (for the technical basis, see also K. Steinbuch 1965 and D. A. Bell 1962).

Cybernetics alone has also revealed the significance of codifying very different external stimuli as similar types of nervous excitations of different frequency. It has shown how important it is that these uniform elementary signals represent "all or nothing" signals (minor neurally conditioned deviations of amplitude are unimportant here). It also explains how it is possible to arrive at an optimum in the selection of information, and so on (see M. Spreng 1964). And cybernetics alone was able to show how a certain frequency pattern in sinusoid fluctuation of light intensity may reverse the regulatory activity of the iris, causing the pupil to dilate in a strong light and contract in a lesser intensity (conditioned by the time lag between the light stimulus and the beginning of the reaction and by incomplete compensation by the reaction of the iris). Finally, we are correct in attaching great importance to model concepts obtained by electronic computing machines which enable us to reach some causal understanding of learning processes, processes of abstraction, and even more complex cerebral activity.

D. The Monistic Principle

Since man began to think along philosophical and scientific lines, he has attempted to reduce the world's diversity to a limited number of basic facts or even to a single principle. It has always been observable that objects differing in quality consist at least in part of homogeneous elements, that processes and corresponding conceptual systems are often arranged in a hierarchical order, and that a succession of stages and superordinate laws can be recognized. So it was natural for some philosophers to carry this unifying tendency further and assume that all matter is ultimately composed of the same elements or governed by a single fundamental law.

Thales of Miletus believed in a primal, material substratum for the universe, namely water or moisture, the various conditions of which accounted for the variety of things. To Heraclitus, the principle was something like fire, and its different manifestations were controlled by a reasoning and lawful order, the logos. Democritus already offered the brilliant explanation, only confirmed gradually more than two thousand years later, that matter is made up of atoms, all alike in quality but differing in size, shape, and arrangement. He taught that all action ultimately springs from these indivisible entities joining and parting by cause and effect. In such hypotheses the sixth and fifth centuries B.C. had already produced *monism*, though this was restricted in the main to the material world.

Monism also developed in India at about the same time, though its roots go back still further; but it was more ontological, or rather epistemological, in character. The multiplicity of phenomena was held to be the illusion of our senses, which, like the "veil of Maya," conceal Brahman, the All-One. And in the sixth century B.C. China too evolved a similarly ontological form of monism in Lao-Tse's doctrine of Tao which he regarded as underlying all being.

Later, European philosophers developed different versions of monism based on materialism, spiritualism, identism, or general ontology. Even medieval dualism, strongly influenced by Christian theory, shows certain monistic tendencies in regard to the material world. And a great many more recent philosophers have upheld more or less consistent monistic views, among them Giordano Bruno, Spinoza, Berkeley, La Mettrie, Holbach, Maupertuis, Diderot, Schelling, Hegel, E. von Hartmann, Fechner, Schopenhauer, Wundt, Lipps, Spencer, G. E. Müller, Erdmann, Mach, Ziehen, Eisler, Haeckel, and Ostwald.

To what degree, however, are monistic ideas supported by present-day science? Has the monistic principle any heuristic value? And can the fact that events and their implications are consistent with a hierarchical, monistic order in nature even be taken as a criterion of truth? These questions are important for philosophic methodology. A few examples will show that a monistic view is in fact justified up to a certain limit.

Restricting our examples at first to the material world, we may begin with the complicated structure and processes in *living organisms*. In every active cell, highly complicated processes are taking place: multifarious metabolic processes, change of energy, synthesis and decomposition of complex proteins, carbohydrates, lipids, mineral salts, etc. This complication is further enhanced by the integration of cells in tissues and organs—in the liver and kidneys, sensory organs, nerves, and brain. And there are also the regulative activities of hormones on metabolism, development, maturation, and reproduction, and the processes of heredity and evolution.

A great many of these complex processes have been ingeniously analyzed in the last three centuries, and especially during the past decades. The result has always been the same: In each case, what is finally uncovered is a causally determined succession of chemical and physical processes. The molecules, ions, and atoms which interact in the vital processes are the same as those in inanimate matter. No "biogenic" molecules with special "vital" properties indiscernible to physicochemical methods have been discovered. By eating and breathing, living organisms are all constantly taking in molecules, ions, and atoms from their environment, incorporating them into their cells, and partly giving them off again. The laws which govern these activities often involve extremely complicated processes, but in the last instance they are chemical and physical laws. Because of the complexity of the processes, these laws are often "system laws." They are considered to be specifically biological, but we shall see (Ch. 6 K IV) that they are conditioned solely by manifold causal relationships.

Investigation of plants and animals had shown that living organisms can be classified in a hierarchically graded system of units. *Individuals*, that is to say individualized biochemical systems showing a relative constancy of their characters during a certain time despite continuous metabolism, constitute the simplest of these units. Many of them are almost alike and belong to the same *species*, which remains constant with regard to the successive stages of the individual life cycle, from the ovum or some asexual initial stage to maturity and senescence, over very long periods. Related species sharing similar structural characteristics can be combined in genera. Higher stages of grouping are represented by families, orders, classes, etc. It is very probable that all living organisms, whether extinct or recent, belong to *one single ramified phylogenetic tree* (see Ch. 3 A IV, V).

Structurally, too, these innumerable species of plants and animals are built up in a remarkable *uniform* manner. All organisms are made up of *cells*, or consist of one single cell; and these cells all possess a nucleus containing chromosomes bearing the main hereditary factors, cytoplasm, a cell membrane that governs intake and discharge of substances, mitochondria for cell respiration (in most primitive organisms, pre-stages of these), and ribosomes where specific nucleic acids are synthesized.

The past years have witnessed another highly remarkable discovery—namely, that *diversity of species rests upon purely quantitative differences in the hereditary factors.* The chromosomes have revealed themselves as coiled systems of filamentous molecules of deoxyribonucleic acid (DNA) enveloped by specific proteins. Each DNA molecule represents a double helix of a regularly alternating phosphate and sugar group held together by four bases which are always alike. Apart from the specific chemical composition of the reacting cytoplasms and the specific mitochondria, all that determines whether the basic hereditary factors shall be those of a tree, a worm, a beetle, a bird, or a human being is the varied sequence of these four bases. In other words, the phylogenetic differentiation of species is mainly based on the sequence of these same two paired bases. Every alteration in the sequences can involve mutation, expressed in some alteration in the organism's characteristics. But such mutations may also result from other purely quantitative changes in the chromosomes. There may be partial interchange of parts of two chromosomes of the same set, or portions of the same chromosome may be reversed (chromosome mutations), or the set may be multiplied (polyploidy).

Recent findings in the fields of *chemistry and physics* speak still more strongly in favor of a monistic interpretation. The same electromagnetic waves apply to radio waves, heat and light-waves, x-rays, and y-rays, the only difference being the length of the waves. Some cosmic rays are only 10^{-13} cm (0,000 000 000 001 mm) in length. X-rays are from 0,000 000 01 to 0,000 000 1 mm; light rays visible to man from 0,000 380 to 0,000 760 mm; infrared rays 0,0008 to 0.4 mm; short waves 10 to 180 m; and long waves 600 to 10 000 m. Thus light rays represent only a minute section of the total range (0,000 000 000 000 1 to 10 000 000 mm).

Molecules have been shown to consist of atoms, and atoms are all made up of the same elementary particles: protons, neutrons, and electrons, combined in different quantities in the different atoms. Matter is also represented by a number of other minute and in part extremely short-lived particles like deuterons, antiprotons, antineutrons, various kinds of mesons, positrons, neutrinos, antineutrinos, and photons. But as these particles are interchangeable, they are probably composed of the same basic components and differ only in mass, charge, spin, and impulse.

As these material particles all have equal spatial and temporal qualities, it was possible to imagine an absolute *space as homogeneous*, three-dimensional, and (because homogeneous) mensurable—that is to say, divisible into constant units of measurement. In the same way too, it was possible to imagine an absolute *time as homogeneous*, monodimensional, mensurable, and constantly shifting. As the time dimension can be conceived as a temporal sequence of spatial extension, there is some justification (within certain limits) for adding it, as a fourth dimension, to the three spatial ones.

Another fact that has given support to monistic ideas is that chemical and physical processes are all governed by one universal causal law. We can, indeed, assume causality in the field of microphysics as well, because macrophysical and chemical processes come about by integration of microphysical ones. The only difference is that in microphysics it is not possible to predict single events (Ch. 6 K III).

It is plain, then, that a great many of the world's facts and processes can be traced back to uniform basic structures and laws. This is the reason why more or less radically monistic ideas have repeatedly been developed in man's attempt to understand the material world. And these ideas have often been successfully applied as heuristic principles. It even seemed to be possible to comprehend all phenomena in a single "world formula" (Ch. 6 I). This could be attempted because in logic and in mathematics, which is based on logical laws, complex relationships could also be traced back to a limited number of basic statements or axioms.

Summing up, we can state that *countless individual observations and comprehensive explanations of structures and processes have shown that up to a certain limit the monistic principle offers an adequate explanation for the "material" world, and that it is of considerable heuristic value.* Within these limits it can be taken as mediating a high degree of truth, and it can be applied—with all necessary caution—as a criterion of truth (on the concept of truth, see Ch. 6 D IV). However, the monistic principle is significant because it has stood the test of proof, and not—as many have suggested—because it affords the most economic explanation.

As we shall see, all monistic reduction has its limits, imposed by the fact that the elementary particles possess irreducible qualities such as mass or energy, charge, spin, speed, and also spatial and temporal qualities as well. Besides, several basic laws are operative in the world—the causal law, universal constants, the laws of conservation, the principles of symmetry, and the logical laws. Despite the relative validity of the monistic principle, then, *the multifariousness of the world ultimately derives from a pluralist source* (Ch. 6, I, K, L).

In the case of psychic phenomena, the monistic principle can be applied only in a very restricted sense. We can probably trace all mental imagery back to sensations and to logical procedures, but the visual, auditive, haptic, and other qualities of sensations and images cannot be traced directly back to uniform basic principles. In later chapters, however, we shall be discussing at greater length how it is possible to identify the diverse phenomena in a panpsychistic, identistic sense with corresponding physiological, i.e., "material," processes in the brain (Ch. 6 H III and K VII). We shall then find that the qualities of phenomena are based on relationships of "matter" determined by systemic laws, and that matter has a protopsychic nature. But though we can study the systematic development of "material" qualities at various stages of integration from the elementary particles to the molecules, we cannot do the same with the phenomenological qualities. They can only be studied in *Homo sapiens*, at the final phylogenetic level and at the highest stage of integration. (We shall be dealing with these important problems in greater detail in Chapters 6 D II; H III, IV, K V; and especially K VII and L.)

Chapter Three

Characteristics of Life

In spite of the enormous variety in plant and animal species, including man, all organisms may be said to belong to one specific class of "things" characterized by certain specific and essential features. Yet the distinction between the animate and the inanimate is not so easy as might at first appear, for almost all the distinguishing characteristics are shared by both groups, and intermediate forms exist as well. We shall see that "life" is not so much defined by certain single characters but by their combination into individualized, purposefully functioning systems showing a specific activity, limited to a certain time span, but capable of reproduction, and undergoing gradual hereditary alterations over long periods.

Although a great deal has been written from classical times to the present day about the uniqueness of life, scientists and philosophers have not been able to develop a generally accepted conception. It is becoming widely recognized that all life processes are governed by chemical and physical laws, but some researchers doubt whether the origin and conservation of purposeful structures and processes, and the progressive evolution of organisms, can be explained in a causal manner. The fact that man and animals are also distinguished from inanimate matter by conscious processes has led to very different philosophical theories of life.

The present work makes no attempt to deal with the voluminous literature on this subject but tries instead to clarify the problems and to develop a general view of life baesd upon our present biological, psychological, and epistemological knowledge. Relevant works on biological theory will of course be referred to. But before considering a causal analysis of the life processes, we must describe life's characteristic features and discuss some of the difficulties in distinguishing between the animate and the inanimate.

A. Characteristics of Living Beings

1. Chemical Components of Life. F. Wöhler achieved the synthesis of oxalic acid and urea in 1824 and 1828. Until then most biologists and chemists had believed that a kind of vital force, a *vis vitalis*, was required for any organic synthesis, and that all organisms must contain specific molecules to accomplish the life processes. With advances in organic and especially physiological chemistry, these conceptions became less and less tenable. We now know that there are no molecules in a living organism which cannot also exist as inanimate components. Concepts of life molecules in the sense of W. Verworn's "biogenes" (1909) are only acceptable insofar as certain compounds are indispensable for the qualities and activities of living organisms—namely, nucleic acids and proteins. *Nucleic acids* (given this name because first found in the cell nuclei and only later also in cytoplasm) make possible specific protein syntheses (thus overcoming entropy), reproduction, hereditary transmission, mutation, and evolution. *Protein bodies*, having large molecules and being highly complex, can combine in a great variety of ways by normal and auxiliary valences and thus cause the extraordinary diversity of microscopic and macroscopic structures, the lability of processes, and the activity typical of all organisms, as well as the development of innumerable plant and animal species including man.

Comparatively few classes of compounds are normally found in animals and plants besides these main ones: carbohydrates, fats, and lipids. We also find isoprene derivatives (e.g., cholesterines) and porphyrin pigments (e.g., blood pigments and chlorophyll) (F. Klages 1961, and others). Other phosphorus compounds (in particular, adenosinetriphosphoric acid = ATP) which can liberate a high amount of energy are essential to life; and so are mineral salts and water, which is the chief solvent and therefore necessary in most biochemical processes. It is not yet possible to state which of these groups of compounds, apart from nucleic acids, proteins, and water, are absolutely indispensable to the simplest organisms. This depends mainly on whether we already ascribe life to such beings as the so-called large viruses and Rickettsiae (Microtatobiotes, Mycoplasmatales) which also contain fats, carbohydrates, and enzymes.

Relatively few *chemical elements* are essential to all plants and animals. Among the ninety-two "normal" elements, not counting the transuranian elements, only the following may safely be termed predominant "constituents": Carbon (C), hydrogen (H), oxygen (O), nitrogen (N), phosphorus (P), and sulfur (S). "True" living creatures, from the bacteria upwards, also contain potassium (K), sodium (N), calcium (Ca), magnesium (Mg), chlorine (Cl), iron (Fe), manganese (Mn) (which may replace Mg in bacteria), cobalt (Co), and often boron (Bo), and copper (Cu) or some other heavy metal. (On bacteria, see, for example, C. Clifton 1957.) These quantitatively predominant elements have a low atomic weight between 1.01 (H) and 40.1 (Ca). A whole range of other elements—silicon, arsenic, iodine, etc.—are present in certain plant and animal groups, but they are not generally characteristic of all organisms.

Compared to inanimate matter, living organisms display an enormous *diversity and complication of chemical compounds.* The multiplicity of *carbon compounds,* of which living substance is largely built up, is based on special characters of the tetravalent C-atom, which other atoms do not show. Two C-atoms may combine as C—C, C=C, or C≡C; this depends on how the outer electron shells unite. They often combine themselves in chains or circles. Double linkages (C=C) in particular lead to easier reactions, because they can split up in compounds in which one C–atom has more electrons than the other one (polarized double linkage). Heterocyclic circles can be formed by the introduction of other polyvalent elements such as nitrogen or sulfur. Chains or circles can also be united in many combinations (see, for example, B. P. Karrer 1963).

In consequence of the multiplicity of carbon compounds, the number of organic substances is virtually unlimited. This is particularly true of proteins. Their basic components, the polypeptides, may consist of an almost infinite variety of some twenty or twenty-five amino acids, tied one after the other in manifold successions. Attempts to estimate the possible variants including isomers (that is to say, the variants composed of the same amino acids but connected in different succession) have yielded an almost incredible total: 20^{150} polypeptide variants are possible, if they are composed of 150 amino acid units on the average. Written out in a normal manner, this would mean a number with 195 figures. These molecules, if tightly packed togther, would form a mass several trillion times larger than the universe, with a radius of one thousand million light-years (F. Klages 1961). The figure for the *nucleic acids*, which play such a vital part in the transmission of hereditary factors, is also extremely high. They can form filamentous molecules up to many centimeters long, with an almost infinite variety in the sequence of the two pairs of bases (see Fig. 2, Ch. 4 E). Clearly, then, every species of animal and plant, and in most cases every individual in these species, has its own nucleic acids and protein bodies; and large, more complex organisms are composed of a very large number of different compounds, though we have seen that relatively few classes of compounds are involved.

Many characteristics of the living organism can be traced back to the particular features of these large molecules (see J. Brachet and A. E. Mirsky 1964; R. Weber 1965; E. Lehnartz 1959). As nucleic acids will be considered at greater length in Chapter 4 E, which deals with heredity, we shall at this point only indicate some distinctive qualities of protein bodies.

Like the amino acids and peptides of which they are composed, these proteins can combine with either acids or bases; in other words, these substances are *amphoteric*, capable of dissociating either in an acid or a basical manner. They may also act as buffers. The globulins and albumins that predominate in the protoplasm are acid, but they largely appear as almost neutral potassium or sodium salts. As cytoplasm and body fluid normally show only a weak alkaline reaction, even slight alteration in hydrogen ion concentration can result in a weak acid reaction. This *labile condition* is of great importance in many vital activities. (Other buffer systems may also play their part.)

Another reason why these relatively stable protein molecules react in this labile fashion is that in water they form *colloidal solutions* whose constituents are relatively large and therefore react in a special way. In addition to the chief valence of the atoms (the normal cohesive power of the atoms which share the same electron shells), there is also some electrostatic attraction. This has only about 1 percent of the cohesive power of the atoms; but with a number of combined large molecules this attraction can exceed that of the atoms. By such van der Waal association other molecules may become loosely attached (adsorbed) to the surface of protein bodies, producing an increase in their reactivity.

Colloidal solution in the cells and body fluids of organisms mostly leads to hydration; i.e., owing to adsorption the protein particles become surrounded by water molecules. If the particles coagulate, the solution (sol) may become a gel, which binds a good deal of water. The degree of hydration can be altered by the ions; that is to say, they can cause swelling or shrinking. These alterations are of great importance both for water transport within the tissues and for the reactivity of proteins and of whole cells. Muscles, for instance, can only function in quite definite conditions of such swelling.

On the surfaces of contact between fluids (e.g., cell fluid) and solid substances (e.g., solid cell components), or between two immiscible fluids (e.g., water and lipids), no reaction between identical neighboring molecules takes place, and this leaves some degree of electrostatic attraction free—what is known as *surface tension*. Here too, molecules are adsorbed, often in a single layer because of the limitations of attraction (in some cases also in a certain direction). *Membranes* are formed in this way, and they play a great part in a number of physiological processes. It is here that substances dissolved in cell fluid or body fluid are selected, and reactivity is heightened in the adsorbed monomolecular layer. This explains what had been revealed by electron microscope analysis; namely, that active cells include many internal membrane systems (the endoplasmatic reticulum, ergastoplasma, Golgi structures, mitochondria, cytocenter, nuclear membrane). The cellular membranes determine the intake and output of substances. In plant cells and animal cells (especially nerve cells) the origin and conduction of excitations are bound to the membranes, in the case of nerve cells by alterations in porosity for sodium and potassium ions.

The *enzymes*, an important group of proteins, are especially active at internal surfaces. They consist of a protein and a simpler prosthetic group, and bring about specific reactions of special compounds. They act as catalysts, causing and/or accelerating certain reactions, but without themselves being consumed. Hence, very small amounts of enzymes are sufficient to bring about reactions of relatively large quantities of compounds. Without them, reactions of the fairly stable proteins, carbohydrates, and fats would run very slowly. *The characteristic intensity and speed with which most life processes take place are thus caused by these biocatalysts.*

Finally, *hormones* are characteristic of many reactions of organisms. Very small amounts of these substances, too, have great effects. They include very different compounds formed within the cells of the organism itself. Body fluids (blood or the water systems in plants) distribute them to different structures or organs where they affect metabolism. Hormones *regulate* the processes of metabolism, growth, and maturation, and are apparently common to all organisms, especially to higher plants and still more to animals. Heteroauxin, the most important growth hormone in plants, is found in bacteria and fungi as well as in flowering plants, Acetylcholine and adrenaline, important for nervous excitation, have also been found to be present in protozoa. Only certain hormones are protein bodies (e.g., insulin and glucagon of the pancreas, some hormones of the anterior lobe of the pituitary body, and the parathormone of the parathyroid gland). Most others are much simpler in structure (those of the posterior lobe of the pituitary body are octapeptides; sex hormones and adrenocortical hormones are steroids; and heteroauxin is β-indole acetic acid).

II. Morphological Basis of Organisms. Life may be said to manifest itself only when the chemical compounds which we have noted as essential combine in organized, clearly confined systems of relative constancy. Animals and plants are *individuals* at every stage of their life-cycle: as ovum, embryo, larva, and when full grown. Yet, though individualized systems are typical of all living organisms, it is also true that clearly confined, organized and relatively stable "individuals" are to be found also in inanimate nature: crystals, sand grains, molecules, atoms, etc. These too can rank as individuals of a higher order of complexity; for atoms consist of elementary particles, molecules of atoms, and sand grains of molecules. Moreover, atoms and molecules, being in almost continuous motion (thermal agitation, radioactivity, and chemical combinations), prove that activity is not an exclusive feature of living matter.

The important point is that organisms—as their name suggests—are organized in a specific and complex way. They represent a *wholeness* made up of specialized parts (structures and organs) adapted to particular functions. The higher plants have roots, leaves, flowers; mammals have bones, muscles, intestines, liver, kidneys, blood vessels, heart, lungs, brain. If such vital organs are removed, the individual usually dies. Most plants cannot survive for long without roots or leaves, nor can mammals continue to live without a liver, heart, or brain. And yet it would be untrue to say that all organisms are indivisible. Protozoa do in fact divide; a whole individual can evolve from only a very small portion of a fresh-water polyp or a planarian, and single separated segments of a tapeworm can remain alive for some time. Many kinds of organic tissue can live and multiply in suitable liquids, though in most cases with certain structural alterations.

Parallels to this condition of wholeness exist in inanimate nature also. A molecule is an organized whole; if divided, it loses its distinguishing characteristics, but its parts may again form a new molecule. Wholeness is therefore an important feature of living matter, but it is not a specific criterion.

Yet in contrast to atoms, molecules, and crystals, the wholeness of the living being is only relatively confined and constant. An organism is continually taking up matter from its surroundings, assimilating it, and giving off other matter. An organism only shows a *relative constancy*; it constitutes an *open system* connected at all times with certain components in its environment (Ch. 4 B). In the realm of the inanimate, a flame behaves in a similar way: it consumes combustible material and gives off gases and light energy.

When characterizing the wholeness of an organism we must also consider that many organisms are *individuals of a higher order*. In the case of siphonophores, the individuals in the colony are functionally specialized to such a degree (as swimming bells, gastrozooids, tentacles, pneumatophores, gonophores) that they seem to be merely separate organs of a single whole. In fact, this was long considered to be the case. From the physiological point of view, the colony functions as a single and uniform individual. Many colonies of bryozoa, too, are made up of similarly specialized individuals. The individual sponge in a colony is mostly indistinguishable. All living creatures, which necessarily live in the closest symbiosis with other organisms, are individuals of a higher order. A typical example is for instance the lichens, a combination of algae and fungi, or the mycorrhize, a fusion between a fungus and the roots of a higher plant. The animal kingdom contains many examples of endosymbiosis with algae (zoochlorella, zooxanthella), as for instance in the ciliate *Stentor polymorphus*, the polyp *Chlorohydra viridissima*, Anthozoa, and the turbellarian *Convoluta*. Cases of symbiosis of marine animals with photogenic bacteria are still more numerous, as in the cephalopods, tunicates, and fishes. Many blood- or sap-sucking and wood-eating insects need symbiosis with bacteria or yeast fungi. The bacteria and protozoa in the intestines of many higher animals are essential to the digestive processes. And we must also bear in mind that the body fluids of most multicellular animals contain cells, leucocytes in the case of vertebrates, which lead an independent life, like amoeba capable of amoeba-like movement, sense reactions, and phagocytosis. Hence humans, too, are individuals of a higher order.

At times this idea of wholeness and the allied conception of "Gestalt" have misled scholars into inadequate and occasionally almost mystical utterance. Many have sought support from Goethe's words:

> Und keine Zeit und keine Macht zerstückelt
> Geprägte Form, die lebend sich entwickelt.

> No time and no power dismembers
> Moulded Form which Life develops.

Those who advocate a theory of idealistic morphology have been led into questionable formulations. L. Wolf and W. Troll (1940), for example, wrote: "A Gestalt has no parts, for it is by its very nature indivisible" (p. 10). It is not easy to see the impossibility of distinguishing such "parts" as leaves, flowers, limbs, or head in a real (or archetypal) plant or animal. The same authors also declared: "At the same time, Gestalt manifests an inherent archetype, though this may not be directly appreciable and usually appears disguised by its diversity" (p. 25). This could simply be a reference to what the biologist terms genotype and phenotype; but if so, it might have been more succinctly expressed. "Gestalt is also unthinkable without symmetry." One is left wondering what symmetry is displayed by an amoeba, a slime fungus, or a snail. "It is the goal of natural order to reveal archetypal unities by means of Gestalt" (p. 51). F. Waaser (1942) has something similar to say on the subject of "wholeness": "This concept no longer lives a merely abstract existence in man's mind; it is formative, creative reality; it is the fashioning principle manifest in everything that lives." This kind of "idea" exerting a formative influence on nature has much in common with Plato's "Ideas," the active force of which seems to have been indisputable since Aristotle's day (*Metaphysics*, Bk. III). I refer the reader in this connection to Chapter 6 K IV, where the philosophical implications of such conceptions (e.g., holism) are discussed.

Leaving philosophical considerations aside, we may say that ideas about wholeness are based on two relatively simple statements. One has been mentioned—that individuals are organized systems. The other is the fact that when constituent components combine to form a larger whole, or when individual structural elements undergo alteration, new relationships come into being, and these may result in the development of new properties.

A few examples may make this clear. When taken singly, carbon, hydrogen, and oxygen (which are gaseous at normal temperatures) have quite different properties from compounds of the same three elements such as sugar, fats, alcohol, and acetic acid. Nerve cells which develop among other tissues have different structures from those grown in isolation in an artificial medium. When the size of a certain breed of dog has been reduced by successive selection, we also find changes in the animal's proportions, because different parts are growing at different rates (allometric growth): the head, brain, heart, liver, and kidneys become relatively larger, and the facial parts relatively shorter. Because the surface of the body is now greater in proportion to its volume, the balance of heat conservation is altered (smaller races are more susceptible to cold). In short, every change in relationships, either within the body or with reference to external factors, creates new conditions and the possibility of new properties.

As every individual, every "Gestalt," represents a distinctive complex of relationships, it follows that structures and functions can only be fully assessed if the coherence of the whole is kept in mind (Th. Ziehen 1930; F. Alverdes 1933, 1935; O. Koehler 1933). Many exponents of the Gestalt or "wholeness" theory would have us believe that before they propounded their views, organisms or their organs had been thought of simply as the "sum" of individual components;

but this is a tendentious imputation. Surely no morphologist or physiologist has ever doubted that structures and functions of organisms come about by complex integrations and not merely by simple summations. Of course one had to begin by analyzing the separate structures and their functions before undertaking a study of interrelations within the wholeness of the organism. But once these analyses had been carried out, it was right to press also for studies of the organism as a whole, because organismic activities are directed to the preservation of the wholeness of the individual. H. Driesch (1901) was one of the first to advocate this method, though his study of regulations in sea urchins led him to the premature conclusion that the forces bringing about the unity of the living organism are beyond the reach of analysis. It was precisely the study of correlative interrelationships and gradients of differentiation that later led research in developmental physiology to important findings, especially to the discovery of inductive processes (H. Spemann 1924).

Many other branches of biology as well owe much to such investigations of wholeness. Research into allometrical relations have proved to be very important for the understanding of individual as well as phylogenetic development (W. d'Arcy Thompson 1917; J. S. Huxley 1932; B. Rensch 1947, 1954, 1958a, 1959; and others). The study of functional anatomy received a fresh impulse (H. Böker 1935, 1936, 1937, and others). A psychology of "wholeness" and a Gestalt psychology developed, opening up new vistas (Ch. von Ehrenfels 1890; M. Wertheimer 1912, 1922-1923, 1925; K. Koffka 1912, 1935; W. Köhler 1920, 1928, 1940; W. Metzger 1940, 1954; W. Ehrenstein 1942; and others). Some exponents, however, have gone much too far in ignoring the established findings of association psychology. But biological theory and natural philosophy in particular have gained a sound basis through closer study of interrelations in the wholeness of an organism (L. von Bertalanffy 1932, 1942, 1945, 1950, 1951; F. Alverdes 1933, 1935; K. E. Rothschuh 1959; and others).

Every true organism is characterized by a specific structure; it consists either of one *cell* or of a combination of cells. This structure distinguishes it clearly from inanimate matter. In 1655 Robert Hooke in his *Micrographia* had stated cork to be a substance made up of small "cellulae." Only gradually, however, did it become recognized that plants, animals, and humans are all made up of such cells. M. J. Schleiden (1838) and Theodor Schwann (1839) were the first biologists to state this important fact. Today we know that *the cell structure of all living beings is basically very similar though often complicated and highly organized.* As already mentioned (Ch. 2 D), they are composed of cytoplasm containing many different kinds of organic compounds and in particular very large protein molecules (Ch. 3 A I); a cell membrane (plasmalemma) enclosing the cell; one or more nuclei; many mitochondria and lysosomes; innumerable ribosomes; a cytocenter; and usually some additional structures such as the endoplasmatic reticulum, Golgi bodies (Fig. 1), cilia, flagella, and plastids in the case of plants. The mitochondria possess a rich inner membrane system and represent the cell's "power station" (J. Brachet and A. E. Mirsky 1959-1964). Within them, respiratory enzymes carry out complicated respiratory processes which produce compounds rich in energy (adenosine triphosphate). The ribosomes, in combination with nucleic acids coming from the nucleus and others from the cytoplasm, govern the protein syntheses appropriate to each specific organism. The lysosomes have enzymic functions; they are so to speak "organelles of digestion." The nucleus contains the long-coiled giant molecules of DNA which transmit the hereditary information. The nuclear membrane largely protects the nucleus from the metabolic processes in the cytoplasm, though it contains pores which open from time to time, allowing exchange of material with the cytoplasm and especially the release of the messenger RNA which transmits the hereditary information to the ribosomes. The complicated cytocenter controls the processes of cell division which normally begin with nuclear division by chromosome division lengthwise, so that the same hereditary units fall to each daughter cell. The wealth of internal surfaces already mentioned is a characteristic feature of both plant and animal cells. The various processes are thus kept separate and in order, and adsorption to these surfaces increases the reactivity of organic materials.

It is only the lowest organisms, the bacteria and blue-green algae (cyanophyta), whose cellular structure is of a simpler kind. Yet even they have a kind of chromosomes consisting of DNA (the nuclear membrane is lacking), small mitochondria or their equivalents, typical ribosomes, a cell membrane, and flagella. They have far fewer inner

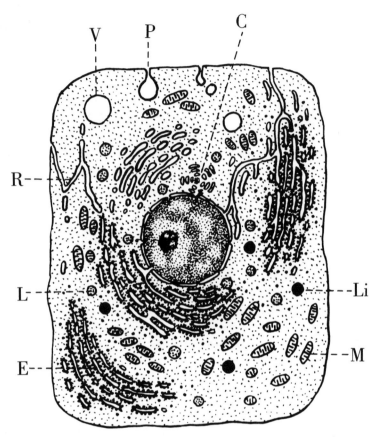

Figure 1. Diagram of an animal cell. In the center, the cell nucleus with darker nucleolus and nuclear membrane with pores. C = cytocenter; E = ergastoplasm (reticulum studded with ribosomes); L = lysosomes; Li = lipoid droplets; M = mitochondria; P = pinocyte; R = endoplasmatic reticulum; V = vacuoles. The larger dots indicate free ribosomes.

membranes, because they have no endoplasmatic reticulum, which is not an important feature in such small organisms, the cell wall being large in relation to the cell volume.

Every cell is functionally independent to a large extent, because it is *a coordinated system of regulatory and operative structures.* Most plants and animals, and hence also man, begin as a cell, usually a fertilized egg cell. The cells resulting from cell division are in general relatively uniform. It is only when various types of tissue are formed that the cell's independent state is more affected by neighboring cells by specialized functioning in the interests of the organism as a whole. Normally, however, all the cells share the same hereditary units; that is to say, the same number of chromosomes—or a multiple of that number (polyploidy).

It is a characteristic of most cells of plants and animals that they are *of microscopic proportions*, although they may vary considerably in size. Bacteria range from 0.15 to 20 μ, mostly between 1.0 and 1.5 μ (1 μ being 1/1000 mm). The smallest protozoa are hardly any larger than this. The microsporid *Nosema bombycis*, an intercellular parasite of the silkworm, is 0.5 to 1.5 μ long; the immature malaria organism *Plasmodium vivax* 1.5 to 3μ; the haemosporid *Theileria brimonti*, a blood parasite of the sloth, 0.5 to 2 μ. Human cells generally range from 10 to 30 μ ($-$ 250 μ); liver cells, for instance, are 18-26 μ. Spermatogonia of the human embryo when $\frac{1}{2}$ cm long measure only 4.5 to 12μ. Red corpuscles in adult man have a diameter of 7.5μ, blood platelets (thrombocytes) only 2 to 3.6μ. *Pelomyxa*, the largest shell-less amoebas, have a diameter of about 3μ. The gregarina *Porospora gigantea*, one of the sporozoa, may be as long as 15 mm. The processes of the nerve cells of larger vertebrates are many centimeters in length. So are muscle fibers, though these are multinuclear and so rather resemble a "tissue without cell walls." The large green unicellular marine algae of the order *Caulerpa* are also multinuclear and can be several decimeters long. The mononuclear cell with the greatest volume is found in the yellow egg yolk of the African ostrich (*Struthio*); but it owes its size not to the cytoplasm but to the yolk stored there as reserve material.

Most of the cells in all multicellular creatures are relatively small. Cells also determine the specific structures of such intercellular matter as coral skeletons, snail and mussel shells, the cartilege and bones of vertebrates, and the like. The reason plants and animals are made up of such small elementary components is probably that their specific needs in regard to food, protection, and reproduction made it essential for them to remain constant throughout a long sequence of generations—in other words, to develop a soundly functioning hereditary mechanism. As the hereditary material in the cell nucleus has only a small range of effectiveness, a construction of numerous small cells was the most suitable.

This relationship also determines *the hierarchical order of elements in organisms*. The many atoms, consisting of protons, neutrons, and electrons, combine to form molecules, and these in turn make up structures and organelles of cells and also whole cells. The cells unite into tissues; the tissues form organs; and the organs, often arranged in organ systems, constitute the individual. In many cases individuals do not yet represent totally autonomous levels of organization but are dependent on a larger group of individuals. This extends the hierarchical framework, when individuals unite in coherent colonies sometimes involving division of labor, or in societies composed of differently functioning castes (ants, wasps, bees, termites).

A few figures may serve to emphasize the very wide range within a single hierarchy. The body of an adult human comprises about one quadrillion (a number with 24 figures) atoms and two hundred trillion (a number with 14 figures) cells—twenty trillion in the blood and some 13,000 million in the brain. And each active cell contains perhaps 100,000 molecules of highly complex respiratory enzymes (after O. Meyerhoff 1937; cf. K. E. Rothschuh 1959). But the microscopic multicellular rotifer *Epiphanes senta*, some 0.4 mm in length, shows only 959 cell nuclei of which 183 belong to brain cells. A bacterium may contain forty to fifty enzymes besides numerous other proteins and organic substances. The threadlike DNA molecules of *Bacterium coli* are about 1 mm long.

Finally, and on a universal scale, *the size of all organisms* is intermediate in a specific order between the large individuals of celestial bodies and atoms and elementary particles. A minimum of several million molecules is required to form a single cell, and the molecules in their turn mostly consist of hundreds or thousands of atoms. Only the level of complication realized by cells is capable of processes which we can confidently characterize as life processes. But organisms vary greatly in size—from 0.15 thousandth of a millimeter in bacteria to a length of some thirty-three meters in the adult blue whale. Man ranges from 0.2 mm in the stage of the fertilized ovum to 1.50–2.10 meters as adult man. A size maximum is mainly imposed for both plants and animals by the chemical and physical characteristics of their organs of support (wood or bones) and their system of food distribution. The minimal size of a true living organism with its own metabolic system is partly determined by the fact that beyond a certain limit thermal agitation of the molecules would disturb the life processes.

It should be emphasized once again that all that has been said about structural features of organisms refers to genuine living creatures alone. Other bodies exist which in many respects can be regarded as intermediate stages between the animate and the inanimate. Examples include the viruses, for instance, and the Microtatobiotes (Mycoplasmatales). These possess a large part of the distinctive structural features and functions of living matter but lack a metabolism of their own. They will be considered in connection with problems of phylogenesis (Ch. 5 E).

III. The Individual Cycle. Every individual among plants, animals, and man has a specific life span; this may vary, but only within the inherited limits for its species. As the organism grows older, progressive degeneration of the tissues sets in, ending in the death of the individual. Plant and animal species can therefore survive only by *reproduction.* This takes various forms, but in each the same process takes place: part components—spores, buds, or sex cells—are set aside, and these transmit the hereditary units to the next generation, while they themselves still retain their "youthful" characteristics. In many unicellular organisms this type of reproduction is replaced by binary fission. But in all cases each new individual begins as a smaller organism, often minute indeed compared to its size at the adult stage. We have already mentioned the size of the human fertilized ovum as having a diameter of 0.2 mm.

The reproductive cells keep their youthful character because they do not lose their developmental powers in the course of differentiation. They thus retain the ability to develop all the structures of an adult individual; they are *totipotent* and in turn can be traced back to earlier totipotent cells. This in fact constitutes *the unbroken thread of life,* connecting the generations by a succession of totipotent cells, in most cases by the "germ tract" of successive immature and mature cells. In this sense the multicellular individuals which emerge from the germ cells, spores, or buds represent so to speak "outgrowths" on the thread of life which ensures the continuance of the species.

Hence, each individual plant or animal is constantly undergoing a change of form. Such an *individual cycle*, which normally can be traced from the ovum through the embryonic, larval, or juvenile stages to adulthood and advanced age, and which is repeated in every generation in the same manner, is typical of all living organisms. But variations of several kinds can be noted. In the case of bacterial sporulation and binary fission in unicellular organisms, the life cycle is very simple. On the other hand, in many parasites and some other groups of animals, the complicated development involves one or more changes of generations, displaying very different shape and structures (as in the malaria cycle or the liver-fluke cycle).

Although adult individuals, except in the case of binary fission, end in death, i.e., as a carcass, the unbroken chain of successive totipotent cells is evidence that living substance does not necessarily die. It may live on in a succession of daughter cells, without a dead body being formed during a long time. This longlasting immortality, another important feature of life, can also be observed in tissues cultivated in artificial media (*in vitro*). If these explantated tissues are divided from time to time and the artificial medium is regularly renewed so that nourishment is maintained and the excretions are removed, tissue can live very many years. A. Carrel, one of the pioneers of tissue cultivation, kept a culture of fibroblasts from the heart of a fowl for thirty-three years—much longer than the bird itself would have lived. Only an accident then destroyed the culture, which might otherwise have existed still longer.

Another characteristic of individual change is that numerous interrelated reactions transform the comparatively simple matter of an ovum, spore, or bud into highly complex and functionally distinct structures and organs like leaves, flowers, sporangia, or limbs, eyes, and brain (see also L. von Bertalanffy 1949).

None of the above-mentioned characteristics of the life cycle is found in the realm of inanimate matter. Metals undergo alterations as a result of oxidation, stones by weather, and colloidal solutions by age. But such changes do not represent a regularly recurrent sequence of the same forms. The so-called fluid crystals, e.g., those of para-azo-oxycin-

namic acid ethyl ester, can continue to grow, and when they attain a
certain size can divide; this may be called a kind of reproduction, but
there is no development from simpler material to highly specialized
structures (O. Lehmann 1907, 1908).

IV. Phylogenetic Change and the Unity of All Life. Plant and animal
species in successive generations remain structurally constant over long
periods, but not indefinitely. *Mutation* occurs from time to time; and
these spontaneous hereditary alterations, together with new gene combinations, natural selection, and processes of isolation, produce changes
in the species. In the course of geological time, species have evolved
into new species or split up into several species. The continuous effect
of such evolutionary alterations gradually enlarged the differences, and
new genera, families, orders, and higher categories came into being
(Ch. 5).

However, despite phylogenetic alterations, individuals of successive
generations were always linked by the above-mentioned chain of totipotent cells. Every individual is bound to its ancestors by this "thread of
life," and all such partially interconnected threads belonging to the same
species lead back to a differently constructed primary form. These forms
in turn have descended or deviated from other species, and so on. The
axiom *omne vivum e vivo* (all life is derived from life), established
some two hundred years ago, has invariably proved true. It has been
definitely proved that many lines of descent are derived from a common
category of ancestors. In most other cases this could be shown to be very
probable. We are justified, then, in concluding that all these threads of
life may ultimately be regarded as ramifications of one great *common
stream of life* to which every living or extinct plant, animal, and human
belongs.

This "stream" of life is usually represented as a *phylogenetic tree*
whose dead branches correspond to extinct groups of organisms and
whose countless living twigs represent the extant species. This picture is
a true one insofar as it portrays life's continuity. And it has proved to
be an accurate representation of the facts for plant and animal groups
of which we have sufficient fossil remains—vertebrates, for instance,
including the line of descent of hominids. In other phyla where fossil
remains are rare or absent, studies in comparative anatomy and embry-

ology have served to determine, at least in a more general way, the phylogenetic origin. We may assume that all living organisms belong to one common phylogenetic tree which branches out near its base into the main division between plants and animals.

We have already mentioned that the genetic substances, the DNA molecules of plants, animals, and man, are composed of the same few components and that the differences among organisms mainly depend on the sequence of two pairs of bases and the length and number of these molecules (Ch. 2 D). This uniformity of substance which produces the basic differences in organisms emphasizes the *unity of all life* in a striking manner.

In the course of phylogenetic development, natural selection has operated to adapt the species of plants and animals structurally to their environment. In most cases the organisms underwent different alterations until an optimum was reached, and then remained stable for some time. But such constancy rarely lasted for more than a few million years; climatic changes, mountain building, erosion, and transgressions of the seas have altered habitats, and so have the emergence of new sources of food, new diseases, competitors, and enemies. Species have also spread to varying habitats and had to adapt themselves structurally to new conditions. Hence in many ways new types of organic construction originated. In some instances organs took over new functions. For example, in the Devonian period, when some Crossopterygii began to live on land and become the first primitive amphibians, their pectoral and ventral fins developed into limbs suited to walking, and their air bladders became lungs. As all structures of plants, animals, and man are determined in such a way by the structures, and therefore the functions, of their ancestors, it follows that some of the essential characteristics are purely *historically conditioned*. Vestigial organs afford particularly striking examples here. These rudiments of once fully developed structures are now carried along through life as useless ballast. Examples include the Greenland whale's rudimentary pelvis and femur; the tiny leg stumps of some reptiles like *Seps*, *Ophiomorus*, and others; the kiwi's miniature wings hidden among its feathers; and man's vermiform appendix. An important characteristic of all living beings, then, is that they are historically conditioned. There are parallels here with some inanimate substances too. The stratification of sediments or the stages of erosion of a mountain also have their historical aspect, and so has a rounded pebble or a piece of rusty iron.

Besides change and adaptation by influences of the environment, however, many lines of descent display an upward trend of evolution. We may disregard this *evolutionary progress* in the present context; more will be said about its characteristics and cause in Chapter 5 C III and D IV. Here it may suffice to note that we are quite justified in saying that mammals are more highly organized than their ancestors among reptiles, amphibians, and fishes; that monkeys are more highly developed than marsupials and Monotremata; and that man occupies a higher place than his simian forebears. This upward trend, noticeable in many lines of descent and in the phylogenetic tree as a whole, and extremely rare and restricted in inanimate nature, can also be reckoned among the distinctive characteristics of life.

V. Metabolism and Activity. Living organisms are normally in a condition of constant activity, as different chemical or physical processes take place. Some of these processes can also take place in inanimate matter, in a test tube, for instance (Ch. 3 A I). The peculiarity of life processes consists mainly in the fact that they all occur within the framework of a definite and very complicated system—an organic unit. Its hierarchical arrangement of structures involves a similarly hierarchical arrangement of processes—often initiated by cell nuclei, ribosomes, or mitochondria. Because of the complication of the chemical compounds, in particular the proteins, and the structures, many mutual relationships exist and most of the processes are correlated and interrelated in countless ways. Hence many of these processes are governed by "system laws" which partly veil the underlying causal basis of individual reactions and have prompted biologists in the past to assume the existence of a *vis vitalis*.

The activity of living organisms is the result of a continuous metabolism. Nourishment and normally also oxygen are taken in, enabling the organism to build up its structures, to maintain the life processes, and to get rid of excreta and excrements. At the same time, energy is usually absorbed in the form of warmth and light or mechanical energy (like sound waves and gravitational force). On the other hand, energy is given off as metabolic warmth and mechanical energy (e.g., by loco-

motion mediated by cilia, limbs, or writhing). In this connection it is of
vital importance that most plants can utilize the energy of sunlight to
transform inorganic compounds into organic ones by photosynthesis,
and these compounds nourish both animals and heterotrophic plants.
The chemosynthesis which occurs in some bacteria (sulfur and iron
bacteria, for instance), by which also organic molecules are synthesized
with the help of an organic process of oxidation, is much less significant,
at least since green plants capable of photosynthesis have predominated
on the earth.

A characteristic feature of the metabolism of all organisms is the
maintenance of a *labile equilibrium*. By the transformation of compounds, and especially by decomposition of adenosine triphosphate
(ATP), the organisms release chemical energy which they need for
all vital processes like cell division, movement (of cilia or muscles), and
neural excitations, and also for other chemical reactions. It follows that
the organism has always to be supplied with energy through its food,
or through sunlight in the case of green plants. But active organisms
never reach a complete chemical equilibrium, for then all activity would
cease. In this sense every living creature may be called an *open system*
which is in a state of *dynamic or flowing equilibrium*—as Von Bertalanffy (1937, 1949) has so well characterized it. The peculiar and
hereditary structure of the organisms and their regulatory mechanisms
ensure that this labile condition is maintained.

It is understandable, therefore, to find that the structures most essential to life processes are not completely static. Especially the cell membrane, cell nucleus, mitochondria, Golgi bodies, and cytocenter all
undergo functional alterations. Von Bertalanffy (1949) has therefore
regarded such dynamic structures as "extended and slow functions."

Many specific functions of active organisms emphasize once again the
unity of all life. Unicellular as well as multicellular organisms, plants,
animals, and humans are all capable of transforming their foodstuff
into substances of their own body, and assimilating them. On the
other hand, they all release the energy required for life processes by
biological oxidation, with the help of breathing enzymes, and partly
also by processes akin to zymosis. All living organisms react to external
stimulations, and plants show processes of origin and conduct of excitations in principle in a manner similar to animals. All normally increase

the number of their cells by mitosis. Only in bacteria and blue algae is the process of cell division simpler, but it follows a similar pattern—first the gene-bearing DNA fibers and then the division of cells. Almost all plants and animals are divided into two sexes, and fertilization takes place in a similar manner. In both groups of organisms the individual's development begins with totipotent cells and ends with symptoms of senility. Processes of regeneration are common to both plants and animals. The laws of heredity, Mendel's laws, operate for all sexually propagating organisms. All organisms undergo mutation, and they show similar abilities of adaptation. Many of them also regulate their processes according to the periodicity of the year and the day by the help of an "internal clock"—a phenomenon which has not yet been fully explained.

Activity and flowing equilibrium in open systems are not characters which clearly distinguish animate from inanimate matter. Springs, rivers, and glaciers are active; they absorb new material which they may then alter and discard (by deposition of sinter, river turbidity, formation and melting of glacial ice). The same holds good for all active machines, and here too an exchange of energy takes place (in a plant powered by water or oil, for example). On the other hand, some organisms can fall into a state in which almost *all activity is at a standstill.* Instances of this include encysted Protozoa, Rotatoria, and Tardigrada, phyllopod eggs, and the spores and seeds of plants. The hard, dessicated seeds of some legumes have remained capable of germination for almost two centuries, and bacterial spores may perhaps do so for thousands of years. Exposure to extreme cold terminates all their activity (anabiosis), but on being slowly warmed they revive again. The above-mentioned multicellular Rotatoria and Tardigrada survived in a dry state for twenty months in liquid air at a temperature of —190 to —200°C and even several hours in liquid helium at —271.8°C (W. Kochs 1890–1895, cited by R. Hesse 1935). So living organisms can exist in an absolutely inactive state without loss of life—a fact that will be of significance when we come to discuss the problem of how life first emerged.

One very important feature of all living organisms has yet to be mentioned. Many of their main processes are not subject to the second law of thermodynamics, the *principle of entropy,* by which order tends to decrease. On the contrary, a specific order in the individual develops and is maintained throughout life, and this remains constant in the succession of generations. In the course of phylogenetic development this order may even lead to higher stages of integration. We shall take no account here of the fact that the principle of entropy is restricted to some extent in inanimate substances also (as in crystallization), because this will be dealt with in Chapter 5 F, in connection with questions of evolution.

VI. Purposiveness. A most salient feature of living organisms is that almost all the structures of plants, animals, and humans are so designed as to function efficiently, with the aim of preserving the unity and activity of the complex system which represents each individual. Moreover, certain structures and processes exist which keep life always "alive" by means of reproduction. In other words, the development of each individual is "purposive"; its aim is to build up this ingeniously constructed unity from a simple ovum, spore, or bud into an adult living organism. This being so, it seems at first almost inconceivable that the origin, development, and functioning of the organisms should be simply a matter of cause and effect. The problems involved will perhaps be most suitably discussed by looking briefly at some examples from among both plants and animals, and considering the extraordinary multiplicity of their harmoniously coordinated structures and activities.

In view of the characteristics we have mentioned in sections I-V, it is clear that the chemical and structural complexity of the organism is such that it can only function properly in an individualized system if all the processes involved are purposefully coordinated and form what K. E. Rothschuh (1959) has so aptly termed "a co-operative action." Even in protists and to a much higher degree in multicellular organisms we can observe the useful principle of *division of labor*. The structure and function of the different parts are always interrelated and adjusted to preserve the whole individual in its active state.

An example may serve to emphasize the complex nature of these interrelations. The human *skeleton* has to develop in conformity with the whole body, and it must be able to function properly. It has to support and carry the body; the bones must provide sufficient surfaces for the insertion of the muscles, and these in turn must correspond to the size and shape of the skeleton and all necessary movements. Now, as the bearing strength of the bones is proportionate to their cross-sectional area, when the body grows larger, this capacity only increases by the square, whereas the volume to be supported increases by the cube. An anatomist would find it extremely difficult if not impossible to work out what stresses the differently structured bones would have to bear and thus what their shape and size should be. Yet nature has solved this task in a masterful fashion, at the same time suiting the developing joints to the muscular system and its functioning.

This functioning of the muscles depends in its turn on an adequate *supply of blood*. Thus there has to be a system of blood vessels branching into such fine capillaries that each muscular fiber receives enough nourishment and oxygen. The scale of these capillaries, too, has to be related to the minuteness of the red corpuscles. The system requires a heart of sufficient volume and power to pump the blood to every muscle and organ in the body and to suck it back again. The blood must circulate fast enough to sustain the intense metabolism in higher animals and man (the approximate five liters in man has to circulate about once every minute). So the finely branching veins must serve the very varied needs of the different parts. The quantity of blood flowing in a certain period (t) through a section (l) is dependent on the volume of liquid (V), the pressure (p), the cross-section of the vessel (r^4), and the degree of viscosity, i.e., the internal friction (K), in accordance with the formula $\dfrac{V}{t} = \dfrac{p \cdot r^4}{l} \cdot K$ ($=$ the Hagen-Poisseuille Law; see H. Davson 1964; E. Schütz 1966). But even with the aid of this formula, no one could work out how an appropriate system should be constructed, allowing for the variation in the dimensions of the vessels and their different elasticity. The calculation is further complicated by the changing requirements of all the organs and structures. Nature, however, has succeeded in creating an extremely efficient and suitable system for animals as well as for the human body.

The blood must also consist of a certain biochemical composition of blood plasma and a suitable combination of red and white corpuscles in sufficient numbers, in order to bring nourishment and oxygen to the tissues and transport decomposition products, hormones, and antibodies. So the tissues in the marrow, lymphatic glands, spleen, and liver have to be so constituted that they function properly. Lastly, several regulatory mechanisms are necessary, so that the circulation alters in proportion to the varying demands of the body as a whole or of individual organs or groups of muscles (see below).

These brief details, covering a very limited area of activity, may be enough to show how complex are the relationships among the structures and functions themselves, and between them and the individual as a whole. More could of course be said on this point, if we were to take into account the histological, physiological, and biochemical adaptation of arterial walls, heart muscles and other muscles, bones, etc.

In addition, almost all organisms possess a number of steering and control mechanisms, operating to suit their varied activities and needs. Some of these are comparatively simple steering structures, like cardiac valves and venous pockets which determine the direction of the blood stream, the nerve fibers along which excitations are conducted, or the micropyles on the ovules of plants, which direct the pollen tube into the embryonic sac and later lead the seedling's root and hypocotelydon out of the seed.

A much wider control over the life processes is exercised by *hormones*, which are extremely effective even in very small quantities. They control cell division and growth in plants (kinetin, heteroauxin), flower formation (florigenes), and in part the special formation of leaf, shoot, or root (e.g., metaplasin). A hormone-like substance also plays a part in the transmission of excitations (E. Bünning 1948 and others). A much larger number of hormones has already been identified in animals, especially in arthropods and vertebrates. They regulate the specific processes of growth (the juvenile hormone in insects, somatropin in vertebrates), and the rate and type of metabolism (thyroxin, parathormone, insulin, glucagon, adrenalin, corticoids, prolactin, etc., in vertebrates). They also control color change in crustaceans, in insects, and in some fishes and amphibians. They determine sexual dimorphism and sexual maturity in vertebrates (estrone, testosterone, etc.) ; the cycle of ovulation (follicle-stimulating hormone) ; pregnancy (luteinizing hormone, progesterone, and chorion gonadotropin) ; and birth of mammals (oxytocin). But hormones are also involved in conduction of excitations (acetylcholine, adrenalin) and they release instinctive reactions (sex hormones producing rut, display, and fighting, and prolactin inducing maternal behavior).

The *nervous system* plays just as large a part in this work of control in multicellular organisms, and it often cooperates with the hormone system by releasing hormones at the appropriate moment. As almost all multicellular creatures (except sponges) are innervated, the nervous system can very suitably operate in adapting individual functions to the needs of the whole and so consolidating the unity of the individual. Even on the level of primitive worms, central nervous systems have developed which direct the movements of organs and the whole body, glandular activity, excitations, reflexes, and instincts. The formation and utilization of memorial traces (engrams, corresponding to mental images in humans) ensure a very finely adjusted control of reaction to complex environmental conditions.

In many of these regulations the fine adjustment is governed by a *feedback system* rather than a simple steering mechanism. A few examples may be sufficient to illustrate this effect. If the *pressure of the blood* pumped by the left ventricle into the aorta rises above the normal, the blood-vessel walls dilate and excite the nerve endings in the walls of the aorta (parts of the tenth cranial nerve, the vagus nerve). These excitations are conducted to the vasomotor center in the medulla oblongata, which causes the peripheral vessels to dilate, so reducing the blood pressure (depressor reflex). If the pressure is below normal, toning impulses from the center of the sympathetic nervous system cause the muscles of the walls to contract, and the pressure rises. The amount of blood received by individual organs depends in large measure on their own activity. If metabolism is speeded up, tension of the smooth muscles of the vessel's walls decreases, the vessels dilate, and their blood content is increased. Adrenalin participates in this regulatory process whereby the blood is directed from the inactive to the active organs.

The *maintenance of a constant body temperature* in warm-blooded animals affords another example. If metabolism in the internal organs and muscular activity fail to produce sufficient warmth, certain centers in the brain stem (in particular in the hypothalamus) are stimulated by temperature-sensitive nerve endings in the skin and also directly through the blood temperature. This causes a reflex action by way of the sympathetic nervous system which increases the tension of the vascular muscles and produces a constriction of the capillaries of the skin, so that less heat escapes. If the temperature drops sharply, the skin muscles are tensed and the hairs of mammals or the feathers of birds rise, thus increasing the warm outer covering (man's "gooseflesh" is a rudimentary form of this). Warmth is also generated by reflexive shivering. If the body temperature rises above normal, however, birds and some mammals react automatically by panting; and ungulates, apes, and humans by sweating. This reduces the blood temperature by evaporation-cold in lungs or skin.

The pupil reaction already mentioned in Chapter 2 C is another example of such a feedback mechanism by which the eye is adapted to differences of light intensity. Others include the depth and rate of respiration, related to the amount of carbon dioxide in the blood, regulation of urine concentration, and instinctive actions which are controlled by sense organs conveying success or its absence to the central nervous system.

The exchange of gases, and transpiration, among plants, is regulated by means of guard cells at the stomata of the leaves. Exposure to light converts starch in the chloroplasts (chlorophyll grains) of these guard cells into sugar, the concentration of the cell fluid rises, and water is taken in by osmosis. As the different walls of these cells have a different thickness, the stomata then open. In darkness, the sap concentration decreases and the stomata close again.

The *embryonic and juvenile development* of the living organisms is also highly purposeful. According to plan, so to speak, the sequence of differentiations leads to the development of the organs and finally to the emergence of the specific structure of the whole organism. Here too we may cite a few striking examples.

In most higher *plants* the first nourishment of the seedling is provided by the cotelydons already prepared in the seed. Positive geotaxis of the root and negative geotaxis of the shoot at once assure that the seedling will receive water and all the materials for photosynthesis and growth. The young plant duly develops vessels for transporting food and water, strengthening tissue, and an adequate leaf system for photosynthesis. Only at a later stage do the organs develop which will warrant the survival of the species—the flowers with all the structures ensuring pollination, and then the fruits and seeds, so constructed that they can be dispersed and can protect the embryo they contain.

In the case of the *human species,* development is ensured by nourishment through the placenta. The amniotic cavity, containing the embryo in its protective fluid, keeps the fragile tissues from being deformed. The *ductus botalli,* linking the pulmonary artery with the main artery of the body, and the open connection of the two ventricles ensure that only part of the blood stream reaches the embryonic lungs which cannot yet function. The size of the child at birth corresponds to the width of the birth canal in the pelvis. At the time of birth, the opening between the ventricles closes, the *ductus botalli* becomes reduced, and normal lung circulation begins. Meanwhile the sense organs, brain, and nerves are so fully developed that the infant is capable of the reflexes necessary to life—clinging, sucking, swallowing, coughing, sneezing, etc. The bones harden only after birth, as the child begins to stir and to move about. All these phases are arranged in a purposive sequence and all the special structures and mechanism of mother and embryo are favorably designed to bring a new individual into being.

Finally, the relations of living organisms to their habitat are also purposefully developed. All plant and animal species are adapted to a specific environment which offers a suitable habitat—fresh or salt water, dry land or air, enough nourishment, adequate protection from inanimate hazards (extremes of climate) or dangers from enemies or parasites, and also the possibility of reproduction.

The structure of the organism is intimately bound up with its special habitat. For instance, higher water plants usually have submerged leaves with a very thin outer tegument to admit gases, water, and salt. They lack stomata, and normally also lack conveying vessels, roots, and a secondary growth of thickness.

In animals, body structure and organic functioning, and especially reflexes and instincts, are all adapted to the world around them. Ungulates inhabiting steppe country have relatively long legs and powerful thigh muscles, few toes, and narrow hard hooves, enabling them to run swiftly. Their young are relatively long-legged already at birth. The adults have teeth suited to chewing the hard grasses and other steppe-land plants. Sight, hearing, and smell, that is to say, all the senses which allow perception of distant objects, are especially well developed. The metabolic system requires little water, and the feces are dry. The gregarious habit and strong instinct of flight are a protection from larger predators. Like almost all organisms, the ungulates form part of a bio-cenosis, a community adjusted to other species, to competitors, and to possible enemies or parasites.

All organisms, then, exhibit not only purposive adjustment of bodily structures and functions, an *entharmony*, but are also adapted to their habitat and to a certain type of life in their biotic community. They exhibit an *epharmony*. Not all structures and processes, however, are harmonious and purposive. As already mentioned in Chapter 3 A IV, their construction is also historically conditioned, and therefore does not represent an optimum. There are many imperfections, unnecessary vestigial organs and functions, as well as what seem to us "evolutionary detours" and exaggerated, "luxurious characters." And we shall of course also have to bear these imperfections in mind when we consider whether a causal explanation can be found for the origin of this pur-posiveness which predominates so generally (Ch. 5).

VII. Psychic Processes. We humans experience perceptions, images, feelings, and volition. As we are a branch of the phylogenetic tree of life, we may assume that at least the higher animals also experience psychical processes of this kind, which correspond to physiological processes in the brain (and possibly also in the sense organs). These represent a further group of characteristics which on biological and epistemological grounds we may consider are possibly shared, at least as prestages, by all living creatures. More will be said about these impor-tant but not definitely determinable questions in Chapter 6.

In this context it may be enough to note that psychical characteristics differ from all other characteristics in that they are normally not equated with physiological processes in the brain but regarded as funda-mentally different (cf., however, Ch. 6 L). The usual opinion is that they run parallel with such processes but without any causal connection between the two. This would mean that psychic qualities were each linked with distinctive patterns of excitation within the system of nerve cells, in accordance with certain "parallel laws" which we should take as datum (Ch. 6 K V). Another feature of psychic processes is their sub-jectivity, which is clearly marked in humans by the connection with our self, but is probably present at a rudimentary stage also in higher ani-mals. Finally, some of these psychic processes within our self seem to be "free," in the sense of being independent of the causal and logical laws generally valid in the universe, even before any humans existed. We shall see, however (Ch. 6 G), that there are grounds for calling this "free will" and also the parallelity into question.

VIII. Summary of the Characteristics of Life. In the foregoing sections I-VII we have endeavored to characterize all the peculiarities which distinguish living organisms from inanimate matter, though we discussed only a few examples in each case. This may have revealed the range of the problems involved. It is clear that a causal explanation of the life process and the emergence of living organisms in the course of the history of our planet has to deal with different and very complicated groups of facts.

As a basis for such discussion (in Ch. 4) it will be useful to summarize the main characteristics and to emphasize those for which there are no parallels in inanimate matter. We shall here consider only genuinely living organisms. Rudimentary forms such as viruses naturally exemplify fewer of such features, though they do possess some highly typical characteristics of life (see Ch. 5).

In regard to its *chemical* basis, life is dependent on the presence of highly complex and manifold carbon compounds, though a relatively small number of classes of such compounds is involved. Proteins and nucleic acids are present and constitute a main criterion of life. Insofar as these compounds are found in an inanimate state in nature, all derive from living organisms or viruses. As giant molecules, proteins have several specific qualities which are important for life processes: colloidal solubility, buffer action, surface activity, facility in changing from a sol to a gel, differences of swelling, and a tendency to form inner and outer surfaces. All living organisms embody special proteins in the form of enzymes which act as catalysts, building up compounds which are characteristic of the species, and directing the decomposition of other compounds. Nucleic acids, which carry the hereditary factors, ensure the constancy of the species and in particular the processes of specific protein synthesis. Spontaneous alterations among them bring about mutation and hence the phylogenetic changeableness which is typical of all organisms. Active organisms are further characterized by their relatively high water content, a necessary means for solution and swelling. Hormones, which regulate metabolism as well as excitation, maturation, and processes of development, are possibly present in all animals, since they occur also in bacteria and protozoa.

Organisms normally remain *individuals* at every stage of their lives. That is to say, they are structural unities showing a temporary constancy and embodying chemical compounds arranged in *hierarchical systems*. One very distinct feature is that all true organisms are composed of *cells* or are unicellular, temporarily (like an ovum) or permanently (like protists). All cells are normally confined by a cell membrane which regulates the intake and output of matter, and they possess one or more nuclei containing nucleic acids, or some equivalent (in bacteria and blue algae), bearing the hereditary units. They have many ribosomes, directing the synthesis of proteins, and mitochondria, where change of energy by cellular respiration takes place. Bacteria show prestages of mitochondria. Systems of internal surfaces in cells and mitochondria ensure separation and intensification of the many biochemical processes.

The *limits to the possible range in size* among active organisms between elementary particles and celestial bodies are mainly determined by physical conditions. The smallest living organisms are still large enough to ensure that thermal agitation of the molecules and corresponding disorder cannot disturb their functions. The upper limit of size is determined by the mechanical properties of organs of support, like bones, shells, and wood.

Disregarding some resting stages which can resist extreme cold for a long time (anabiosis), all organisms are characterized by *metabolism and change of energy*—that is to say, a typical *activity*. They are open systems in a condition of dynamic equilibrium. An important point is that despite the principle of entropy the individual is preserved, both as such and as a link in the chain of successive generations, because order in each individual is transmitted from order of the generation before.

It is an especially characteristic feature of all living organisms that their structures and functions are predominantly *purposive*. The individual development, too, is purposive. This *entharmony* of every species corresponds to the *epharmony*, i.e., to its normally efficient adaptation to the conditions of its habitat and its biocenosis.

Organisms are also characterized by their ability to *reproduce* by means of *totipotent* cells. This means that all present life is bound by an unbroken chain of such cells (normally in the sense of a germ tract) to that of former generations. Unlike inanimate substances, each living organism has a *rhythmically repetitive life cycle* from the single cell (spore, ovum, or daughter cell) or bud, to the adult and then the aging individual.

One decisive point is the unity of all life. Living organisms can only spring from other living organisms, and it is very probable that all belong to a single *stream of life*—in other words, that all are branches of the same phylogenetic tree. It is characteristic that all living species undergo gradual alterations through *mutation* and natural selection. This means that their characteristics are adapted to the conditions of their life, and are also *historically* conditioned. Another important fact is their capacity for *progressive evolution,* which has also led to the emergence of man.

Lastly, humans and higher animals at least are distinguished by some characteristics which differ fundamentally from the rest: *perceptions* and *mental images,* corresponding to physiological processes.

After enumerating all these characteristics it might be desirable to attempt a *definition of life.* But this will not be possible in a single phrase. Leaving aside occasional exceptions and borderline cases like viruses, we can characterize living organisms in contrast to inanimate matter in the following terms.

Living beings are hierarchically organized, open systems of predominantly organic compounds. They usually constitute clearly delineated cellular individuals showing a temporary constancy. Their cells are morphologically characterized by specifically functioning organelles (cell membranes, nucleus, chromosomes, ribosomes, mitochondria, or their prestages). With regard to their chemical constitution, they are characterized by specific proteins and nucleic acids. Metabolism and exchange of energy give rise to activity and maintain the organism in a state of dynamic equilibrium, determined by complex structural and functional interrelationships and controlled by particular steering and feedback systems. They show specific reactions to external stimuli (light, gravity, currents, contact, chemical matter, etc.). All their structures and processes are mainly purposive, serving a rational functioning of the organs and the maintenance of the individual and the species, but historically conditioned by the structure of the organism's phylogenetic ancestors. Reproduction through totipotent cells is linked with changes of form in the course of the individual's life. Organisms undergo phylogenetic alteration through mutation of the hereditary factors. They are links in the continuous chain of cells that constitute the stream of life to which probably every species of organism ultimately belongs. Progressive development in many lines of descent made the emergence of complicated psychological processes possible.

As a number of these characteristic features are limited to living organisms alone, even one or two would be enough to distinguish them from inanimate matter. Some definitions of this sort are often aptly phrased, as the following examples show.

Life is "the continuous adjustment of internal relations to external relations" (H. Spencer 1898).

"In contrast to all inanimate substances, all organisms are characterised by certain highly complex chemical compounds and in particular proteins" (M. Verworn 1915, p. 151).

"Life is not a property of a homogenous matter; it is the achievement of a system" (H. J. Jordan 1929, p. 333).

"A living organism is a gradation of open systems whose functioning sustains it while its constituent parts undergo changes" (L. von Bertalanffy 1949, p. 124).

"Life seems to be orderly and lawful behaviour of matter, not based exclusively on its tendency to go over from order to disorder, but based partly on existing order that is kept up" (E. Schrödinger 1944, p. 73).

Naturally, all the authors quoted here have also referred to other features that distinguish life. But if we are to discuss the important scientific problem, whether or to what degree life processes and their emergence can be causally explained, we must base our considerations on a really comprehensive general definition.

B. Plants, Animals, Humans

In characterizing organisms, we have already mentioned the *unity of all life*, and how plants, animals, and humans are all phylogenetically related. We stated many important characteristics common to all true living organisms: these include complex organic compounds like proteins, nucleic acids, and lipids, which combine with water to form protoplasm; cellular structures with typical organelles like nucleus, ribosomes, mitochondria; processes of mitosis, assimilation, cell respiration, change of energy, regulation by hormones, excitation, sexuality, reproduction, change of form in the sense of rhythmical life cycles, capability of regeneration; the same basic structure of genetical substance (DNA); the same genetical rules; the same types of mutation; adaptation to the habitat, and so on. In order to understand how three quite heterogeneous categories—plants, animals, and humans—could originate in spite of these conformities, we must briefly discuss the *differences* as well.

It is impossible to draw a sharp line between *plants* and *animals*, for unicellular groups exist in which some species are more plantlike and others more animal-like. But the multicellular chlorophyll-bearing plants fall clearly into place, because of their method of nutrition and the structural and functional peculiarities involved. These plants are *autotrophic,* that is, they are capable of building up organic compounds, at first glucose, from carbon dioxide in the air and water, with the help of enzymes. The necessary energy for these processes is provided by sunlight which they absorb through pigments, chlorophylls a and b, xanthophyll, and carotene, and convert it into chemical energy. This *photosynthesis* is replaced in several bacterial groups (sulfur bacteria, for instance) by chemosynthesis, using energy from inorganic oxidation processes to synthesize organic compounds. Nitrogen compounds provide the plants with the means of building up specific amino acids and proteins. These compounds are absorbed by most higher plants in inorganic form through their roots and by heterotrophic higher plants in organic forms.

Animals, on the other hand, are *heterotrophic* and cannot synthesize organic compounds from inorganic ones. They feed on organic compounds which are rich in energy, mainly carbohydrates, proteins, and fats, all of them built up by plants or other animal species. Hence solar radiation remains the chief though indirect source of the energy needed for the animal's intensive metabolism.

However, there are also parasitic or semiparasitic flowering plants living wholly or partly heterotrophically—for example, dodder (*Cuscuta*), toothwort (*Lathraea*), *Orobanche*, some orchids (*Neottia*), and mistletoe (*Viscum*). Other heterotrophic plants include the saprophytic

or parasitic higher fungi (Eumycetes), the Myxomycetes, the Phyco-
mycetes, and the great majority of bacteria. Their relatively firm cell
walls and manner of reproduction place these groups more or less among
the plants. The myxomycetes, like amoebas and many flagellates, live
chiefly heterotrophically throughout the main stages of their life cycle,
and might therefore also be termed Mycetozoa and rank as animals.
Bacteria, too, may be reckoned as a basic group of either plants or
animals. Most flagellates are autotrophic, but other species closely allied
to them are heterotrophic. *Euglena,* for example, has chlorophyll,
whereas *Astasia,* another of the Euglenoidea, has none. If some chlorophyll-bearing plantlike species of *Euglena* are kept in darkness, within
several generations they produce purely heterotrophic animal-like forms.
Thus nutrition is not an unambiguous criterion for distinguishing plants
from animals, nor is there any other characteristic which makes this
distinction possible. The flagellates and the myxomycetes belong in fact
to both categories.

Multicellular plants, however, differ from animals in other respects
also. As their nourishment, deriving largely from air (CO_2) and soil
(water, mineral salts), is available in many places, they can remain
fixed to one spot. Animals, on the other hand, normally have to seek
their nourishment, which comes from plants or other animals—all the
more since they are often adapted to a special kind of food by their type
of mouth, intestines, enzymic system, physiological needs, and specific
instincts. Thus they have to be mobile, and must have more elaborate
and responsive sense organs and faster reactions than plants. Exceptions
here include the sessile aquatic animals like sponges, hydrozoa, corals,
and bryozoa, which swirl their food toward themselves, and some parasites which live in their hosts' alimentary systems, as tapeworms do in
the intestines, or in nutritive tissue. The females of the cirriped *Saccu-
lina,* for instance, have rootlike processes ramifying into all parts of
their host's body.

Because it was advantageous to take up large quantities of carbon
dioxide from the air and to absorb as much solar energy as possible,
multicellular land plants developed large surfaces by a system of
branches and flat leaves. They also developed a similarly branching root
system, in order to take up enough water and mineral salts, and to
anchor themselves firmly in the soil.

Animals, on the other hand, do not need such an enlargement of outer surfaces. The necessity to move about caused them to develop a compact body shape. And as their organic nourishment needs a certain time to be broken up by enzymes, they require an inner cavity. This usually consists of an intestinal canal with a stomach or only a simpler sac, as in coelenterates, turbellarians, and trematodes.

The tissues of branching plants which are rooted in the soil are supported by firm cell walls. Any excitations which are conducted pass comparatively slowly. The cells of most animal tissues, on the other hand, have only a very thin cell membrane. This allows them to take up food molecules, water, and oxygen more easily, and ensures the intensive metabolism necessary to free-moving organisms. Firmer cell walls would also prevent the tissues from making the necessary rapid response to stimuli and excitations. So animals alone were structurally capable of acquiring nerves and a central nervous system, allowing the development of reflexes, instincts, memory, and actions of choice.

Because plants are rooted in the soil and branching, they have a terminal growth which often continues until death. In larger plants therefore, especially in shrubs and trees, many parts of tissue remain totipotent, so that new shoots, leaves, and flowers or sporangia may originate. It is remarkable that some sessile animals such as corals and some hydrozoa behave in a similar manner, though here terminal growth applies to the colony and not to the individual, and it is the tissue uniting the individuals which is totipotent. In free-moving animals, on the other hand, with more complex organs, growth has to take place at many parts of the body, and it ceases after a time and is succeeded by a fairly long and more constant period of adulthood. Continuous growth would have seriously interfered with the way in which the animal's method of movement adapted itself phylogenetically to its environment.

Hence we see that in spite of the many characteristics already mentioned as shared by both groups, green plants differ from animals in several decisive characters. As the autotrophic method of nutrition is the prime cause of these differences, it is clear that despite their common phylogenetic source, plants and animals were bound to develop in fundamentally different directions in regard to shape and function. We shall be returning to this point in the section on evolution (Ch. 5 D).

Man, however, owes the direction of his phylogenetic development toward his unique and dominant place as the most highly integrated living organism to quite a different kind of specialization. The morphological and physiological differences between him and the anthropoid apes are more or less quantitative ones. Humans, chimpanzees, and gorillas have the same number and kind of bones, their organs show the same arrangement and histological structure, and they function in the same ways. Man differs from these animals only in the relative size and particular shape of certain parts of the body and, in some cases, in their relative positions. This makes it difficult to draw a firm line of distinction, on skeletal evidence alone, between the oldest human types and their apelike ancestors.

The Australopithecines, known to us from a large number of skeletal fragments from South and East Africa, walked erect, and the pelvis and limbs were shaped like those of man. Some features of the skull, especially the type of teeth, were also human. But in addition they had markedly apelike characteristics—a sloping forehead, a snoutlike prognathous jaw, and especially a relatively small cranium, the cavity varying from 400 to 600 cm³, hardly larger than that of the present-day anthropoid apes. The frontal part of the forebrain, which plays so important a role in advanced mental processes, was not much more developed than that of a chimpanzee. It is worth noting that some authorities did not at first class this early type with the Hominidae. But more recent research has shown that the Australopithecines—or at least *"Homo"* (= *Australopithecus?*) *habilis*—were capable of making very primitive pebble tools. However, their way of life seems to have differed only little from that of the animals.

So we must disregard these intermediate forms of some 1.0 to 2.6 million years ago if we want to state the characteristics which distinguish man from the animals The early human types of the level *Homo erectus* (= *Pithecanthropus*) also had prognathous jaws and a smaller cranial capacity (about 800 to 1200 cm³) than that of present-day man (1000-1800 cm³). Judging from their still very primitive stone tools, their mental powers bore little resemblance to what we think of as typically human. Only *Homo sapiens*, who dates from some 70,000 to 100,000 years ago, developed those qualities which place man above the animals in so radical a fashion.

The characteristics which distinguish the whole family of Hominidae from the higher animals and from extant and extinct anthropoid apes are few in number—erect stance, particular structure of pelvis and legs, typical dentition with poorly developed canines, some minor skeletal features, and also the ability to make tools. But gorillas and chimpanzees also walk erect at times, and anthropoid apes not only use tools but can make rudimentary ones (by fastening sticks end to end, piling up boxes, and stripping off the leaves and twigs from branches). It is only when we compare present-day man, *Homo sapiens*, with the anthropoid apes, that the essential differences become apparent.

The main morphological and anatomical characteristics of *Homo sapiens* are the greatly reduced body hair, relatively long legs with well-developed calf muscles, the wide scoop-shaped pelvis, the upright posture leaving the hands free for carrying and tool-making, and the flattened thorax. The head is better supported because its center of gravity lies above the body's center of gravity, thus allowing a relatively larger skull to develop. The jaw protrudes only slightly, and there are no large canines. The brain is both relatively and absolutely large, with more highly developed frontal and temporal lobes. *Homo sapiens* alone is distributed over all continents, but is divided into races which have interbred to a remarkable degree, though they differ greatly. Typical developmental features include a long embryonic period, a relatively high birth weight, a prolonged juvenile phase, the late onset of maturity, and a comparatively extended life span.

The enlargement of the associative regions in the forebrain made it possible for a motor area for speech to develop in the temporal lobes of the forebrain, and a new center at the base of the frontal part is important for associative processes. These special features and the larger number of ganglion cells in many regions of the forebrain greatly increased the capacity for memory and ability for actions of choice as a result of experience. In consequence of these alterations, instinctive behavior became less important for ensuring the life processes.

The increasing development of speech allowed the formation of more abstract concepts, logical and causal thinking, and planned actions. Communication with others made it possible to build up a body of tradition, the basis of rapid material and cultural advance. But it is only in the last ten to twenty thousand years that man has grown autonomous, learning to rule the world and to alter it by means of agriculture, animal husbandry, the construction of houses and roads, villages and cities. The gap between man and the animals became increasingly widened by the development of ethical and religious codes and ideals, science and arts, the increasing prevalence of a spiritual life, and the extension of thought into the future and the past. At present the phylogenetic connection between the two groups may appear scarcely credible and even "degrading."

So the most important characteristics which distinguish man from other animals are not found among all hominids; they belong to *Homo sapiens* alone, and particularly to late stages of more advanced cultures. This raises the significant question whether and to what extent the successive development of these decisive differences can be causally explained. (On man's evolution, see more detailed treatment by W. Gieseler 1957; G. Heberer 1952, 1956, 1959b; W. E. Le Gros Clark 1954; G. R. H. von Koenigswald 1953; B. Rensch 1965a; and others.)

Chapter Four

Causal Analysis of Life Processes

A. The Problem

If we are to arrive at a view of life based on natural philosophy, we must first establish whether or how far its many specific processes can be causally explained, and to what extent they can be traced back to chemical and physical processes—to processes, in other words, which are also a feature of inanimate matter. We shall defer consideration of the emergence and phylogenetic development of organic systems to Chapter 5.

Aristotle, whose *Historia Animalium* contains a wealth of acute biological observations, had already occasionally established causal relations. He recognized, for example, that fishes' eggs fail to develop if unfertilized; that several human defects such as blindness, deformation of the legs, and birthmarks are inherited; and that as a result of castration in youth the voice does not break, body hair does not appear, and baldness does not occur. More and more causal connections of this kind were observed as science and medicine advanced. But it was not until the seventeenth century that anatomy and physiology reached a stage when more far-reaching generalizations could be made. Inspired by Harvey's discovery of the circulation of the blood (1628), Descartes declared that an animal's body functions like a machine, in accordance with physical laws and in response to unconscious reflexes. This view was expanded by G. A. Borelli, S. Santorio, and other "Iatro-chemists" who made a study of movements, metabolic phenomena, and other physiological processes in animals. Descartes had also conceived the human body as a machine, but a machine guided by a soul capable of reasoning. La Mettrie went still further. In his famous *L'homme machine* (1748), a work largely misinterpreted and reviled, he stressed the dependence of psychic phenomena on physiological processes, and declared (1909, p. 66) : " Let us boldly conclude that man is a machine, and that the whole universe contains only a single substance varied in a number ways."

All these admirably consistent hypotheses were much in advance of their day. Even at the beginning of the present century, biological research had not reached a stage when a more general causal interpretation of life processes could have been made. Change of energy by cell respiration, the processes of protein synthesis, the regulation of life processes, and the transmission and realization of hereditary characters must have appeared marvelous events to biologists of that time, when respiratory enzymes, regulatory hormones, and the character and mode of action of nucleic acids in transmitting hereditary information were all as yet undiscovered.

And what is the position now? Are we at this date in possession of a satisfactory causal understanding of metabolism, of excitations in the sense organs and nervous system? Do we know enough about reproduction and sexuality, hereditary processes, and individual development to offer a causal and in part chemical and physical interpretation of the phenomena of life—an explanation which shall at least seem plausible? With the great and accelerating progress in biology, it appears that we can. The causal method in research has stood every test without exception, and we may safely assume that this will continue to be true. All kinds of vitalistic interpretations seem less and less tenable, and so does an "unknown form of determination," independent of normal causality, postulated by N. Hartmann (1950, p. 689).

Again and again, seemingly insoluble puzzles have yielded to the use of newly devised apparatus and methods; and this gives grounds for hope that in the foreseeable future we shall have unraveled all the problems of life processes. Invention and improvement of microscopes and microtomes, histological techniques, and new methods of elective staining have rendered research into cell and tissue structure possible, and so laid the basis of modern biology. Techniques of transplantation and tissue culture in artificial media have resulted in crucial discoveries in developmental physiology. With the aid of micromanipulators we can now make very minute operations on cells and tissues. The analysis of metabolic processes has made dramatic progress, following the introduction of radioactive tracers in the organism and the development of spectroscopes, the ultracentrifuge, and infrared spectroscopy. It had long seemed impossible to analyze proteins, until they could be examined by paper chromatography, electrophoresis, and mass spectrography. Cathode ray oscillography has enabled us to make a detailed study of excitation processes in sense organs, nerves, and brain. Even rapid movements can now be studied by films using the slow-motion technique. Many other appliances and methods could be mentioned which have opened up new fields of research and led to astonishing new scientific knowledge .

In the ensuing sections we shall show how far the analysis of life processes has been able to proceed. But we can only give a relatively cursory survey and illustrate with a few typical examples. Any comprehensive treatment of the present state of our knowledge in this field would require several volumes. For the reader not specially versed in biology, the examples may, however, indicate the chief methods, the validity of causal analysis as a heuristic principle, and to some extent also the present limits of our knowledge.

B. Analysis of Metabolism and Change of Energy

In certain biological processes analysis has now extended to the microphysical sphere. This has already prompted discussion of the concept of causality with reference to some phenomena of life. We may leave this question aside here; it will be treated in greater detail in Chapter 6 K III. We are now considering cause and effect in the sense of a universal causal law acting without any interruption, which has been found valid for every macromolecular process in physics, chemistry, and biology.

The first general fact to establish is that the structures and functions of the living organism obey *physical laws.* In consequence of gravity the body presses onto the substratum, one organ presses upon another, and organs draw down from structures from which they are hanging (fruit from trees, for instance, or freely suspended forelegs from the trunk). The thrust needed for running, jumping, climbing, swimming, digging, and flying obeys the laws of mechanics. The specific structure of joints determines the specific ability of moving the limbs in accordance with the same laws. Flight is governed by the laws of aerodynamics. The production of animal sounds is determined by the laws of vibration. The processes in the semicircular canals of the labyrinth of vertebrates—stimulating sense cells which inform the brain—are based on the principles of inertia (the endolymph at first lags behind the movement). The blood moves in accordance with the laws of hydrodynamics. The tissues take up oxygen by diffusion, water is passed from one cell to another by osmosis, transpiration in the plant takes place by cohesion of the water molecules and evaporation at the stomata. Perspiration cools by the thermal principles of evaporation. The body conducts and radiates heat according to physical laws. Optical laws govern the refraction of light rays in lenses and other dioptric structures of the eyes of animals. One could easily multiply these examples of processes, all controlled by physical laws; but these few may suffice to show that bodily activities are subject to physical laws in the same way as are equivalent processes in inanimate nature.

The same is true of *chemical laws*. The laws of affinity govern chemical reactions in organism and test tube alike. In Chapter 3 A I we have already noted several biochemical processes of particular importance for the phenomena of life, including colloidal solution, van der Waal's association, surface tension, the action of surfaces of contact, buffer action, and catalysis by enzymes; but none of them is confined to living organisms. The efficiency of an acid or a base is determined by the constant of dissociation in the same way in a plant or an animal as in inanimate matter. The law of conservation of energy holds good for all living substances as well. Chemical energy is transformed into work or heat, or is used to synthesize other chemical compounds. The mechanical equivalent of heat holds good, which allows us to express work (i.e., power which accelerates mass) in units of heat, in calories (1 erg $= 2.389$ g $\cdot 10^{-8}$ cal.) * Living organisms are also subject to the second law of thermodynamics, the law of entropy, whereby disorder in matter normally tends to increase. Hitherto compact well-ordered molecular groups in a crystal split up in disorder in a solution. But the law of entropy applies less universally here than in inanimate nature, because organisms represent highly complex systems of a peculiar kind,

* 1 erg $= 1$ dyn $\cdot 1$ cm; 1 dyn $=$ the force required to accelerate 1 g by 1 gal; 1 gal $= 1$ cm/sec^2; 1 cal $=$ the amount of heat required to raise 1 g water from 14.5° C to 15.5° C.

in which order is continually determined by order. This important fact will be further dealt with in Chapter 5 F.

Causal analysis of metabolism and the exchange of energy is made more difficult by the fact that all the processes take place in a highly complex system in which many components are constantly changing. So it is essential to begin by studying certain processes as far as possible independently of the whole system—for example, in separate tissues or by homogenized cell material—and only then to investigate the relations to other processes within the organism. On the other hand, one can sometimes make an assessment from some general processes—for example, by establishing the basic metabolic rate, which indicates the intensity of the whole metabolism of the body. The combination of numerous physiological and biochemical methods has already permitted a far-reaching causal understanding of plant, animal, and human metabolism.

It is possible to mention only a few of the more important recent advances in the causal analysis of life processes. The biochemical basis of *photosynthesis*, which is vital for the emergence and maintenance of all life, has been uncovered in its chief features. A series of complicated reactions is initiated by the absorption of photons by chlorophyll molecules. Thus an electron is raised to a higher level of energy and is transferred to a cyclic redox system, causing the formation of adenosinetriphosphoric acid (ATP), which is rich in energy. The effect of light is also necessary for photochemical decomposition of water (photolysis). Then a series of complicated reactions runs off, only understood after experimental introduction of radioactive carbon dioxide ($^{14}CO_2$) by which carbon dioxide is converted into a carbohydrate (a sugar). (See C. L. Prosser and F. A. Brown 1950; H. Davson 1964; P. Karlson 1966).

Another important discovery has been that of the presence, action, and chemical composition of *vitamins*. These are made up of very different classes of compounds, and are mainly necessary to build up vital enzymes, particularly those needed in cell respiration, visual purple, and mucopolysaccharides in the connective tissue; and in small quantities they also participate in other reactions. Plants and lower animals can build up their own vitamins. In the course of phylogenetic development, higher animals and man have lost this faculty, probably because they derive a sufficiency of vitamins or pro-vitamins from the plants and animals they eat.

The isolation and chemical characterization of *hormones* have been no less important. Hormones are extremely effective even in very small quantities. They are not taken up with food but are built up within the plant or animal; and they are mostly transported to the appropriate organs by the body fluids and particularly the blood stream. Numerous experiments in the past few decades have shown that they act as sensitive regulators throughout the life of the organism. Thus thyroxin, the hormone secreted by the thyroid gland, steers the entire metabolism, and adrenalin, produced by the adrenal medulla, converts the stored glycogen into blood sugar. Insulin, a product of the pancreas, reduces the blood-sugar level by causing more glucose to pass through the cell membranes. The parathormone in the parathyroid glands controls the blood calcium and phosphorus levels, and the formation and metabolism of the bones. The sex hormones secreted by the gonads determine male and female characteristics, as well as reproductive and aggressive instincts. A follicle-stimulating hormone of the pituitary (FSH) regulates ovulation and, once fertilization has occurred, a luteotropic hormone causes the growth of the corpus luteum in the ovary. The corpus luteum then produces another hormone, progesterone, which in mammals prepares the mucous membrane of the uterus to receive the embryo.

In addition to their physiological effects, many hormones of higher animals and humans arouse certain instincts and govern psychological behavior, which of course we can establish with certainty only in man. The luteotropic hormone, also called prolactin because in addition it stimulates the mammary glands, releases maternal behavior in higher animals and humans. Progesterone in mice fosters the instinct to build large nests for breeding. Testosterone, a male sex hormone, induces "manly" and also aggressive behavior. The intensity of thyroxin production influences the temper, and excessive adrenalin increases excitability.

Hormones also control active color changes in arthropods and in fishes and amphibians, and molting in arthropods, as well as pupation and development of adult insects (imagines). They direct nervous processes, intestinal peristalsis, the movements of the nephridia in lower animals, and the growth and flowering of plants. It is only fairly recent research into hormones that has brought us to a causal understanding of the delicately balanced phenomena of life, and of how the various phases that lead up to maturity are regulated.

We have already mentioned in Chapter 3 A I that hormones are composed of very different classes of compounds. The chemical constitution of many of them has now been established. Insofar as they are proteins, this has been possible in some but not all cases. Very recently the structure of insulin was established, and this important hormone has even been produced synthetically—a notable achievement, for this giant molecule is made up of fifty-one amino acids composed of two chains linked by sulfur bonds (see H. Zahn 1966).

The last few years have also brought us nearer to a definite solution of the central problem—how the living organism acquires the energy it needs for its varied activities. It has long been recognized that the decomposition of foodstuffs in the active cells of animals and heterotrophic plants normally takes place with the help of oxygen, and that energy is acquired and stored by these processes which can be used for contraction of muscles, movements of cilia, production and conduction of excitations, or for further biochemical reactions. For a long time, however, it was not known in which cell structures this "cell-respiration," this "biological oxidation," occurs and how the reactions run off. Research has made great strides since 1944, when with the help of ultracentrifuges (which can rotate at over 60,000 per minute) it became possible to separate the cell organelles in homogenized cell material and to examine the different sediments obtained.

We now know that the *mitochondria*, discovered as early as 1886, are present in fairly large numbers in every active cell of plants, animals, and man, and that they represent the "powerhouses" in which this vital exchange of energy takes place. Electromicroscopical investigations have revealed that these tiny ellipsoid filaments or globules (only some 0.5 to 10 μ in length) have double walls, with cristae or tubules on the inner membrane. Studies of separate parts of fractionized mitochondria have been carried out, and it now appears probable that the material which will be used for cell respiration, phosphorus compounds, and molecular oxygen penetrates the outer lipoid membrane, and reaction sets up in the cristae or tubules by a regular arrangement of enzyme systems. Each mitochondrion contains a series of several thousand checkered arrangements of twenty to thirty enzymes. The most important enzymes belong to the citric acid cycle and to the chain of respiratory enzymes. As the materials cannot become directly oxidized, and as a rapid decomposition into carbon dioxide and water would release too much heat, only those chains of reaction could be phylogenetically developed by which the whole process became divided into several steps, so that the build-up of excess heat was avoided. The reactions take place with the help of the yellow respiratory enzyme which passes hydrogen from the substratum and an electron transport chain of the cytochromes (biological oxidation). This process is coupled with the formation of the highly active adenosinetriphosphoric acid (ATP), providing energy for many reactions within the mitochondria and mainly after it has been set free in the cell plasma itself (see, for example, H. Davson 1964; M. Klingenberg 1963; W. Vogell 1963; E. Grundmann 1964; E. Lehnartz 1959; P. Karlson 1966).

These briefly noted examples demonstrate that it has been possible to make a causal analysis of important metabolic processes which characterize life. At the opening of the present century these processes were still a "marvel," and the existence of a "vital force" was an obvious presumption. The building up of organic compounds from inorganic substances by photosynthesis, the conversion of organic foodstuff by enzymes and assimilation, the decomposition and the exchange of energy, and their utilization for life processes are more or less completely recognized today as links in a chain of cause and effect. We now know how hormones control the changing requirements of metabolism; the periods of growth and maturity and the reproductive cycles are also controlled by hormones. It has also been possible to analyze many other complicated phenomena, such as muscle contraction, the formation of blood, the formation and excretion of urea, uric acid and related compounds, the origin of plant and animal pigmentation. (For more detailed treatment, see C. L. Prosser and F. A. Brown 1950; E. Lehnartz 1959; H. Davson 1964; P. Karlson 1966; and the more specialized books on biochemistry.)

Many questions still need to be answered, partly because the very heterogeneous lower and higher plant and animal species have developed such different structures and functions. However, there is no longer any doubt that metabolism and exchange of energy are causally conditioned chemical and physical processes. This conviction has proved to be a valuable heuristic principle at each stage of investigation. What all these analyses do *not* yet explain is how these living physicochemical systems could originate and become so constructed that the processes follow one another in a purposeful manner and combine to function efficiently and in so finely adjusted a manner that the individual system and the species survive, and the continuation of life is ensured. This crucial problem can only be considered in connection with the problems of evolution (Ch. 5).

C. Analysis of Nervous Processes

Studies in neural physiology, which we shall now consider briefly, have established a succession of fairly exact statements of lawful causal processes. In 1793 Galvani first demonstrated the origin of electricity in muscular activity. In 1840-1843 Dubois-Reymond and Manteucci, working independently, succeeded in measuring the resting potential of nerves and muscles. Soon after this (in 1850) Helmholtz determined the speed of the conduction of excitation along the nerves. In 1902 J. Bernstein developed the membrane theory, and in 1949 A. L. Hodgkin and B. Katz showed that excitations come about by movements of ions. All these findings form the basis for the present theories of the *origin and conduction of excitations*.

The basic discovery was that nervous excitations arise on the membrane of the nerve cells, where sodium ions predominate on the outside and potassium ions on the inner side. This distribution causes a resting potential which is kept constant by oxidative processes (phosphorylations in the mitochondria) preserving a dynamic equilibrium. Mechanical, electrical, and chemical stimuli produce an abrupt change in the permeability of the membrane, which then permits the flowing-in of about five hundred times the number of sodium ions as in the resting stages. The membrane thus becomes depolarized and electrically negative. When the electric field reaches a certain threshold, the excitation passes to the neighboring parts of the membrane, and a rapid exchange of ions occurs along its entire length. (The speed remains constant in well-insulated myelinated nerve fibers.) In peripheral nerves, the sudden change of permeability is caused by acetylcholin, which also transmits the excitation from the endings of the nerve fibers to a neighboring nerve cell or muscle fiber, thus altering the permeability of its membrane as well. After the excitation, the flowing-out of sodium ions is then initiated by the help of energy from phosphoric compounds, and the resting potential is restored. (Specific details of this process are still debatable.)

We shall not go into the many particular problems of nervous phenomena—problems concerning the significance of calcium and chloride ions, carbonic acid, the specific activating substances, aneurin (Vitamin B 1) and inhibiting substances, or the special features of saltatory conduction and the origin of excitations in the different types of sense cells. In the context of our philosophical study it may be enough to state that the investigations of nervous processes have proceeded to the analysis of the physical and biochemical phenomena involved. Here too, the laws of chemical affinity, the physicochemical laws of the origin of electrical potentials, the extension of the electric fields, electric currents, resistance, and so on are valid (see, for instance, A. von Muralt 1945, 1958; J. C. Eccles 1957; H. Lullies 1959; H. C. Lüttgau 1960, 1963). The same holds good for the sense organs (see Ch. 4). We do not yet know the exact details of frequency modulation by which, as they pass into the nerve fibers, the generator potentials in the sense organs are transformed into potentials of nerves; but it is clear that the processes of excitation are all causal in character and that at every single stage they obey the law of causality, presuming that this law also governs microphysical events (see Ch. 6 K III). Any questions not yet adequately solved are unlikely to yield different conclusions in this respect.

The analysis of the more complex nervous processes, in particular those in the brain, has led to a great advance by numerous experiments including extirpation of brain parts as well as biophysical and biochemical studies. For many animal *reflexes* and almost all reflex processes in man, we already know the releasing factors, the course of the excitations to the reacting muscles or glands, and the possible inhibitions. Detailed studies of *conditioned reflexes*, in which the reaction can be released by an inadequate stimulus, have grealty helped to explain associative learning. Nervous processes in sense organs have been elucidated through using microelectrodes and cathode ray oscillographs by which the potential oscillations of excitations can be enlarged and recorded. In such a manner the reaction of the eye to light rays could be analyzed. Using corresponding methods, electroencephalograms have been plotted, to show the activity of the various parts of the brain. Injections of radioactively tagged substances have allowed examination of the metabolism of the brain and the paths of different metabolic elements.

The study of *instincts* has also made considerable progress. These processes too had seemed inexplicable marvels a few decades ago. Without having learned it or thinking about it, an animal is capable of carrying out at the right moment a succession of complicated actions which are advantageous for the individual and for the survival of the species. A garden spider goes through an identical series of separate actions to weave its typical web, and then reacts in a constant manner to the vibrations caused by an insect caught in it—running up to its prey, biting it, spinning round it, bearing it off to its den and devouring it (= instincts of food-getting for purposes of self-preservation). When a male wren has marked out its breeding area, it builds several of its typical nests with a side opening. When a female arrives, attracted by its song, it chooses one of the nests and lines the interior. The two then mate, and the eggs are laid in the nest (= instincts related to the survival of the species).

All these instinctive actions which tend to preserve the animal's course of life have been examined by means of numerous observations and experiments, in particular by using successively simpler dummies to release the instinctive processes. Other methods include electrophysiological investigations, the removal of portions of the brain, and also drugs and hormone treatment. This has already led to a causal understanding of some phases of these processes.

The releasing mechanisms of the instinctive behavior are almost always found to be comparatively simple signal stimuli (a certain coloration, a succession of sounds, smells, movements, etc.), to which some innate nervous structure responds by reflex action steered by stimuli of the environment (K. Lorenz 1943, 1957; N. Tinbergen 1951; F. A. Beach 1962; D. S. Lehrman 1961; G. Tembrock 1964; and others). The "mood" that leads to reaction to a particular signal stimulus is usually determined by metabolic conditions. The animal when hungry responds to a nutrition stimulus; if its sex hormones are active, it reacts to a releaser provided by its mate; if cold, it seeks shelter and sunshine. Normally the mood causes appetitive behavior; a hungry animal moves about in an undirected manner, and this will give it a chance to meet a signal stimulus (the scent of a flower, in the case of a bee), which will release the reflexive instinct response of settling, stretching out the proboscis, and sucking the nectar. In this way an animal, without "knowing" why it acts, is automatically led to satisfy its requirements.

When microelectrodes were introduced into different parts of the brain stem of cats or fowl, and in particular into the hypothalamus, different instinctive behavior could be artificially induced by electric stimuli. In this way it was possible to localize the structures involved to some degree (see W. R. Hess 1954, E. von Holst 1957). It was also possible to localize the centers responsible for the different mating calls and "song" of locusts and crickets (F. Huber 1965).

Thousands of experiments have been conducted to examine the *learning process* in multicellular lower and higher animals, and to determine how far this depends on the intensity and complication of the stimuli received by the different sense organs. In higher animals, capacity and length of memory were found to be largely independent of any particular brain structure. Although the brain of an octopus, a bee, a fish, and a mammal are quite different in structure, all these animals can learn similar optical tasks and retain a number of them for some time. The ability to *abstract*, to *generalize*, and to act "as if" they had formed abstract averbal concepts (not connected with words) is more developed in birds and mammals than had been supposed. Apes and monkeys can also act according to *plan and with foresight* (Ch. 6 F). For the moment we are concerned with physiological aspects and may ignore any corresponding psychological processes which we may assume in animals (Ch. 6 C).

However, we do not yet know very much about the physiological basis of these higher accomplishments of the brain. We do know which parts of the mammalian brain and which regions of the forebrain are most involved in the learning process, though we are no longer so confident of a more precise localization as formerly. It is more probable that learning always involves physiological and cytological changes in a large range of neurons and in different parts of the brain. Despite many decades of research, it has not yet been possible to identify the histological or physiological bases of *engrams* (Ch. 6 C). But their formation leaves no reason to doubt that they are causally determined.

D. Analysis of Reproduction and Sexuality

Sooner or later every organism must die; it displays progressive and irreversible signs of age and at last it ceases to function. So living organisms could only originate when they developed the ability to reproduce themselves. Most unicellular creatures do this by bipartition, and this involves rejuvenescence, because each daughter cell is smaller and has a larger surface in relation to its volume and so can take in more oxygen and nourishment. In multicellular plants and animals certain cells or cell groups remain juvenile and totipotent, i.e., capable of producing the entire organism. Spores, buds, or sex cells are reproductive cells of this kind. Their being small in relation to the adult individual represents a metabolic advantage and also a juvenile phase during which they start to develop a new organism. These cells produce a new individual before they themselves would also finally age.

The living substance, then, is potentially immortal insofar as an unbroken chain of reproductive cells forms the thread of life on which the succeeding generations are strung. In multicellular organisms these cells also produce other cells which do not remain totipotent but become successively differentiated and produce the tissues and organs of the individual. The discovery of this continuity represented a great advance over older concepts whereby every individual had been regarded as an entirely new creation.

The simplest and most primitive form of *reproduction* is *asexual.* It occurs in prestages of true life, in viruses, as well as in higher plants propagating by live buds, runners or slips, and in bud-producing multicellular animals like sponges, polyps, corals, lower worms, and bryozoa. Identical twins or quadruplets from one single fertilized ovum and cases of polyembryony may also be taken to represent a form of asexual propagation.

But why did *sexual* reproduction also develop during phylogeny? Several hypotheses have been put forward in the past to solve this problem, but most of them were not adequately supported by relevant statements. It is only within the last decades that a satisfactory explanation has been found.

First, it must be noted that fertilization, the fusion of two cells, is not an absolutely necessary precondition of sexual reproduction. Doubtless in many multicellular organisms fertilization initiates the emergence of new individuals. However, unfertilized egg cells can also develop, as rotifers, water fleas and aphids show, which multiply by parthenogenesis; and it is also possible to initiate the development of unfertilized eggs of sea urchins and amphibians by different artificial methods.

Originally, sexuality has quite a different significance. When the germ-cell nuclei fuse in the act of fertilization, *new gene combinations* are formed; in other words, the variability of the species is increased. This is a great advantage, since it means that as natural selection always operates very intensively and in many ways, the species has a greater chance of survival. Even when conditions become extremely unfavorable, it is less likely that *all* the individuals will perish; for some variants may survive (for instance, those immune to a disease or individuals resistant to extreme cold). Besides, if there are more variants, selection can be more versatile, and the organism can adapt itself more readily and quickly to the different living conditions of all stages of its life cycle (Ch. 5 B). As characters which are not kept efficient through selection tend to dwindle or disappear in the course of phylogenetic development, we may also argue as follows: Sexuality which is present already in some bacteria owes its survival to a continuous process of positive selection, for it has maintained itself through the hundreds of millions of years that elapsed in the evolution of present-day higher plants, animals, and man. It is also probable that this selection operated in favor of a *clear bipolar pattern of sexuality* which led to the development of unambiguous male and female forms. Some of the more primitive organisms still show an incomplete stage in this respect. Some algae (e.g., *Ectocarpus*), fungi, and flagellates (e.g., *Trichonympha*) have sex cells of different power (probably because of variously developed fertilizing material); not only do male and female cells fuse, so do strong with weak male cells and strong with weak female cells. Here we are concerned with *relative sexuality*.

Bipolar differentiation has steadily intensified during the course of phylogenetic development. In some protists the sex cells alone differ morphologically or only physiologically (at least by producing different fertilizing material). However, in multicellular organisms also the primary male and female genital organs are differently constructed, and in most higher animal groups and some plant groups, this difference extends to other organs. Secondary sexual characters and differing sexual instincts originated, by which the sexes attract one another, mate, and in some instances share in rearing their young. Since the possibilities of sexual differentiation and reproductive processes are almost unlimited, an enormous variety in the sexual life of animals and plants has been developed. In some classes of animals the processes of fertilization are simple and direct, in others highly complex and even grotesque. The only condition for the survival of the species was the certainty that reproduction would take place. The sexes come together in response to very different stimuli—chemical, optical, acoustic, or tactile. Varied preliminaries to mating and forms of courtship—for example, song and display —keep the pair together for longer or shorter periods. Fertilization may be external, as in most fishes, or extremely complicated copulative organs and actions may be developed. In flowering plants, very varied means exist to bring the male pollen to the female ovary. Often the same individual has both male and female sex organs. Hermaphrodites of this kind are found among many groups of worms, snails, and flowering plants. They have usually developed structures or processes which prevent self-fertilization (often by successively maturing male and female germ cells). We can see this as another argument that sexuality came into being because an increase in genetic combinations was advantageous.

Although the sex cells and their nuclei unite on fertilization, the chromosomes do not; so the body cells get double the number of chromosomes and in mature sex cells this diploid set must be reduced to a haploid set. Hence, this process of *reduction-division* (meiosis) must have been phylogenetically developed parallel to the origin of sexuality. But reduction can take place at any stage before fertilization occurs, usually only when the germ cells mature.

In animals, sexual reproduction is best assured when the sexes are equally represented. The discovery of how this is brought about constituted an important step forward in our knowledge. In many cases specific sex chromosomes could be discovered which are unequally distributed in both sexes. Usually (in humans also) all the mature ova in the haploid set contain one X-chromosome, but one-half of the spermatozoa have one X-chromosome and the other half one Y-chromosome.

On fertilization, therefore, two equally frequent possibilities exist: a spermatozoon with an X-chromosome may unite with an egg with an X-chromosome from the female, resulting in a female (XX); or a spermatozoon with a Y-chromosome may unite with an egg with an X-chromosome to form a male (XY). In some organisms (in birds and butterflies) the combination XY gives a female and XX a male; or a Y-chromosome is lacking, so that one sex is XX and the other X(o). However, this scheme is too much simplified. The sex is realized by the balance between the effects of the X-chromosome, normally determining female characters, and several genes in the other chromosomes (autosomes) which determine male characters. In several cases the factors are unbalanced at first and the sex will be changed after some time (cf. G. Bacci 1965). According to the theory of O. Bütschli, C. Correns, and M. Hartmann all organisms with sexual differentiation have a *bisexual potency*; that is to say, each individual can develop as either a male or a female. This is determined through certain realizing factors laid down at the moment of fertilization (genotypically) or through later physiological influences (phenotypically). In vertebrates, the hormones act as these realizing factors. As each individual of these animals also produces the hormone of the opposite sex, though to a limited extent, mixed types (intersexes) occasionally arise. In consequence of bisexual potency it also became possible to change the sex by the administration of the hormones of the opposite sex. It was found possible to transform embryonic cocks, whose sex had been hereditarily determined by the sex chromosomes, into adult hens, by repeated injection of the female sex hormone (oestrone) into the allantois of the embryos. In mammals on the other hand, only hereditarily determined female embryos could be transformed into males by injecting male hormone (testosterone). In such cases it is only the chromosome set that reveals later on which sex was determined by fertilization.

Also in snails and worms it was possible to change the sex by physiological methods. In some cases this could be done by fairly simple methods. The marine polychaete worm *Ophryotrocha puerilis* becomes a mature male as soon as it reaches a length of fifteen to twenty segments. When it continues to grow up to thirty segments, it becomes a female. If the last two-thirds of these segments are cut off, the anterior part reverts to a male again in only two days. A similar change of sex can also be brought about by starving a female until it shrinks and becomes sufficiently small.

It was a great scientific success when the fertilization processes could also be analyzed biochemically in unicellular algae as well as in fungi, annelids, molluscs, sea urchins, and fishes. It was found that the male and female sex cells produce certain materials, so-called *gamones*, which cause the germ cells to come nearer and unite. The female cells of lower plants and animals usually produce compounds which attract the spermatozoids or spermatozoa chemotactically. The male cells provide the matter which causes them to adhere to the ovum and dissolve its membrane. Some of these latter gamones have been analyzed. Feather-moss spermatozoids react chemotactically to saccharose (cane sugar), those of liverwort to proteins, and ferns to malic acid. The pollen tube of flowering plants is guided to the ovum not only mechanically but also chemotropically, by sugar for instance. In some flagellates (*Chlamydomonas*), sea urchins, and fishes, the female gamones which cause the male and female sex cells to adhere and unite proved to be glycoproteids (see M. Hartmann 1956).

Thus we can see that during the last decades a considerable advance in the understanding of reproduction and sexuality has taken place. Although several questions have still to be answered, we can claim to have arrived at a causal understanding of the basic processes. The discovery of the potential immortality of the living substance, the genotypic and phenotypic determination of sex, the general bisexual potency of organisms, and the relative and transformable nature of sex—all these discoveries have proved to be of crucial importance. And the study of sex hormones and gamones has already led to a biochemical understanding of sexual processes. (For fuller treatment, see, among others, J. Meisenheimer 1921, 1930; V. Dantschakoff 1941; M. Hartmann 1943, 1951; J. Hämmerling 1951; G. Bacci 1965).

E. Analysis of Hereditary Phenomena

Research into the problems of heredity has rapidly expanded since the beginning of this century into an independent branch of biology. It will not be possible here to discuss all the varied aspects of genetics; we can deal only with its most important findings, indicate the sequence of the main discoveries, and then touch upon some of the chief results of biochemical research in this field.

Modern genetics can be said to have sprung from Mendel's conception of hereditary factors as separate units (as "genes," one might now say) and his experiments of their transmission from generation to generation. He established certain laws (1865), but these failed to find proper recognition, probably because the constant ratios which he found could not yet be interpreted at that stage. This only became possible after Mendel's laws were rediscovered independently by C. Correns, E. Tschermak and H. de Vries in 1900, because by that time germ cells and chromosomes had been more closely studied. W. S. Sutton and Th. Boveri (1902-1904) provided a solid basis by propounding the *chromosome theory of inheritance*. This combination of cytological and genetical research proved very fruitful and quickly produced many important results.

It was now realized that many characteristics must be inherited *linked*, because the relevant genes lie on the same chromosomes and the chromosomes are normally passed on to the daughter cell as whole units when cell division takes place. It also became clear that during meiosis, when the chromosome number becomes halved, the hereditary factors are divided by "random distribution" of chromosomes from the father and the mother among the daughter cells. An important finding was that *the laws of heredity all apply equally to plants, animals, and man*. Thus hereditary phenomena of universal application could be studied in organisms where one generation rapidly succeeds another, as in fruit flies of the genus *Drosophila*, mice, and later also in bacteria. Studies in human genetics were possible by investigating relation groups and especially twins (comparison of identical twins who have the same genes with nonidentical twins who have more different genes). These studies led to the important discovery that *psychical characteristics*, or their histological bases, are inherited.

Th. H. Morgan and his co-workers succeeded in localizing the hereditary factors on the chromosomes of *Drosophila melanogaster*, so that a fairly accurate *chromosome map* could be produced. His deductions were based on the hypothesis that occasional interchange between portions of the homologous chromosomes (from the father and the mother) must take place at the stage of reduction division. The correctness of the localizations could be confirmed later on, by a totally different method, after the giant chromosomes of the salivary glands of diptera had been discovered in 1933. These giant chromosomes, where chromatids are multiplied up to 1000-2000 times but do not become separated, show the structural features as if correspondingly enlarged. They are therefore extremely valuable in the investigation of many genetic problems (see below).

The study of *mutations* became another very important field of research. Although the hereditary factors remain constant over a considerable period, after hundreds of thousands or millions of generations they fairly regularly undergo a spontaneous alteration. This can happen in various ways. Either a single gene-locus is altered or a small particle at the end of a chromosome is broken off, or portions from two chromosomes of the same chromosome set may be interchanged, or a chromosome section may be reversed through 180°, or the whole chromosome set is doubled or multiplied. Only crossing experiments with such mutations have made it possible to study the genetic behavior of individual genes. The mutation rate can be speeded up by exposure to hard rays (ultraviolet, x- and γ-rays), by temperature changes, and by chemical methods. The importance of these studies for the problems of hereditary alterations in the course of phylogenetic development is obvious (see Ch. 5).

The hereditary factors have a dual function: (1) They control the specific differentiation of the cells, as the organism develops, i.e., they govern its metabolism. (2) They then play a major part in determining the function of definitely differentiated (adult) cells. These genetic effects could not be understood until the chemical nature of the gene and the chains of reaction were discovered, which occur between the chromosomes in the nucleus and the cell plasma where specific syntheses take place and structures are formed. It was a turning point in the history of genetics when O. T. Avery and his co-workers established in 1944 that the hereditary information is not transmitted, as had been supposed before, by qualitatively different protein bodies but by *nucleic acids*. And the new molecular genetics was definitely founded in 1953, when J. D. Watson and F. H. C. Crick succeeded in elucidating the molecular structure of deoxyribonucleic acid (DNA).

Avery and his co-workers had shown that bacteria can take over

characters of a mortified genetically different strain and that the transforming substance is DNA. After enzymes had destroyed or removed
the proteins, lipids, and ribonucleic acids (RNA) in the dead strain,
transformation was still possible. But it did not take place if the DNA
had been destroyed by an enzyme. Investigations by A. D. Hershey and
M. Chase (1952) on bacteriophages confirmed these findings. As these
viruses (visible only with the electron microscope) show reproduction,
hereditary constancy, and mutation (Ch. 5 E), they can be considered
as models of prestages of primitive living organisms. Hershey and
Chase were able to show that these phages infect bacteria in such a
way that the protein coat of the phages remains on the bacterial cell
membrane and only the nucleic acid enters the cell. The nucleic acid
then determines the bacterial cell plasma to build up the characteristic
proteins which are typical of the phages' protein coat.

The Watson-Crick model (later confirmed and somewhat modified)
shows the DNA molecule composed of nucleotides. It forms an extremely
long thread consisting of a double helix of equal strands with regular
alternation of a sugar (pentose) and a phosphate group. Each sugar
group is connected with a base which is linked by a relatively weak
hydrogen bond with an opposite complementary base of the other strand.
With very few exceptions, the two strands of the DNA molecule are
linked by only two different pairs of bases: a purine in one chain is
paired with a pyrimidine in the other, adenine with thymine, and
cytosine with guanine. Ribonucleic acids (RNA) have ribose in place
of deoxyribose and uracil instead of thymine. I have already referred
(p. 31) to the great theoretical significance of this totally unexpected
finding: the hereditary information handed down from one generation
to the next in all living organisms is passed on by substances composed
of fairly simple structural elements, differing for each species only in
the sequence of two pairs of bases (Fig. 2). As the filamentous molecules are very long, with molecular weights of some millions, the number
of possible sequences is virtually unlimited. Thus *the qualitative variety
of organisms rests ultimately upon quantitative differences in the structure of the hereditary substance.*

The nucleic acids can annex different proteins at certain points
determined by the different sequences of bases. And their arrangement
varies with the functional stage of the germ cell. Present studies are

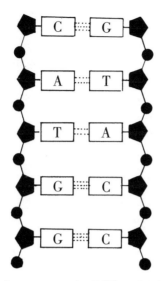

Figure 2. Diagram of a segment of a DNA molecule. The black dots are phosphate groups, the black pentagons sugar (pentose). The two chains form a double helix, linked by four bases: A = adenine, C = cytosine, G = guanine, T = thymine. The dotted parallel lines indicate hydrogen bonds. (After M. Florkin and E. Schoffeniels)

proving more and more successful in determining some base sequences of DNA. This new science of *molecular genetics* is beginning to yield more precise information about individual hereditary factors. The usual terms "gene" and "allele" are not now fully adequate, for they stand for units whose functions alone were characterized by genetical experiments, but whose morphological basis was not sufficiently known. The biochemically characterized hereditary units representing distinct segments of the DNA thread are therefore called cistrons, operons, mutons, and so on.

Once the DNA structure had been revealed, it was possible to arrive at a causal understanding of an important characteristic of all life: the *identical replication of the hereditary factors* in the course of all cell divisions. Before mitosis begins, the chromatids split lengthwise in such a way that half is later assigned to each daughter cell. As the DNA molecules of each chromatid are in the form of a double helix, whose strands are held together by fairly easily soluble hydrogen bonds, the helix can divide into two separate chains (Fig. 2). Then each chain can build up the complementary chain with the help of an enzyme by taking up successive nucleotides after the manner of a zip fastening. A purine becomes connected with a pyrimidine, or vice versa. A. Kornberg and his co-workers have clearly demonstrated this to be so; for they achieved such a synthesis in nonliving material, using an enzyme (a polymerase) obtained from coli bacteria. (For fuller treatment, see J. H. Taylor 1967; Cold Spring Harbor Symposia of Quantitative Biology, vol. 33, 1968.)

Genetic experiments on bacteriophages and bacteria, radioactive tagging of nucleotides, and many cytological and biochemical investigations then allowed well-established conceptions to develop about *how the hereditary factors cause both the process of differentiation in the individual's development and also the cell metabolism of the adult organism* (see J. H. Taylor 1963, 1967; B .C. Bresch 1965; W. Beermann 1962, 1965; F. Kaudewitz 1964; J.-P. Changeaux 1965). When segments of the DNA thread become active, messenger RNA is synthesized. This builds up the complementary sequence of bases (for one strand only), and reaches the cytoplasm by way of pores of the nuclear membrane. There it meets the many ribosomes contained in every cell, where it synthesizes the proteins appropriate to the particular stage of development or phase of metabolism. This takes place by the help of the RNA of the ribosomes and a number of soluble transfer RNA. With the help of enzymes and adenosinetriphosphate furnishing energy, the transfer RNA are linked to a particular amino acid, providing the material for the protein bodies. The messenger RNA attaches itself to a chain of ribosomes, each of which has a triplet of three nucleotides (a codon), which is arranged in such a manner that a complementary triplet of the transfer RNA (anticodon) can be inserted. The messenger RNA then successively moves on, and the next transfer RNA fits its codon to its anticodon. Enzymes and guanosinetriphosphate providing energy help to unite the amino acids contributed by the transfer RNA. In this way many amino acids combine to form polypeptides, the basis of protein bodies (Fig. 3). If, for instance, a codon contains a uracil triplet (U-U-U), this determines the connection with the amino acid phenylalanine. The codon uracil-uracil-cytosine determines the connection with leucine. A whole series of such "three-letter code-words"

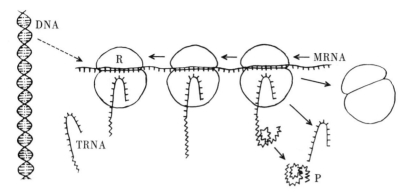

Figure 3. Diagram illustrating specific protein synthesis through the action of genes. DNA = double-spiral DNA chromatid molecule. MRNA = single-stranded messenger RNA molecule. R = ribosome. TRNA = transfer RNA. The same ribosome is represented from left to right at four successive functional stages. The messenger RNA moves along the ribosome, as indicated by the arrows, combining its codon (triplet) with an anticodon from a transfer RNA. The zigzag lines in the transfer RNA indicate the growing chain of amino acids forming a polypeptide. P = polypeptide (protein).

has already been worked out. There are as many types of transfer RNA as there are code words for about twenty different amino acids. These types of TRNA "read off" the code of the messenger RNA, so to speak, and fit their amino acids in the correct order into the polypeptide molecules to be synthesized.

Hence we see that the causality of this whole process of specific protein synthesis is already conceivable to a far-reaching degree. It is a succession of information of growing complexity. The DNA chromosomes "code" of four letters (the four bases) is taken up by the messenger RNA, matched for a time with the "anticodes" of the transfer RNA, which transform it into a code of twenty "letters" (the different combinations of the twenty amino acids). Inumerable "words," i.e., polypeptides or proteins, are then formed. These first protein bodies are enzymes, and they in their turn synthesize or break down other specific substances. It is essential to grasp this "stream of order" leading to more and more complex systems, in order to understand the causal nature of the development of organisms (Ch. 5).

When a *gene mutation* occurs, i.e., when the sequence is altered at one point in the DNA thread of the chromosomes, the messenger RNA transmits a somewhat altered pattern, and so when a polypeptide is synthesized on the ribosome, an "incorrect" amino acid is incorporated. This has already been analyzed in a human mutation which alters the constitution of the hemoglobin, the pigment of the red bood cells, and causes sickle-cell anemia. In this case, in one spot on a DNA thread a thymine takes the place of an adenine, and abnormal red corpuscles are formed because the code words in the combination of the hemoglobin molecule's 150 amino acids have substituted a glutamine for a valine.

As far as we can judge at present, *virtually the same process of transmission of hereditary information to the cell plasma holds good for all plants and animals including man, and even for prestages of organisms like viruses* (in the case of phages, transmission to the plasma of the infected bacteria).

We are also approaching the causal solution of the problem how the different hereditary factors receive the information which activates certain segments of the DNA molecule at the correct time. Recent experiments have shown that various hormones can promote or hinder the functioning of the genes. Estrogen, for instance, stimulates the formation of messenger RNA and transfer RNA (E. H. Davidson 1965). And the study of the giant chromosomes of certain midges has also yielded some important findings. A correlation was found between the process of differentiation in the tissues and the behavior of the genetic structure. The pattern of the DNA threads composing the salivary gland chromosomes shows at time swellings called "puffs" at the points where the DNA segments become active, i.e., where certain genes are affecting the cell metabolism. This finding received confirmation when it was shown that these puffs are associated with intensive RNA synthesis (W. Beermann 1962). It was also found that only a certain percentage of DNA molecules produce puffs, and that probably these alone impart information for protein synthesis while others regulate gene activity. Experiments with bacteria and yeasts have yielded similar results.

Monod and his co-workers have recently worked out the following conception (B. J. Monod 1959; J. H. Taylor 1963, 1967; C. Bresch 1965; J.-P. Changeaux 1965) : An operon, that is to say a DNA segment which becomes active, consists of structural genes on which messenger RNA is synthesized, and also an operator gene which regulates the activity of the structural gene. It may interrupt this activity by means of a regulator gene, which causes the cell plasma to form repressor molecules, and these combine with the operator gene (allosteric control). The repressor molecules act, for instance, when the structural genes have produced a sufficient quantity of a certain material. When enough isoleucin (an amino acid) has been synthesized, for example, the repressor is brought into action. It conveys a message, to stop or start the action of the structural genes from the cell plasma to the hereditary units on the chromosome. The inactive genes are most probably blocked by histones. C. Bresch (1965) has used the following comparison to describe the process:

"The 'genome' [the hereditary substances, DNA] sends as it were an unadressed bottle-post [which reaches the ribosome in the cell plasma], containing a long letter [the messenger-RNA]. The plasma answers simply Yes or No. But it must address the reply to a particular operon. As the plasma cannot write, however [only nucleic acids can do this], nature employs the trick of making the genome also provide all the addressed envelopes (repressors), but each envelope only contains a message concerning the presence of a specific molecule."

Although some links in the chain of evidence of these processes are as yet based on indirect conclusions, it is clear that causal analysis of hereditary processes and specific cell metabolism is rapidly advancing. Cytogenetics and molecular genetics—what we may call *chemocyto-genetics*—are rapidly leading to the causal understanding of processes which only a few decades ago had seemed impossible to explain.

F. Analysis of Individual Development

The adult body of every higher plant or animal is made up of many tissues and of organs whose structures and functions show enormous differences. The cells themselves which characterize the different tissues are very unlike in type. Animals have very different cells in skin, muscles, intestine, kidneys, bones, nerves; different sense cells in eyes, ears, and tongue; different cells in red and white blood corpuscles; in spermatazoa; etc. Plants have parenchymatic, epidermic and sieve cells, sclerenchymatic, hair cells and sensory cells, tracheids, sieve tubes, etc. So research into individual development is faced with the difficult problem of explaining by what causal sequences a comparatively simple structure—an ovum, a bud cell, or spore—differentiates into an adult individual of such complexity, including the astonishingly complicated structure and versatility of a brain such as that possessed by higher animals and man. The problem is made all the more difficult because every part of the organism is correlated with other parts and does not develop independently, so that at each embryonic and juvenile stage, other factors come into play and affect the gradual process of differentiation.

In view of this interplay of varied genetic, histophysiological, and histochemical factors, affecting every stage and every part, it is not surprising to find that the causal analysis of ontogenetic processes has made less progress than research into types of metabolism in the mature organism. Yet the past eighty years have seen considerable advance, recently extended even to biochemical and biophysical processes. Some examples may indicate the state of research today. In view of the multifarious experimentally obtained results in different groups of organisms, a wider survey would be impossible, and furthermore unnecessary in connection with our philosophical questions. The following brief survey of the chief methods and results is for the benefit of those less familiar with recent work in biology. (For fuller treatment, see H. Spemann 1936; A. Kühn 1965; F. Seidel 1953; F. E. Lehmann 1945; C. H. Waddington 1956, 1957, 1962; J. Brachet 1960; M. Abercrombie and J. Brachet 1961-1967; E. Bell 1965; R. Weber 1965; and others.)

After Wilhelm Roux had initiated the experimental analysis of embryonic differentiation in 1865, research into individual development led rapidly to some remarkable discoveries. That portions of the germ are capable of *regulation* which can lead to the development of a complete individual was a discovery that aroused heated discussion. Experiments with sea urchins and vertebrates showed that artificial division of the germ at the 2-cell stage produced two normal individuals. Isolated cells at the 4-cell stage in sea urchins and at the 16-cell stage in the hydroid polyp *Clytia* also produced complete individuals. The reverse process yielded a still more surprising result. Two germs at the 2-cell stage of newts, when pressed together crossways, formed a single normal individual. All these findings indicated that each of these cleavage cells has still all the developmental power to build up a whole individual, that it is *totipotent*. But this power of regulation does not extend to all animal groups. When one of the pair at the 2-cell stage of an ascidian was destroyed, only half an embryo developed. In these creatures the different regions of the egg skin already have a more definite assignment, and therefore regulation to a whole individual cannot take place in the separated half.

Many adult coelenterates and worms also possess certain totipotent cells which remain embryonic and can help to form complete individuals out of minute parts of the animal: 1/100 of a turbellarian or 1/200 of a fresh water polyp can produce all the structures and organs. Germ cells, of course, also remain totipotent. And the same is true of the spores of fungi, mosses and ferns, and the plant cells which remain meristematic, as with plants reproducing by means of runners, slips, and live buds.

Many other *methods* have been employed in the study of development. Plasma regions of different weights were displaced by the centrifugalizing of ova. Such experiments proved the significance of the egg membrane for the initial processes of differentiation. By tying up portions of an ovum or developing egg, the potency of the different regions could be ascertained. By uniting germ portions belonging to different species or genera, chimeras could be produced which showed not only the powers of differentiation of each species but also how the heterogeneous elements might react and combine with one another. Explantation of cells or tissues and cultivation in artificial media gave information on autonomous potencies. Transplantation of embryonic or juvenile tissue or of entire buds (Anlagen) of organs into different parts of the same or some other species proved autonomous potencies as well as influences on neighboring parts. When different portions of eggs or germs were marked by vital staining, the emergence of structures and organs from such regions could be traced. Regeneration experiments also revealed how structures or organs can be re-formed. Biochemical analysis of the phenomena of development has been furthered by changes in environmental factors and the introduction of radioactively tagged molecules.

At the outset of developmental physiology, the problem already arose, *how far the differentiation process is autonomous*—that is to say, controlled by the cell structures and in particular by the hereditary units in the nucleus of the germ cell—and how far it is affected by the neighboring tissues and by environmental factors. Numerous experiments have established that all three factors are relevant. The initial differentiation of the germ cell into several similar cells is probably not, or at least not primarily, caused by the genes but largely by the egg membrane and the centrosomes, for it has been established that even after the nucleus of a sea urchin's egg is removed, the cell often goes through processes similar to cell division, with cleavage centers and radiating fibers, and forms spherical morula-like "cell" masses. Apparently the hereditary units of the chromosomes are in many cases only involved later in the processes of differentiation, when cleavage divisions are finished and invagination, excrescence, and the development of specialized cell structures begin.

The differentiation of the first two germ layers ectoderm and endoderm is also largely autonomous. J. Holtfreter (1939) illustrated this by the following experiment. Cleavage cells from newt germs at early stages were isolated. When placed in contact, they recombined (positive affinity). But if they were kept separate long enough to have developed ectoderm and endoderm, they became two types of cell which no longer recombined when placed in contact (negative affinity), though cells of the same type continued to do so. This shows that during the period of isolation these cells had undergone autonomous differentiation into ectoderm and endoderm cells which do not adhere to one another. In a corresponding manner cells of later embryonic stages kept in artificial media show autonomous differentiation. Embryonic nerve cells, for instance, form typical nerve fibers. Isolation experiments with cells of embryonic organs of vertebrates showed that the separated cells of the same type not only adhere once again but may even develop into normal organs. Isolated kidney cells, after being united again, developed kidney tubules and glomerules, and isolated cells of embryonic skin taken from birds formed complexes leading to the development of normal feathers (P. Weiss and A. C. Taylor 1960). It is clear then that the cell surfaces contain specific structural features which guarantee the right combination of cells for correctly functioning tissues and organs (see also G. Gerisch 1965).

But cell differentiation is also *affected by neighboring tissues*. This has never been in doubt for purely histomechanical effects. Isolated cells without rigid membrane or filamentous inclusions mostly take on a spheroid shape or an amoeboid shape with lobelike processes. If such cells grow within the body, their position in or on organs determines their particular structures and shape: prismatic, cubic, flattened, spindle, or filamentous.

H. Spemann's discovery that environmental influences include chemical ones, and that these may affect the development of the organism in a far-reaching manner, has been of great importance. Such *inductions* were first observed in experiments with newts. When the upper lip of the blastopore at a late stage of gastrulation was cut out and transplanted to the future belly region of another younger gastrula, it not only continued to undergo autonomous differentiation to chorda mesoderm but also induced the ectoderm above it to form a new spinal cord. In a corresponding manner, an embryonic eye vesicle, when transplanted below the belly skin of a newt larva, induced the skin to form an eye lens.

Similar organizing effects on neighboring tissues were established in other animal groups and in many organ systems, and they could also be shown when different tissues were cultivated in artificial media. J. Holtfreter (1934) was able to show that embryonic nerve tissue grows into a spheroid form, with the cells lying thickest at the periphery. But if nerve-cell masses were grown enclosed by embryonic connective tissues, the latter had an inductive effect causing the nerve cells to tend away from the connective tissue toward the center of the sphere, producing a cavity there. If chorda tissue was then attached to the nerve-cell mass at one side, this cavity narrowed to a slit with its main axis toward the chorda tissue. Many further experiments proved that the correct regeneration of parts of the body is induced by the neighboring intact tissues.

The effects of induction are brought about by chemical substances, as proved by the fact that even tissues killed by freezing, heating, or alcohol can still direct differentiation processes in neighboring tissues. Tissues implanted in structures induced in this way, moreover, may themselves become inductors. If, for example, a piece of future ventral skin from a newt germ (marked by vital staining) is implanted in the roof of the primitive gut (the archenteron) and is removed twenty-four hours later, it has acquired the faculty of inducing an extra spinal cord when implanted in the "ventral" region of a young gastrula. Research is now going on to discover the chemical nature of these induction processes. With the brain and spinal cord, this is probably being carried out by a gradated effect of a ribonucleotide.

Another statement of great importance for the causal analysis of development concerns the part played by *physiological gradients*, in most cases corresponding to initial differentiation processes of germ cells as well as organs. Sea-urchin eggs exhibit a gradient from the animal to the vegetative pole (so-called because the characteristic animal organs such as the nervous system develop at the former and the more vegetative organs of nourishment at the latter). The yolk material increases toward the vegetative pole, respiration intensity decreases, and the skin structure alters as well. Later, as a result of more accumulation of yolk material on one side, differentiation between dorsal and ventral regions develops. (This does not become recognizable, however, until the rudimentary intestine begins to form.) The brilliant experiments conducted by S. Hörstadius (1939-1949) have shown that these gradients may be more significant than initial morphological differentiation. A sea urchin at the 32-cell and the 64-cell stages shows three different types of cleavage cell: larger ones in the vegetative half than in the animal half, and very small ones (micromers) at the vegetative pole. Yet in spite of this differentiation, normal larvae are produced even if the larger vegetative cells are removed, and the micromeres joined to the animal half. Normal larvae can also be produced, however, if at the 32-cell stage the topmost animal and lowest vegetative cells including the micromeres are removed. Thus it appears that all that is required for normal larval development is a balance of the animal-vegetative gradient; strong animal plus strong vegetative cells (micromeres) from the poles can produce a larva as well as weaker animal plus weaker vegetative cells of the equatorial region.

Gradients were also found when the very first buds of organs appeared or regeneration buds began to form. They can originate by the different speed and sequence of self-differentiation within the tissues in question, or under the influence of inductors from neighboring regions. It is often difficult to distinguish between these two possibilities. In the limb buds of amphibia, the primary cause is apparently self-differentiation of the mesoderm. This mesoderm acts in a gradated manner as an inductor on the skin above it, centered in a sector of the leg bud facing upward and backward. The skin is induced to adapt itself to the configuration of the developing limb. This even occurs if the organ rudiment is transplanted below a different portion of skin. As a gradient is often present from the onset of organ differentiation, a general *theory of physiological gradients* has been established. First suggested by Th. Boveri, it was developed by C. M. Child (1914, 1928, 1941) and, on the basis of various other statements, by A. Dalcq and J. Pasteels and by J. Runnström, S. Hörstadius and P. E. Lindahl.

Growth gradients can also be established at later stages of organ development. They remain constant for a certain period insofar as the increase of a single organ remains constant in relation to the increase of the whole body. Such *allometric growth* can be quantitatively expressed by calculating an "allometric exponent" (see J. Huxley 1932; B. Rensch 1954; etc.).

As several embryonic structures always develop simultaneously, the differentiation processes are generally *geared to the organism as a whole*. In some cases this can be directly observed in the mutual compensation of material. Continuous reference of the development of structures and organs to the growth of other parts and the whole organism increases the variety of influences at every stage, and so it often becomes difficult to assess the causality of the processes with sufficient clarity (see, for instance, C. H. Waddington).

A further complication is the fact that *external factors* can influence the development at different stages. Usually they are necessary for normal differentiation and especially for the final differentiation of organs. Sea-urchin germs, for example, must take up sulphate ions from sea water, and these, combined with polysaccharides, control normal protein metabolism. If these ions are lacking, the animal tendencies predominate. Harmful metabolic processes are then developed, and the larva fails to achieve bilateral symmetry. An insufficient supply of oxygen can disturb the harmony of development in amphibian embryos—that is to say, the constantly altered balance of the successive stages (what Waddington terms homeorhesis)—and lead to deformities (H. and H. Tiedemann 1954). The neuron structure in the mammalian brain only develops fully if the relevant region begins to function.

A great deal of study is being carried out on the *chemical analysis* of developmental phenomena, which we have already touched upon in these examples. But we are still at the beginning of a full understanding of these complex processes. One important finding, for instance, has been that during the development of a sea-urchin germ the ribonucleic acids, the protein synthesis which depends on them, and the oxygen consumption in successive periods increase in proportion as the glucosamid content decreases. Special phases of activity in the differentiation process can be marked out in this way. Specific proteins appear during the cleavage divisions and the development of the endoderm, i.e., during gastrulation. The continuous decrease in glutathione during cleavage is probably due to the part this tripeptide plays in the protein synthesis required by the gastrula. Such descriptive findings, outlined here by a few examples, form at least a good basis for wider biochemical causal analysis in the future. This will no doubt also be helped by molecular genetics, which is beginning to reveal the basic processes at the initial phases of all differentiation.

This brief survey of some of the more important results of developmental physiology may be enough to show how far we still are from any sufficient causal understanding of the complex processes of differentiation. However, some important findings have been made which may be summarized as follows: control of the initial differentiation by the egg membrane; chemical analysis of gene action and early metabolic processes; processes resembling cell division in ovum fragments without nucleus; proofs of autonomous differentiation processes by different methods; release of certain trends of differentiation by inductors; the high degree of regulative powers after division of germs or other interferences; concerted action of autonomous and inductive factors in regeneration; existence of physiological gradients in the ovum, during gastrulation and initial organ formation; temporary constancy of growth gradients; influence of the structure of the cell membrane on tissue building; effect of positive and negative affinities; the significance of external factors. All these results contribute to a causal understanding and a histomechanical and biochemical explanation of developmental processes. In some cases indeed, they have already achieved this. We may expect further research to lead us to a final causal understanding of individual development. One question, however, remains. We have to ask how in the course of phylogeny these chains of developmental stages following the same sequence from one generation to the next could first originate—reactions leading "purposively" and "meaningfully" from the ovum to the vast and intricate complexity of the mature individual (Ch. 6 K IV).

G. *Present Situation of the Causal Analysis of Life Processes*

The examples briefly considered in Chapter 4 A-F show the very rapid and significant advances made during the twentieth century in biological research. We have come nearer to a causal explanation of phenomena in all fields, phenomena which had previously been regarded as incomprehensible effects of vital forces or even simply as "marvels." Often, new and ingenious methods have combined to extend the analyses to the fundamental biochemical and biophysical processes.

Among the metabolic processes which are especially important for the general problems of life, photosynthesis and energy change in the mitochondria, and the chemical structure and significance of vitamins in synthesizing enzymes and other vital material could be elucidated. The structure and function of hormones in regulating metabolic processes, sexual maturation, and instinctive actions could be analyzed. Light- and electron microscope studies have revealed much about the function of the sense organs; and some surprising discoveries have been made in this field—for example, that bats and some birds steer their flight by echo sounding, that insects and some crustacea react to polarized light, and that some fishes have electric sounding devices. The chemical and physical bases of the origin and course of nervous impulses have been uncovered. Analysis of instincts has led to an understanding of appetitive behavior and "release" mechanisms and to such unexpected discoveries as the "language" of bees and how animals use an "internal clock" to guide them by the changing position of the sun. Experiments into animal psychology have revealed that vertebrates have a wide capacity for abstraction and generalization, that birds and mammals can comprehend causal connections, and that apes and monkeys, at least, can plan a course of action.

It has been of great importance for the understanding of life that the potential immortality of living substance could be established. The significance of sexuality became more or less clear. Further problems of sexuality have been solved through an understanding of the bisexual potential of the living organism, experimental changes of sex, and the chemical nature of the fertilizing material (the gamones). Mendel's rules

could now be satisfactorily interpreted by the chromosome theory of inheritance. The study of mutation has provided a precise basis for in-
terpreting change in hereditary characters. Elucidation of the structure
of nucleic acids has yielded quite unexpected information concerning the
hereditary process. On the basis of these discoveries molecular genetics
could develop and the self-replication of the hereditary units and the
action of the genes could be explained. The very complex processes of
the individual development of organisms could not yet be studied in
quite so much detail. Great progress has, however, been made by the discovery of inductors and the analysis of their interaction with autonomous developmental factors. The statement of positive and negative cellular affinities has been important for the understanding of gastrulation
and tissue formation. Chemical and functional gradients in the ovum,
early embryonic stages, and initial organ formation have been established, and chemical and physiological processes in different parts of the
germ could be analyzed.

This brief résumé of some main results cannot of course cover the complexity and variety of the newly acquired knowledge; but it may show
that each advance has been made by following the causal chain further
and further back, in several cases even into the molecular processes.
*Causal analysis has proved itself as a heuristic principle at every stage
of research and without exception.* Yet there are still many questions to
be answered, especially in regard to the highly complex biochemical
processes of giant molecules in the active cells. However, the increase in
research institutes in many countries and the growing expenditures for
science, which have made possible a rapid intensification of biological
studies, justify the optimistic assumption that we may finally solve the
many outstanding problems and establish a satisfactory causal explanation for all the phenomena of life.

The processes to be studied are, as already mentioned, particularly
complex because each represents only a partial activity within the wholeness of an individual. So the analyses of particular processes must always
be supplemented by investigation of their correlations with the overall
system and by studies on their effect in other structures and functions.
The comprehension of these mutual relations is a difficult task because
organisms are "open systems" of only relative constancy (Ch. 3 A II,
V and VIII).

Although doubts have been expressed from time to time, especially by the "vitalists" (see Ch. 6 K IV), we are justified in saying that in principle there is no limit to causal analysis either at the specific or the general level. It is only when we come to the realm of atoms that what W. Heisenberg has termed the "uncertainty principle" imposes limits to biological and physical research.

Summarizing our present knowledge and leaving aside the psychological aspect which will be treated later, we can define the phenomena of life as a system of complex and interrelated reactions governed in the last instance by chemical and physical laws of a causal nature. But the fundamental biological processes are so complexly integrated, especially in their interposition in individualized systems, that certain *specific laws relating to living organisms* originated. We shall be examining to what degree they can be regarded as complications of causal processes (Ch. 6 K IV).

A special feature of all living organisms, mentioned several times already, is the fact that biological processes in general appear to be "meaningful." They are not only appropriate to the immediate conditions but also seem to be directed to some purpose which in individual development is only achieved at a relatively late stage and after many modifications of form. We must therefore ask how such purposive living systems could evolve, and how they could be sustained. These are problems of evolution which will be discussed in the following chapter.

Chapter Five

The Phylogenetic Evolution of Organisms

A. The Problems

Even a reader who is only slightly versed in biology will have gathered from the foregoing résumé of the present state of our knowledge that in principle there are no limits to the rapidly advancing causal analysis of the structures and functions of living organisms. Despite the complexity of the interrelated factors, a firm basis has undeniably been laid down. Some biologists and many philosophers, however, find it much harder to answer the question posed in the preceding chapter: How could living organisms come into being—organisms that are highly complex, individualized, open systems with purposeful structures and functions, which reproduce themselves and continue to build up the same purposeful structures? Some scientists also doubt whether we have already solved the problem of how in the course of long geological periods not only new species but also new organs have originated, and new types of organization and higher organisms have evolved, culminating in man himself. While they recognize that research into evolution has made a start in solving these problems, they question whether our present knowledge is sufficient for us to assert that we have arrived at a satisfactory explanation.

Some philosophers are particularly skeptical. They accept the origin of new species, the existence of a phylogenetic connection among all organisms, and the progressive development of many phyla; but they believe that the explanations hitherto offered for the process of evolution are insufficient. Thus Nicolai Hartmann has written (1950, p. 637): "The theories put forward in explanation of this problem have all erred in going too far on this point, and so they have led us astray." He later added that the "categories of organic matter . . . are as inevitably irrational as the body of facts for which they stand. . . . But the purposiveness of the evolutionary processes is so bewilderingly complex and intricate, and yet also so patently obvious, that one cannot help feeling there are some underlying immanent purposes guiding its course." Several passages in his chapter on phylogenesis, however, make it clear that Hartmann was far from sufficiently conversant with the state of research into evolution. Although hundreds of special investigations and numerous comprehensive treatises had been published by 1950, he wrote about the problem of descent: "At present science has little to say on the matter." Ignoring the many studies of such well-established lines of descent as foraminifers, ammonoids, brachiopods, horses, deer, and bears, he remarked: "The species known to us by chance from fossil remains represent more or less isolated members." Despite the great number of carefully documented monographs on the evolution of species published between 1920 and 1950, he declared: "But no attempt is being made any longer to tackle this problem for its own sake."

Some contemporary biologists have also published skeptical remarks. The zoologist A. Remane (1956, p. 140) sums up his views thus: "The result of hitherto existing attempts to unravel the driving forces of phylogenesis has been rather slight. No doubt phenomena of combination and mutation play a part in transforming the organism; we know very well the mechanisms that led species to split up; we know the bases of the origin of many races, of regressive phenomena, changes of symmetry, etc.; but all this does not bring us much nearer to understanding the transformation of organization in organisms." The physiologist K. E. Rothschuh (1959, p. 127) expressed a similar opinion: "The mechanisms so far suggested as producing phylogenetic alterations are not sufficient to explain the very points that strike a physiologist most forcibly—the remarkable degree of correlation, the multifarious correspondence, the harmony of the various parts, the interweaving of mating-partners, etc." The zoologist A. Portmann has voiced a similar view on many occasions. On the other hand, a much greater number of experienced students of evolution who have analyzed not only the factors of species formation but also the complex processes of development of new organs, types of construction, and progressive evolution have been quite successful in explaining these processes. Several such works containing hundreds of special references are W. d'Arcy Thompson (1917); A. N.

Sewertzoff (1931); J. Huxley (1932, 1942, 1963); J. Huxley, A. C. Hardy and E. B. Ford (1954); Th. Dobzhansky (1937, 1951, 1960, 1962, 1967); E. Mayr (1942, 1963); G. G. Simpson (1944, 1949, 1953); G. Heberer (1943, 1959, 1965); B. Rensch (1947, 1954, 1959, 1960b); S. Schmalhausen (1949); W. Zimmermann (1949); G. L. Stebbins (1950); the Cold Spring Harbor Symposium of 1959 (Vol. 24); and the symposium volumes edited by S. Tax (1960). These writers are all aware how many-sided and difficult the problems are; but most of them are convinced that in principle a causal explanation is possible for species formation, the development of harmoniously interrelated structures and functions in emergent higher categories, the appearance of new organs, and the progressive evolution of many lines of descent. We are even beginning to frame more precise conceptions of how living organisms first came into being, although we are still far from understanding, for instance, how chromosomes, mitochondria, or whole cells were first formed in the course of phylogenetic evolution.

It will be impossible in the following chapters to discuss all the facts of evolution as fully as they deserve. We can only provide a few typical examples and a general survey of the present state of knowledge. We shall show the complications and characterize the causal connections in such a way as to make it possible to understand the main problems. These include the evolution of purposefully structured living systems with purposeful individual development; the division of such systems into new species; the origin of new types of construction; and evolutionary progress. We shall also discuss whether such an understanding can be reached without recourse to some hitherto unknown "autonomous evolutionary forces." The consideration of psychological properties will not be treated until Chapter 6 H.

B. Factors and Ways of Species Formation

The process by which hereditary change brings about the emergence of new species has been studied by different methods. In relatively complete *lines of descent* of fossil animals found in more or less unbroken sequences of deposits, the change of characters could be examined. Where possible, changes of environmental factors have also been taken into account. A great number of such investigations could be carried out in groups which yield well-preserved fossil material, such as foraminifers, corals, brachiopods, snails, ammonoids, trilobites, and mammals. But the results are only partly satisfactory, because while we can estimate morphological change, we know very little about the factors which cause it. Only by analogy with known present-day forms can we try to determine which features were hereditary. As hereditary and nonhereditary morphological characters can look completely alike, in cases of racial variation it is often impossible to speak of origination of new species. However, in longer lines of descent and especially in sequences of genera exhibiting more pronounced morphological change, analogy with recent forms is a reliable guide to whether a certain feature is hereditary.

The study of *polytypic recent species* affords sounder evidence on genetic differences for races as well. Many animals and plants can be arranged in large-scale species composed of a number of well-defined subspecies representing one another geographically or ecologically. Many of these polytpic species or "Rassenkreise," which have been studied (B. Rensch 1926, 1929, 1954; E. Mayr 1942), can at least serve as models of speciation, as they have gradually developed from a basic type by successive spreading in comparatively recent geological time. Their geographically distinct and graded stages illustrate the process of speciation. Experiments in many such cases have shown racial characteristics to be hereditary. The increased variability of hybrid populations (caused by secondary overlapping of territories) also proves the heredity of the distinguishing characters. In other cases the characteristics have been shown to be independent of external factors and are therefore probably hereditary. In certain borderline cases, when more differing races have been geographically separated and have later come in secondary contact, it is found that no interbreeding occurs. Artificial crossing of such types often shows that their fertility is diminished. This indicates that geographical races may become species.

A third method of analyzing the process of speciation is the *study of mutants and breeds of domestic animals* whose genetic base, fertility, viability, and selective advantages or disadvantages can be assessed. The many investigations of this kind which have been carried out range from bacteria to flowering plants, from protozoa to insects, particularly *Drosophila* species, and to mammals. They have been especially informatory in the prestages of organisms, in viruses. Sometimes it has even been possible to create new plant and animal forms which produce no fertile offspring when crossed with the parent stocks, and may therefore be claimed as artificial species and not merely races.

All these paleontological, taxonomical, genetical, and population analyses have yielded the following results. (1) Mutation and natural selection are the decisive factors in the formation of species. (2) Other factors are also involved in speciation. Many new gene combinations, made possible by sexual processes, arise in each generation. (3) The spread of genes in a population (gene flow) and changes in predominance or occurrence of certain genes or gene combinations brought about by alterations in population size influence the speed of speciation. (4) Further factors include isolation of populations or entire races, annidation (when certain variants establish a foothold in a niche), and secondary hybridization of existing races or species. In some cases only one or two of these factors may be operative in producing new subspecies, but more often many factors have combined to produce new races. There are, then, *different ways of speciation*. To provide a causal explanation of the process, we shall now give a brief account of these factors and the resulting types of speciation in order to show how far a causal explanation of these processes is already possible.

All hereditary change ultimately depends upon mutation and new combinations of genes. *Mutations* are "spontaneous," "random" in occurrence, and affect the gene content of individual cells. With regard to race and species formation in organisms with sexual reproduction mutations are only significant if they affect germ-tract cells, that is to say, cells which will finally produce germ cells. If a mutation occurs early in the germ tract, it will be reproduced in all subsequent germ cells affected. Mutations are very rare, but they occur with some degree of regularity, i.e., proportionate to time. The mutation rate for the fruit fly *Drosophila melanogaster*, whose genetical basis is fairly well known, is 0.0001 to 0.005 percent of the germ cells, though for some labile genes it is as high as about 1.0 percent. As the numbers in most species often run into many millions, mutations have to be reckoned with in each generation. They may affect any kind of hereditary unit.

Generally speaking, mutations are disturbances in a long-established and harmonious phylogenetic development of structures. Many of them (lethal mutations) put a premature end to individual development, while most others have the effect of reducing viability (fitness and survival prospects) or fertility. Only a few of the better-known mutations are neutral or produce some advantageous change. But advantage or disadvantage often depends upon how they combine with other genes or are related to external circumstances. Experiments with whole populations have shown, for example, that at normal temperatures the white-eyed mutant "white" of *Drosophila melanogaster* has much diminished viability compared with the wild form, but it has a better prospect of survival when the temperature is either very low or very high. Mutants with multiple chromosome sets (polyploids), which occur in many of the higher plants, are generally less viable at first than the normal (diploid) forms. However, they prove more resistant in extremes of climate.

The different types of mutation vary greatly in their impact on the formation of species. There are gene, chromosome, genome, and "plasma" mutations, and in plants also plastid mutations. *Gene mutations,* occurring in an undirected manner, may alter morphological, physiological, and psychological characters. One point should be stressed here which, in spite of its philosophical significance, is often underestimated: It is the structural and chemical characters of the germ cells, mainly the DNA structure in the cell nucleus, which govern man's hereditary factors, and among these characters they also determine the structure of the later nerve cells and sensory cells, and therefore the corresponding psychological phenomena. So mutative change in the DNA bases may produce alteration in hereditary mental characters.

Another important point is that most mutants are recessively inherited. They produce no alteration in the heterozygous condition, because of the nonmutated "healthy" gene of the homologous chromosome deriving from the other parent, which determines development along normal lines. As populations are normally fairly large, most of the mutants, appearing in single individuals, pair off with nonmutated normal forms; so most recessive gene mutations do not become visible. It is only in the homozygous condition, when both parents carry the same mutation, that the alteration becomes effective. The same holds true, of course, in man. A single human, therefore, may be carrying on a number of harmful genetic factors to the next generations, though this will only become apparent if both parents have the same factor. Dominant mutants, on the other hand, are always visible whether homozygous or heterozygous. This difference of recessive and dominant heredity has a very significant bearing on the formation of races and species. If the mutant is dominant, each individual is exposed to natural selection. Thus harmful dominant mutants are mostly eliminated fairly rapidly, while advantageous ones spread. Recessive mutants, on the contrary, are not affected by natural selection unless they occasionally appear in a homozygous condition.

In discussing the action of the genes (Ch. 4 E), we noted that often the primary effect of mutations is to alter or inhibit the synthesizing of a protein, probably an enzyme. A particular reaction in the process of development may therefore be prevented, and a certain character fails to emerge. For instance, an oxidase may be lacking which would complete the synthesis of a melanin, so only the colorless prestages of the pigment are formed. In many cases, however, several chains of reactions are involved, which depend upon the primary alteration. The frequency of such effects is already shown by the fact that most of the morphological mutants in *Drosophila* were found to diminish their viability and fertility. These pleiotropic effects are of course very important in the formation of species, because several of the organism's characters are then exposed to natural selection. Alterations of the bristles of *Drosophila* (the mutant "bobbed" in *Dr. funebris*), for example, may be relatively unimportant for natural selection; however, the correlated reduction in viability and fertility may have a clearly negative effect.

With regard to the philosophical evaluation of evolution, it is now important to consider whether or to what degree it is possible to analyze the conditions which cause gene mutation. It has sometimes been assumed that since microphysical processes are involved in which causal connection cannot be established, there can be no causal explanation. Referring to mutations induced by high-energy radiations (x-rays), P. Jordan (1945) even went so far as to say that in such cases the macrophysical hereditary process is a continuation of the microphysical acausality. But the acausality which he alleges is open to doubt. We shall return to this question in Chapter 6 K III.

Some progress is already being made toward a causal explanation of the origin of gene mutations. Mutation can be induced mainly through those ultraviolet rays which are best absorbed by DNA and by mutagenic substances. The base thymine, for instance, can be replaced by bromine-uracil. If in the normal keto-form this will pair with adenine. But it can also change in the enol-form and will then pair with guanine. Several alterations of this kind can already be produced with a certain degree of probability. Mutagenic substances such as nitrous acid can deaminate adenine to ,hypoxanthine, guanine to xanthine, and cytosine to uracil. But hypoxanthine can pair like guanine and uracil like thymine and so gene mutation can be brought about (C. Bresch 1965). In such cases some main causal connection becomes clear. When the hydrogen bridges in the double helix of DNA are blocked by acridine compounds, replication becomes altered: one or more bases may be passed over or they may be read off twice. In this manner deletion mutants or insertion mutants originate. The action of other mutagenic substances probably also has an inhibiting effect on certain genes. All these findings mark a distinct advance toward elucidation of the causal connections.

Chromosome mutations of various types also contribute to speciation. They come about in different ways. A small part of a chromosome may break off; having no spindle, it may be lost when cell division takes place (= deletion). Or a part of a chromosome may become attached to another chromosome (= translocation); or a part of a chromosome may turn over through 180°, most often as a result of the formation of loops (= inversion). As the genes of each chromosome are all inherited linked, translocation brings about new linkage groups. This is important for the process of speciation, because natural selection having an effect on one inherited character will operate on the whole group of linked characters. The importance of inversion lies in the fact that on reduction (Ch. 4 D) exchange between the inverted portions from the father's and mother's sides (crossing over) does not take place. This means that the genes from those parts remain associated and unaltered and so the characteristics they determine are kept constant for a longer phylogenetic period. This is particularly important when advantageous structures are involved.

Most *genome mutations* appear as a multiplication of the chromosome set. Such polyploidy takes place when the chromosomes divide without the nuclear membrane disappearing. As it is possible to induce this process experimentally by the use of colchicine, this has opened the way to an understanding of causal connections here. Polyploidy disturbs animals to a much greater degree than plants, particularly because of disturbances in the development of the sexes. Thus it probably affects species formation only to a minor degree. It is most evident among those animals which reproduce parthenogenetically. In plants, on the contrary, many races and species owe their origin to polyploidy.

Rather less is known about mutations of the cytoplasm. In general, mutation can take place outside the hereditary structure of the chromosomes only when nucleic acids capable of self-replication are present. Hence mutations are known in plastids (changes in the chlorophyll), and they may also possibly occur in mitochondria, kinetosomes, and centrosomes.

All the other factors of species formation can be dealt with more briefly, for their causal explanation is easily conceivable, their importance is obvious, and their effects have been fairly well analyzed. In the context of the philosophical implications of evolutionary processes it is chiefly important to show how the varied interplay of evolutionary factors determines the type of race and species formation, and how these processes then lead to the development of new organs and new anatomical structures.

As mutants are continuously produced and a small percentage of them always remain, it follows that every species of organism consists of a large number of different hereditary variants. Sexual reproduction, already developed in protists and predominant in all higher animals and plants, has always led to *new combinations of genes* in every generation. Because of the large total number of genes and continuous mutation, the variety of possible combinations is almost limitless. So we may assume that almost every individual in a population is genetically unique (Th. Dobzhansky 1951, p. 108). The only exceptions are the individuals produced by asexual reproduction; identical twins, triplets, etc.; or cases of polyembryony in invertebrates.

In the course of time, any surviving mutant may spread to the limits of its own population or those in contact with it. This *gene flow* is intensified by the tendency of expansion, especially by the younger individuals among higher animals, owing to overproduction of offspring.

If a population is relatively small, containing only some hundreds of individuals, even a few mutated individuals may constitute a considerable percentage. Such variants or certain gene combinations can spread more easily in smaller than in larger populations. It is equally possible, however, that the few mutants, independent of their selective value, may diminish or disappear. As most populations increase very markedly and are often multiplied after the reproduction period and then reduced to about their former number of individuals, these *variations in the number of individuals* speed up the evolutionary process. (For mathematical analyses, see S. Wright 1932, 1940; O. Mittmann 1940; etc.) This is illustrated very clearly among wall lizards, as in races of *Lacerta lilfordi* on some of the islets in the Mediterranean. The formation of distinct races is also intensified by the fact that the gene set of the founding individuals represented only a part of the total genetic material of the species.

All living organisms have an overproduction of progeny. This means that if the numbers are not to increase by geometrical progression, most individuals must perish before maturity. For example, nearly all the many eggs and young produced by a pair of song birds which would live eight years can perish, and only two individuals must survive if the size of the population is to remain constant. With other organisms which do not rear their young, such as the larger fishes and also trees, which are much longer-lived, up to many millions of the offspring must perish before reaching maturity. When the destruction of such vast numbers takes place, it is probable that normal or advantageous variants of whole populations have a better chance of survival than disadvantageous, weaker, or less fertile ones. Hence, *natural selection* is always operating, eliminating all structures or functional characters of the organism which display some weakness and preserving normal and advantageous structures.

Selection can affect the individual in various ways during the whole course of life: (1) by environmental factors such as variations of climate (for instance, extreme cold, heat, drought), or changes in oxygen or salt content in water, which are important for aquatic animals; (2) by enemies which eliminate animals lacking protective colors or protecting instincts, or having a weaker constitution or defective sensory organs, etc.; (3) by superior competitors, of the same or other species, for food or habitat; (4) by parasites or diseases; (5) by mating rivals or choice of those mating partners which usually react best upon those individuals of the other sex whose visibly acoustic or olfactory sexual characters are most conspicuous. Selection also operates to some extent with regard to the choice of the habitat; for variants whose corresponding instincts have undergone a change do not generally seek the specific habitat of their species, nor do they avoid unsuitable ones. In fact, every individual goes through a whole series of "selection tests," normally with very little chance of success in surmounting them. Selection may indeed be said to be "omnipotent."

The quantitative evaluation of the selective effect is very important. It has been calculated that a variant with even a small selective advantage, perhaps only 1 percent, may increase so greatly in number that in the course of many generations it replaces the form completely. In a small or medium-sized population a dominant mutant with a 1 percent advantage, representing no more than 0.1 percent of the population and producing fifty offspring each year, by the end of some 420 generations would be present in 99.9 percent of the population. If the mutant were recessive, this proportion would only be reached after about 25,400 generations (O. Mittmann 1940). This tempo of evolutionary change corresponds to observed facts; the origin of new subspecies (races) requires some thousands or tens of thousands of years or generations, whereas new species require a minimum of some hundred thousands or millions of years. During all these phylogenetic changes selective processes work on several different characters simultaneously. And selective advantages often amount to more than 1 percent, especially if there is competition between species.

The conceptions about selection have been confirmed by a great number of experiments. In particular, the selective advantage of protective and warning coloration has been proved in many cases. Furthermore, generations of related species, races, or different mutants have been studied in population cages, to determine the effects of selective advantage and the way in which a more vital or more fertile form can successively supersede a weaker one. Such investigations on mixed populations have also been successfully carried out in the field (see Th. Dobzhansky 1951, 1960). However, inferior or disadvantageous mutants are not always eradicated. A small percentage may survive, especially in "niches" free from competitors (annidation).

Selection operates on single disadvantageous or weaker characters. But it affects the whole individual, that is to say, the whole gene set. Wthin a population whose total gene combinations are decisive for survival and further phylogenetic development of the race or species, an increase of individuals with relatively advantageous traits normally takes place; but some or all individuals may also have single defects (often correlated to advantageous features). The result is a more or less balanced polymorphy, which makes survival possible in spite of selection of various kinds. As the selecting environmental factors always fluctuate and alter in consequence of extreme climatic phases, the appearance of new enemies or competitors, and so on, every plant or animal species must provide a great number of variants, only a few of which may be able to withstand extreme conditions of selection (preadaptation). This fact emphasizes once again the great advantage of sexual reproduction, because of the enormous variety of genetic combinations it makes possible. It also explains why *sexuality*, developed so early during phylogeny, has been retained from unicellular creatures right up to the highest plants and animals including man. If the process had not had this vital importance for the survival of the species, it would have disappeared long ago, like so many other structures and functions. A certain selection pressure has obviously operated in most organisms in favor of retaining sexuality.

Thus we see that natural selection has a stabilizing effect (see I. I. Schmalhausen 1949). The same applies to many other characters of organisms. "Successful" structures are usually retained over long periods in the course of phylogeny. Since their first appearance, mitochondria, for example, have continued to be an important feature of every organism. After the coelenterata developed nerve cells and muscle cells, and higher worms blood vessels, these structures which enable the organism to function more efficiently have been retained and further developed in all the higher animals (for other examples, see B. Rensch 1947, 1954, 1960a; Ch. 7 B). Wherever mutation and selection have led to the development of advantageous structures and functions, selection has had a progressive or creative effect. As mutations are undirected, and as many mutations and processes of selection must be gone through before a new type of cell, tissue, or organ is formed, development of this sort has proceeded step by step. Hence, selection has also had a canalizing and normalizing influence (see C. H. Waddington 1957). And it has often led to greater adaptability (I. I. Schmalhausen 1949), to increased plasticity of structures and functions (B. Rensch 1947, 1954, 1960), and to homeostasis (J. M. Lerner 1954).

With regard to our philosophical considerations we must now ask whether the processes of race and species formation can be sufficiently explained by undirected mutation and natural selection, gene combination, gene drift, gene flow, and isolation. Most researchers would certainly agree that they can. The idea first put forward by Charles Darwin that the marvelous purposefulness and harmony which we find in all living organisms can be explained by causal processes (Ch. 3 A VI) has proved to be a reliable conception. There is no need to assume the existence of any purposive, harmonizing "evolutionary force" beyond the scope of biological analysis. The only process not yet fully explained in a causal context is the release of mutations. And even here some of the chemical and physical factors behind it have been discovered, and we may assume that further progress will be rapid.

It is important to bear in mind, however, that with so many different combinations of factors at work, there are different types of species formation (B. Rensch 1947, 1954, 1960a). "Mutation pressure" may prevail and cause undirected race formation independent of external factors, as shown by the many races of birds and butterflies in tropical archipelagos. "Selection pressure," on the other hand, is often conspicuous in areas of temperate and cool climate, causing different species inhabiting the same area to develop the same characteristics. For example, races of warm-blooded animals usually attain a larger body size in regions with low winter minima (Bergmann's rule). But in small isolated populations, like those on the Mediterranean islets (scoglia) already mentioned, "population waves," that is to say, fluctuations in the number of individuals, can already lead to race formation. So too can isolation based on sexual physiology, when variants have developed a different reproduction period from the rest. Sometimes, especially with higher plants, new races and occasionally species as well have come about through secondary hybridization among existing races or species.

Geographical races are the chief prestages of new species. Most species of mammals, birds, reptiles, amphibians, freshwater fishes, several orders of insects, and land snails are made up from a mosaic of geographical races—or rather, many forms formerly described as species must now be recognized as geographical races of polytypic species ("Rassenkreise"). But geographical variants are also found among most other animal groups and many groups of plants. Such races are not merely *secondarily* developed components of species, and speciation is not different in principle from race formation, as R. Goldschmidt (1935, 1952) and some others have supposed. This is proved by the presence of many borderline cases appearing in all geographically varying groups: some forms may "still" count as geographical races or may "already" be regarded as species. It is probable and in some instances certain that such cases have been isolated for longer and are thus phylogenetically older than normal geographical races (observed, for example, on islands isolated for different periods). New races and species of many parasites have developed as a result of purely ecological isolation in or on different animal or plant hosts. Here too, many borderline cases between races and species exist.

The knowledge of mechanisms leading to natural race formation has also allowed the breeding of many artificial races with specific or desired characteristics. In many plants especially, but also in the fruit fly *Drosophila*, it was even possible to breed new species, i.e., new forms, which are infertile when crossed with parent forms.

Summing up, we may state that various methods of speciation have already been analyzed, and a fairly complete explanation has been established.

C. Transspecific Evolution

1. The Problems. Mutation, as we have seen, is typical of all living organisms, and the new gene combinations which it brings about are continually subject to natural selection because a surplus of progeny is always being produced. Paleontological evidence shows us that species formation has always taken place, though the rate of alteration has varied greatly. Knowing these facts, however, does not solve all problems of evolution, for the lines of descent display other pecularities and rules, which are not yet observable in the process of speciation. In many cases phylogenetic development maintained a certain trend, sometimes leading purposefully to a different kind of structure or a particularly advantageous one. We may also establish other rules of phylogenetic development. After a new structural type had appeared, a period of rapid radiation of new species and genera set in, followed by a period during which evolution slowed up and fewer new forms emerged, and leading to a period of temporary stability. Many groups finally died out and some of them developed excessive or degenerate forms before doing so. Moreover, not only has the phylogenetic tree continually thrown out new branches, species, genera, and families; both animals and plants have produced new types of constructions—new, highly complicated and purposefully structured and functioning organs. Besides pure ramification or *kladogenesis* (phylogenetic branching) in many lines of descent, evolutionary progress—*anagenesis*—also occurred (B. Rensch 1947, 1960a), leading in one line to the emergence of man.

Can we attribute these transspecific alterations and rules to the same factors as those which underlie species formation? In particular, can undirected random mutations—which, so to speak, represent "accidental" failures of the replication process of the hereditary units—and selective factors produce such regular and meaningful changes, and can they lead to progressive evolution? Several biologists have doubted this. They have assumed that autonomous directive forces or other purposive and creative evolutionary principles must exist. But such factors, which should be discernible also at the level of speciation, at least to a certain degree, could not be proved. This assumption was only an expression of imperfect comprehension of what lies behind the facts.

It is however possible to explain new organs and types of construction, the development of phylogenetic rules, and progressive evolution in terms of mutation, selection, and other well-analyzed factors. Evidence in favor of such causal explanation comes from widely differing fields of research which are independent of one another: paleontology, comparative anatomy, taxonomy, and genetics. It is impossible to discuss adequately the immense number of relevant facts in the context of our philosophical considerations. We must restrict ourselves to the most important findings and conclusions, and refer the reader to those writings which have laid new foundations of a causal explanation of the whole phenomenon of evolution: Th. Dobzhansky (1937, 1951); J. Huxley (1942, 1963); E. Mayr (1942, 1963); G. G. Simpson (1944, 1949, 1953); B. Rensch (1947, 1954, 1960a); I. I. Schmalhausen (1949); and

G. L. Stebbins (1950). Recent studies also include *Die Evolution der Organismen* (ed. by G. Heberer), the lectures delivered at the *Cold Spring Harbor Symposium on Genetics and Twentieth Century Darwinism* (Vol. 24, 1959), and the lectures of the symposium *Evolution after Darwin,* edited by S. Tax (1960).

II. Continuance of the Factors of Speciation. Detailed studies of many plant and animal groups have shown that the undirected character of mutation primarily led to undirected formation of races and species but that this process has been controlled and directed in certain lines of development by natural selection, according to the conditions of life. Yet the primary undirected character is still often conspicuous, even in the development of new genera, families, and other higher categories. The supply of new gene combinations, i.e., variants with alteration of many characters, which appeared in each generation, has not only produced advantageous types from time to time but has also ensured that everything that was biologically tolerable survived. In many cases even, *all* possible directions of structural and functional alterations have been evolved, so far as they were viable.

A few examples may illustrate this. Alterations in the development of single organs or structures in genera of frogs sometimes mainly affected early embryonic stages (for example, the yolk sac in *Eleutherodactylus*); at other times late embryonic and larval stages (for example, clinging organs of larvae living in rivulets); but usually the juvenile or adult stages. Among many animal groups, changes in growth ratios of various body parts have led to the development of different genera and families with very heterogeneous structural characters. This is particularly well illustrated among crustaceans by the genera of the isopods (B. Rensch 1947, 1954, 1960a; Fig. 15), amphipods, and decapods; among insects by the genera of beetles; and among vertebrates by bony fishes, especially grotesque deep-sea fishes. Certain plants produced forms of flowers in an undirected manner, especially the Orchidaceae and Scrophulariaceae; and many groups of plants, such as ferns, have developed nearly every possible type of growth: herbs as well as shrubs, climbers, tree, and aquatic genera. Butterflies may hibernate as eggs, young or old caterpillars, pupae, or in the adult state.

In addition, undirected evolution is proved in several animal genera and families by the development of structures of so exaggerated a size or shape that they can only be described as "luxurious" features. Certain beetles, for instance, have excessively long feelers, mandibles or legs, or protuberances on the thorax; hornbills have excessive beaks. Other developments, too, appear quite purposeless: regressive and useless rudiments such as the vestigial pelvis and legs of certain whales and snakes and the tiny stunted wings of flightless insects. There are also unnecessary "détours" of individual development retained from ancestral types. The armadillos, anteaters, and whalebone whales, for example, developed vestigial tooth buds which became reduced later on. All these examples can be regarded as "construction failures" which have only survived because they were biologically tolerable.

Summing up, we may say that undirected mutation and natural selection affect the evolution of higher categories in the same way as the formation of races and species.

III. Particular Features of Transspecific Evolution. The conception that evolution is mainly caused by the combined effect of mutation and selection has long been regarded skeptically, because many lines of descent which are well established by fossil remains show that certain evolutionary tendencies have persisted over considerable periods. In many animal groups, especially in mammals, body size always continued to increase (Cope's rule). At the same time, the facial bones increased in proportion to those of the cranium, and the forebrain became larger in relation to the whole brain, and also more convoluted. In the well-known horse line of the Tertiary period the number of toes has been reduced successively. It was tempting to assume that some autonomous and directive evolutionary factors were at work.

But most of these lines of development, these "orthogeneses," can be explained in terms of stable selective factors. During the late Tertiary and the beginning of the Pleistocene, the annual minimum temperature continued to drop. This favored larger variants: because they had relatively less body surface, they did not lose as much heat as did the smaller variants. The larger variants (or species) in competition with smaller species in every generation also enjoyed further advantages; they were stronger, swifter in aggression or flight; and their absolutely larger brain contained more complex ganglion cells with richer ramification.

The increase in some relatively fast-growing structures (i.e., structures growing mainly with positive allometry) was linked with the general increase in body size. Hence the facial bones increased faster than the skull as a whole. Variants or species with a larger forebrain (or regional increase) and therefore a more convoluted cortex were capable of better and more versatile performances. Horses with fewer toes could run better, so the number was reduced by selection from four on the forelegs and three on the hind legs to a single toe on each. But such tendencies were neither as invariable nor as constant as those who favored a theory of autonomous and directive factors supposed. Basing his findings on an immense amount of Tertiary fossil material from America, G. G. Simpson (1959) was able to show that these tendencies were subject to considerable changes in intensity in many branches of the horse family. In some cases there was even a decrease in body size, in other words, a reverse direction of development.

The conditions under which many other lines of descent lived—as far as we can estimate these for epochs in the geological past—indicate that the retention of evolutionary tendencies was primarily due to stable selective factors, to orthoselection. Among well-known lines of descent of mammals one might cite the increasing differentiation of the canines and the scissorlike molars in the families of cats (Felidae) and dogs, the development of successively more flattened molars among elephants, the gradual disappearance of the pelvis in sirenians, and the relative increase in brain size and particularly the relative enlargement of the forebrain in many mammalian orders. Besides, if the main selective factors remained the same, this can explain how adjustment to very different conditions could be made so quickly and many structures showed alterations at the same time. When, for instance, certain crossopterygians changed over to life on land in the Devonian period, all those variants were favored by selection which had a rather tougher skin and better organs for moving about on land, for breathing air, and for seeing objects at a distance.

Something further seemed to point to autonomous directive tendencies—the cases of "overspecialization," when some meaningful development appeared to overreach itself. Natural selection did not seem to account for the huge, top-heavy antlers of the Ice Age Giant Elk (*Megaceros*), the mammoth's tusks curving backward, or the giraffe's excessive length of neck. It was not realized, however, that exaggerated structures like these came about by positive allometric growth in proportion to the body. As this allometric tendency persisted when body size increased, it was bound to lead to excessive growth of this kind. The mammoth's tusks, for example, simply continued the process of curved growth. Selection obviously operated on body size; for the ancestors of the Giant Elk, the mammoth, and the giraffe were much smaller. The giant animals "put up with" these excessive structures for the sake of the advantages of being larger.

Another argument put forward against an explanation of evolution in terms of mutation and selection is that related lines of descent often show parallel development, that they underwent similar structural changes quite independently and reached a higher level by parallel routes. E. Dacqué (1935) used the expression "time-signatures" in this connection. A typical example is the parallel development of characteristically mammalian features in five lines of descent deriving from the reptilian family Therapsidae (G. G. Simpson 1959). The chief features were the development of a secondary maxillary joint (the squamosum-dentale joint), the reduction and successive transformation of two jaw bones (quadratum and articulare) into auditory ossicles (malleus and incus), differentiation of teeth (these three alterations probably connected with the change-over to chewing), and a change in the proportions of the pelvis. To this extent it is possible to speak of the class of mammals as having a *polyphyletic origin*, though the various lines all ultimately derive from a single reptilian group (see also A. S. Romer 1966; H. Frick and D. Starck 1963). In the reptile class, the characteristic features developed in a corresponding manner in probably seven distinct lines (E. Kuhn-Schnyder 1963) and in amphibians in at least three. But each of these polyphyletically developed classes ultimately derived from one ancestral group. The later lines of descent had brought along with them certain inherited characters which allowed the parallel emergence of new advantageous characters by natural selection. More efficiently functioning structures were selected, and this led to progressive development. Another instance of parallel evolution, already mentioned, is the relative enlargement of the mammalian forebrain and especially of the cortex, proved by casts of the cranial cavity of fossil skulls. Such enlargements were successful in competition because a larger forebrain generally leads to more complex performance and more versatile reaction.

Among more nearly related groups, parallel development of less significant features is fairly common. They are based upon the fact that related species or genera share some hereditary units, leading to the same or similar gene combinations and to homologous mutations. V. Haecker (1925) called this "paripotency." It is clear, then, that parallel development also can be explained in terms of mutation, gene combination, and selection.

Another problem for which an explanation by random mutation and selection seemed to be unsatisfactory is that of the *emergence of complicated organs*. The eye of the vertebrate has often been used as an example with such objections. Its parts are made up of very many different types of tissue: the firm sclera, the nourishing choroidea, the transparent cornea, the iris, the lense, the vitreous tissues, and the retina with its three layers of different nerve cells and sense cells. It seemed inconceivable that such an organ could have developed as a result of utterly "random" mutation of the many genes and gene combinations necessary for these several structures, and of natural selection. We cannot trace the phylogenetic development of the eye of vertebrates. We only know that its structure in the different classes of recent vertebrates shows an increasing degree of perfection. Compared to birds and mammals, the lower vertebrates have relatively fewer eye cells and their mechanisms of accommodation are not so perfect. The majority of warm-blooded animals alone have a fovea centralis which eliminates most of the distortion caused by intervening layers of ganglion cells and fibers.

Certain invertebrate species, especially some marine polychaetes and many snails, possess similarly structured vesicular eyes with a cornea, lense, retina, and a protective pigment layer, whereas other species within these groups have more primitive eyes. Some have only shallow or deeper cup-shaped eyes allowing an accumulation of photosensitive cells, which are protected by pigment cells. Others have spherical vesicular eyes with a larger number of visual cells behind a transparent skin layer; others again also have a ball-shaped lense where the light rays are concentrated (see Fig. 101 in B. Rensch 1947, 1954, 1960a). These various stages can be regarded as models which illustrate the probable phylogenetic development of complicated vesicular eyes. The important point is that random mutation could bring about each of these higher stages. It could produce variants with increased sensitivity to light; concentration of such cells and combination with pigment cells; increase in the numbers of visual cells leading to a depression or ball-like cavity of the sense region; increase in light effect through a ball-shaped transparent substance (a lense) with improved refractive exponent. Every one of these stages could come about by mutation, and each offers some selective advantage: simple discrimination of light and dark by means of light rays; indication of the direction of light by a cup-shaped depression; retinal image produced by the ball shape of the depression and development of a transparent pupil; greater effect of light through a lenselike structure; and eventually, some kind of picture seeing, that is to say, reaction to restricted stimulus patterns. It is possible that the ancestors of lower vertebrates may have developed their eyes by corresponding stages.

In order to understand the evolution of complicated organs it seems preferable, however, to avoid speculation of this kind as far as possible, and to concentrate on examples of development for which evidence exists in the form of fossil material. In most of the relevant cases we do in fact find that new organs emerged step by step, and that random mutation and natural selection provide an adequate explanation for every stage. Besides, the succession of changes did not always lead in one direction; there were often alterations of different kinds, and only the more advantageous of these led to progressive development. There was also frequently a complete *change of function*.

Let us take, for example, the phylogenetic development of the human arm and hand, organs which have been vitally important for man's physical and mental evolution because of their versatility and adaptiveness. The evolution began when primitive fishes developed ventral fins for steering and also partly for support. In the crossopterygians these fins already contained some bones which were homologous to arm bones of higher vertebrates. In the Devonian period, when the stegocephalians, descending from a group of crossopterygians, became the first land vertebrates, their limbs supported by hinged bones proved useful for moving on land. Some of the fin tips developed into fingers. The arms and fingers of reptiles and of the lower mammals which evolved from them during the late Triassic period showed structural improvement. Later, at the stage of the primates, the opposable thumb gradually developed, allowing the animal to grasp branches and hold its food securely. Natural selection in these climbing animals also caused an improvement of the shoulder, hand, and finger joints. The innervation of the hands became finer, and impulses could pass more directly from forebrain to fingertips by way of the pyramidal tract. All these stages involved structural alterations in the fore limbs; and they could all have taken place as a result of random mutation and—because of their functional advantage—natural selection.

A double change of function took place in the course of phylogenetic development of the auditory ossicles in mammals and man. These are three small bones—the hammer, the anvil, and the stirrup—which transmit sound waves from the eardrum to the inner ear. The stirrup derives from the columella, the only auditory ossicle in the mammals' reptilian ancestors; the anvil from the Os quadratum, an upper jawbone in reptiles; and the hammer from the Os articulare, a lower jawbone. In reptiles, the quadratum and articulare had formed the joint between the upper and lower jawbones. As mammals emerged, both these bones became successively smaller and were transferred to the middle ear, while a new joint developed in the jaw. This remarkable change in function was probably caused by the intensification of chewing in mammals. The columella itself goes back to the hyomandibular bone in fishes, which had a different function: it acted as a "mandibular shaft" between the cranial and facial bones. Earlier still, it had derived from one of the cartilaginous gill plates of the Agnatha, the most primitive of all fishes. Random mutation can quite well have been responsible for these changes in size, shape, and function which altered the growth gradients, that is to say, the allometric tendencies. And natural selection was operating at the successive phylogenetic stages in different directions determined by the gradually changing functions of the relevant bones. It must of course be remembered that our real knowledge is limited to the bones themselves and their correlations. We are not in a position to estimate sufficiently the particular selection conditions at early phylogenetic stages.

The *emergence of new types of construction*—a typical feature of higher categories, whole orders, classes, etc.—is closely connected with that of the development of new organs. In many cases we can trace clearly the course of the gradual alterations of structures in the lines of descent. Reptiles and mammals, for instance, certainly exhibit great anatomical and physiological differences. Yet the main steps of alteration by which these two classes became differentiated are well illustrated in some of the therapsids of the Upper Trias and the first true mammals of the Jurassic and Cretaceous. Such types show the gradual specialization in dentition, the evolution of the three auditory ossicles already mentioned, the development of a double (in place of a single) Condylus occipitalis, the adoption of erect bearing on the soles of the feet or the toes, and so on.

As it was an advantage for mammals to be warm-blooded, viviparous, to rear their young, and to have a relatively larger and better-constructed brain than the reptiles, they could quickly come to predominate and to adapt themselves to different habitats and ways of life. This rapidly led them to split up into many different orders. Here too we have fossil evidence of these gradual changes. Intermediate forms show how the insectivores of the late Cretaceous led on to the carnivores, ungulates, and lemurs of the early Tertiary period. The evolution depended mainly on the alteration of the proportions of many structures, i.e., their growth gradients, and this involved physiological changes. Such alterations of growth gradients could always have come about through random mutation and natural selection, in the same way as breeding has produced races of domestic animals with very different proportions; for example, greyhounds, terriers, dachshunds.

For some specimens, however, we have little or no fossil evidence for the gradual transition from one type of construction to a radically different one. These *gaps in our knowledge* have led to a good deal of speculation about unknown evolutionary processes. But the more probable explanation is that alterations occurred relatively rapidly, because selection operated very intensively as the animals adapted themselves to entirely new conditions. So transitional forms only existed in a relatively short period, and the chances of finding such fossil remains were correspondingly remote. For instance, selection must have been rigorous when some of the crossopterygians became the first land vertebrates. The change-over to life on land involved a number of simultaneous alterations. Variants had to alter their skin, their organs of locomotion and breathing, their sensory organs, etc. It is only by a particularly happy chance that examples of transitional forms from this short-lived phase have been found, such as *Elpistostege*, a link between the crossopterygians and the earliest amphibians.

If the entire site in the lithographic limestone near Solnhofen had been exploited before the fossil remains were found there, we should be relatively uncertain about what the earliest birds looked like. But the three *Archaeopteryx* specimens found there have thrown light at least on certain essential points. We may be happy that intensive paleontological research during recent decades has been rewarded by the discovery of a number of such transitional forms that establish a link between orders and classes of both animals and plants.

Summing up, we may state that transspecific evolution, that is to say, the emergence of higher taxonomic categories with novel organs and new types of construction, can be regarded as a continuation of infraspecific evolution. It too is due to mutation and selection, and there is no need to assume processes involving unknown evolutionary principles (Ch. 6 K IV). All evolution is ultimately based upon causal laws, though it is difficult to grasp their complicated interactions. As the laws of infraspecific and transspecific evolution also include the laws of progressive evolution, they are of great importance for our philosophical considerations. So we must now discuss them in more detail.

D. Lawfulness of Evolution (Bionomogenesis)

I. The Particular Nature of Biological Laws and Rules. The biochemical and structural complexity of all organisms including even the most primitive bacteria makes it difficult, and up till now at times impossible, to trace life processes back to their ultimate causes. A great deal of examination has convinced us that these processes are causal, but because of mutual influences they cannot be as clearly analyzed as chemical and physical processes of the inanimate. It follows that there are exceptions to almost every established biological law. Such laws are therefore generally described only as *rules*. We must bear in mind, however, that we have to do with ultimately lawful events partially crossing with one another (see Ch. 2 A). Even what we might describe as a "random" happening in the sphere of biology is bound up in causal relationships, though their analysis may be impossible (Ch. 6 K IV).

So the laws of evolution which we are about to discuss are in fact rules, in the above sense of the term. It is impossible to follow every hereditary alteration within a natural population, for the gene combination of almost every individual is unique. Nor can the co-activity of many changing selective factors be sufficiently analyzed under natural conditions, though in every case such action is causal in origin. Yet in spite of these difficulties certain rules can be drawn up; and their number has gradually grown to a point where the lawfulness of evolution becomes conspicuous (B. Rensch 1960b). The implications of such lawfulness vitally affect our philosophical picture.

A biological rule cannot be established simply by collecting a number of relevant examples. One has to prove that these examples predominate and that exceptions are comparatively rare. Rules of evolution can be proved by exact calculations of significance in only a few cases. But it is possible to evaluate their validity fairly well, and for this reason the rules have become more or less generally accepted. We also have to include here a number of primarily physiological laws and rules which apply to phylogenetic processes, because they impose certain limits on the possibilities of evolution.

This is not the place to discuss or confirm all these rules in detail; for that the reader must consult the specialized literature on evolution, genetics, and the relevant branches of ecology and physiology. In a symposium lecture the author set out sixty-three such rules (B. Rensch 1960b); and in another lecture (1964) suggested how they might be classified. A number of those rules are enumerated here, to show their importance for both infraspecific and transspecific evolution and because their implications have a bearing upon natural philosophy.

II. Rules of Infraspecific Evolution. 1. Proportionate to time, individual "spontaneous" and nondirected mutation occurs in all organisms. These mutations may be recessive, or more or less dominant; they are in most cases harmful and often reduce vitality or fertility.

2. Mutants contribute to evolution only if their effect on the individual development of the normal structures and functions is not such as to prevent the individual from reaching the age of reproduction.

3-5. When mutants of sexually reproducing species lead to the establishment of new gene combinations by fertilization, Mendel's three rules will hold good. The third, however, applies only if the genes in question are not situated on the same chromosome.

6. In the course of the generations, mutants spread throughout the population, race, or species (gene flow).

7. Species are limited to particular habitats by their specific requirements of food, shelter, and opportunities for reproduction. Natural selection is continually operative among individuals, populations, or competing species, through their varied ability to adapt themselves to their habitat.

8. All organisms produce a superabundance of progeny, of which the majority (often over 99 percent) die before reaching sexual maturity. This overproduction causes selection to become especially intensive.

9. As a consequence of the steadiness of mutation and natural selection a continuous transformation of races and species takes place, though at very different and changing speeds.

10. Mutation, gene flow, natural selection, expansion of the species in consequence of overproduction of progeny, and isolation lead to splitting of races and species and hence to a ramification of the phylogenetic tree.

11. There is more rapid emergence of new races and species among small, isolated populations, and those which derive from relatively few individuals, than among very large ones.

12. Less mobile forms, whose populations are more sexually isolated, split off into races more easily than do more mobile or passively spreading forms.

13. Geographically isolated races composed of a relatively small number of individuals tend to become homozygous.

14. In marine species, race formation is less common and the tempo of species formation is slower than in land animals, because their populations are usually larger and more uniform, environmental factors are more stable, and therefore selective conditions are less varied.

15. In general, races and species are differentiated in proportion to the length of time during which they have been isolated.

These rules all apply to the majority of plant and animal groups. In addition, there are a number of less general rules which govern race and species formation in certain phyla, classes, or orders. These include the following:

16. Geographical races of warm-blooded animals inhabiting cooler regions, especially those with low winter minima, are generally larger than those inhabiting warmer regions (Bergmann's rule).

17-22. Races of mammals inhabiting cooler regions have relatively shorter ears, tails, and feet (Allen's rule), relatively longer hair, and more underfur; and they produce more young ones per litter than those races of the same species which live in warmer regions.

23. The underplumage of birds in colder regions is denser than that of related races in warmer areas.

24-26. The hair or feathers of warm-blooded animals inhabiting warm, moist regions contains more melanin than that of races which live in drier and cooler regions. Melanin-pigmented races of warm-blooded animals in dry, warm zones mostly show light yellowish or reddish brown coloration. In colder regions the reddish and yellowish phaeomelanins are more or less absent, and also the dark eumelanin is reduced (Gloger's rule).

27-29. Geographical races of birds in cooler regions have relatively shorter beaks and feet than races of the same species in warmer regions, and they produce more eggs per clutch.

30. Migrant birds have developed more pointed wings than non-migrant races of the same species.

31. Among bird species which are distributed from cooler or temperate to tropical zones, those races which inhabit the warmer regions show little or no inherited migrating instinct.

32-36. The heart, stomach, liver, and kidneys of tropical birds are relatively smaller, and the intestines relatively shorter, than in races or nearly related species in temperate or cold zones.

37-38. Tropical finches have a slower basic metabolism and pulse rate than related races or species living in cooler regions.

39. Races of poikilothermic animals are on the average hereditarily larger in habitats with optimum conditions than in regions where the conditions are unfavorable.

40. Races of ground beetles inhabiting warmer regions have rela-

tively longer legs and feelers than those of the same species living in colder regions.

41. Insect races which have survived the last glacial period in refugial areas display greater hereditary variability than races in areas settled in the postglacial period.

42. Races of butterflies in regions of cooler climate produce fewer generations in the course of a year than do races of the same species in warmer regions.

Similar rules govern the formation of races and species in other animal groups. Many such rules could also be established for plants; they would be concerned with dependence on special conditions of the type of site, as well as adjustment in the shape of the flower to possible pollination by birds or insects, or adaptation of the color of flower to the kind of vision possessed by pollinating animals. The fact that more polyploid forms are found where environmental conditions are difficult could also rank as a rule. In general however, many further rules governing formation of races and species in plants and animals have not yet been adequately established or formulated.

III. Rules of Transspecific Kladogenesis. A much larger number of rules can be deduced only from the study of longer lines of descent, or by comparing extant species, genera, or families deriving from a common ancestral group. Here too we shall simply refer to a number of zoological examples, in order to indicate the multifariousness and diversity of these rules.

43. Continuous natural selection normally leads to progressive adaptation to environmental conditions in the specific habitat (adaptiogenesis).

44. During all phylogenetic changes the tendency has prevailed to maintain sexuality, for by fertilization a sufficiently large number of very different gene combinations were made available for natural selection, and the preservation of the species was better ensured. Moreover, diploidy rendered most of the harmful characters of the largely recessive mutants ineffective.

45-46. The emergence of new advantageous types of structure is usually followed by a period when many new species, genera, or families arise. Later on, this form radiation tends to decline in the lines of descent, in the same measure as available habitats are occupied by specifically adapted forms belonging to the animal groups in question.

47. Natural selection operates continuously toward a perfect adaptation, and then toward maintaining this condition for a relatively long period (stasigenesis).

48. Early ontogenetic structures are generally less affected by phylogenetic alterations than are later stages (= law of conservative first stages). The sequence of ontogenetic stages therefore often illustrates the course of phylogenetic development (biogenetic rule).

49. Phylogenetic development tends to abbreviate successively the détours in the course of ontogenetic differentiation which result from the retention of conservative phases.

50. There is a tendency—most marked in larger flightless land animals—toward a progressive increase in hereditary body size (Cope's rule).

51. Large species, especially those of vertebrates and arthropods, have relatively smaller heads and brains than related smaller species (Haller's rule).

52. In tropical regions a larger number of species of land animals and plants exist than in cooler ones, but in the latter regions there are a very large number of individuals within each species.

53. In ecologically comparable habitats of different continents (for example, damp tropical forests) the percentage of large, medium, and small species of birds—as well as the percentage of trees, shrubs, lianes, epiphytes, and ground plants—are similar.

54. Animal groups forming the principal prey of enemies which hunt by sight, such as insects devoured by birds, or smaller birds and mammals eaten by birds of prey, have evolved protective coloring or shape, threatening coloring, or mimicry.

55. Birds which hatch their young in holes almost always lay light-colored, usually white, eggs; eggs hatched in the open mostly show protective coloring.

56. There is a higher percentage of brightly colored species among tropical birds than among those inhabiting temperate or cold zones.

57. If several similar species of birds of the same genus which recognize their sexual partners by sight live in the same habitat, the sexes

tend to differ more markedly in coloring or call than if only one such species is present.

58. Animal species whose young receive some degree of care normally produce fewer offspring than those without these instincts.

59-60. The respiration rate and pulse rate are slower in large, warmblooded animals than in smaller related species.

61-63. The embryonic period is longer in large, warm-blooded animals than in nearly related smaller species; they take longer to reach sexual maturity, and they live longer.

64. Large animal species more frequently develop excessive structures —excessively long legs or necks, bony processes, extremely long antennae, processes on the head or thorax in insects—than related smaller species.

65. Large beetles show more marked morphological sexual differentiation than related smaller species.

66. Giant species of land reptiles (giant saurians) and mammals tend to develop columnar legs.

67-68. Animals of the steppes and the savannahs which are specialized for running tend in the course of phylogenetic development to evolve longer legs and fewer toes.

69. Insects as well as vertebrates which developed locomotion by jumping evolved powerful femurs, long and relatively thin tibiae, and narrow feet.

70. In the course of phylogenetic development, fast-swimming vertebrates evolve a slender, drop-shaped body.

71. Gliding birds developed long, narrow wings in the course of phylogeny.

72-75. In the course of phylogenetic development, insects that live in caves show an involution of eyes and skin coloring; they evolve longer antennae and a relatively soft chitinous covering on the body.

76. The brain of larger insect species contains a greater number of neurons than those of related smaller species.

77. The brain of larger insects shows mushroom bodies with a larger upper surface, and more globuli cells, than those of related smaller species.

78. Large vertebrates produce larger motor neurons than related smaller species.

79. Larger species of bees and wasps have more complicated instincts than related smaller species.

80-81. Large vertebrates have a greater learning capacity and a longer span of memory than related smaller species.

82. The eyes of large vertebrates have relatively smaller lenses than those of related smaller species.

83-84. The eyes of large mammals have relatively smaller pupils and a flatter lense than those of related smaller species.

85. Large vertebrates have a relatively thinner retina than related smaller species.

86-87. Social animals developed more pronounced instincts of precedence and communication than related nonsocial species.

With our present knowledge we could add greatly to the number of these rules. Every biologist could supply some from experience in his own field. It must also be borne in mind that all cytological, histological, anatomical, physiological, and developmental rules, and ultimately the corresponding physical and chemical laws as well, have a limiting effect and thereby also determine the evolutionary processes. Thus mammals cannot normally develop any system of blood vessels that is not in agreement with the Hagen-Poisseuille law, because such a system would not supply the needs of the organs. Nor could the shape of the lense of the eye in vertebrates alter in such a way as to comply with the laws of refraction. Similarly, every phylogenetic alteration in the epithelium of the lung is restricted by the laws of diffusion. If a bird exceeded a certain body size, it would cease to be able to fly, for while the effectiveness of its two-dimensional wings would only increase by the square, its three-dimensional body would increase by the cube.

IV. Rules of Anagenesis. As already mentioned, the phylogenetic progress of organisms has been put forward in the past as one of the chief arguments for assuming that unknown, purposive developmental factors are at work. So it is important to state that anagenesis also is governed by certain rules which are causally conceivable. The following rules, which apply generally to all animal groups, or at least to higher animals, may be sufficient to exemplify this.

88. Most lines of descent in which progressive evolution successively

took place derived from relatively unspecialized forms (Cope's "law of the unspecialized").

89. There is a tendency among most main lines of descent to retain any particularly advantageous structures, organs, and functions that have been developed. The result is that many such features accumulate in the course of phylogeny.

90. An increase in structural complexity, especially an increase in the number of cells, usually led to greater differentiation of labor among the tissues, and therefore to more efficient functions.

91. Those variants or species whose structure and function were more efficient were more successful in the competitive struggle. This contributed to progressive development.

92. Functions tended to become centralized during the course of phylogenetic development. This also contributed to progressive development (emergence of the brain, heart, liver, kidneys, etc.).

93. In the course of phylogenetic development, organic functions and behavior gained versatility as more and more control mechanisms and regulatory feedback systems came into play.

94-95. The eyes and ears of land vertebrates tended to improve in efficiency as they developed more sensory cells and mechanisms for accommodation and adaptation.

96. The major lines of descent of higher land vertebrates show a fairly general tendency to improve the efficiency of the brain, especially the capacity for learning and the ability to generalize.

97. Lines of descent of mammals show a phylogenetic tendency to display a successive relative increase of the forebrain.

98-100. In the phylogeny of mammals the tendency prevailed to enlarge the number of cortical neurons and to develop more specialized functional areas and associative regions.

As all these rules of anagenesis led to structural and functional advantages, they can be regarded as caused by natural selection.

V. Conclusions Drawn from the Rules of Evolution. The reason for enumerating so many rules was to show their wide scope and to emphasize their theoretical importance, which has hitherto been largely underestimated. And we have noted exactly one hundred rules in order to indicate by this round number that many more could have been formulated, especially for smaller groups of organisms, for families, and for genera embracing many species. In addition, the already mentioned indirect "regulatory" effect of many physiological and developmental laws and correlations which restrict the directions taken by evolution is most important. Some of them are ultimately laws of biophysics and biochemistry.

Even at the present stage of our knowledge, then, it is clear that evolution is a largely lawful process, and with regard to the effect of continuous mutation and natural selection it is also a *determinate process*. The evolution of organisms already began in the prestages of life, and it automatically led the phylogenetic tree to branch out more and more. This bipartite tree includes all former and extant species of plants, animals, and man; and its live branches represent the species of our own day. Moreover, it stands "erect," that is to say, many of its branches show anagenetic development.

If we want to evaluate to what extent it is already possible to recognize phylogenetic alterations as causally determined, the best method is to apply the rules to certain individual cases, and to estimate the *degree of predictability*. It may be sufficient to illustrate this by the following two hypothetical examples, already published elsewhere (B. Rensch 1960b, pp. 110-111; 1961b, pp. 303-304).

1. If a species of songbird from a temperate zone spreads to a tropical region with a warm, damp climate, and there develops a new geographical subspecies (or a new species) adapted to this different environment, then we may expect with a probability of 70-95 percent the following new hereditary characters: smaller body size; relatively longer beak and feet; relatively larger head, brain, and eyes; less pointed wings, the second and third primaries in particular shorter in proportion to neighboring ones; relatively shorter contour feathers; sparser down feathers; feathers richer in melanin; heart, liver and kidneys relatively smaller; intestines relatively shorter; fewer eggs per clutch; no instincts to migrate; lower basal metabolism; reduced heartbeat frequency; shorter life span.

These characters have been quite well established whenever a tropical form has been recognized as belonging to a polytypic species otherwise found only in a temperate region.

2. If late Tertiary sediments were to yield fossil remains of a carnivore belonging to a line of descent hitherto known to exist only in the Eocene or Oligocene, we might expect that the skeleton would differ from those of its ancestors in the following respects: body size larger; skull relatively smaller; facial bones relatively longer; cranial cavity relatively smaller; forebrain relatively larger and more convoluted (casts will reveal this); parietal crista more marked; canines relatively longer; molars more specialized; and eye sockets relatively smaller.

In all such cases, then, we possess a good deal of detailed information. Of course some quite different shape may have developed, but in most cases this will not have happened.

We have discussed the rules of evolution to show how far phylogenetic alterations may now be recognized as causally determined. Much detailed study still needs to be done, but it has become increasingly probable that *evolution is a determinate process,* in complete conformity with the unbroken chain of causal events which governs and has probably always governed the history of our planet and of the universe as far as we know it.

The *rules of evolution,* fashioned by the interplay of various particular laws, *have manifested themselves successively in the sense of an epigenesis.* Mendel's rules could not be formulated until sexuality and therefore the reduction division of the chromosomes had developed. The rules governing improvements in the brain could only come into being after the emergence of central nervous systems. Bergmann's rule only became effective after warm-blooded creatures had been developed. *Like all biological laws, the rules of evolution can be regarded as potential effects of the universal law of causality, and they existed implicitly, so to speak, before they became manifest as specific complexes of causal relationships.* (For the present, as before, we shall leave the psychological laws out of account.)

This idea, that despite its enormous complexity evolution is a determinate process, has far-reaching philosophical implications. It means that all future evolution is also determined. *The whole phylogenetic development has a predestined spatial and temporal pattern which manifests itself successively along the time coordinate* with its constantly advancing point of actual presence. Though this conception is borne out by all that we know of evolution, it can of course only be expressed as a hypothesis. We shall return to the problem when we discuss causality (Ch. 6 K III).

E. *The Origin of Life*

Studies in paleontology, comparative anatomy, and comparative embryology have shown that higher plants and animals have developed from primitive organisms and probably from very simple unicellular organisms. Already a century ago this well-founded conception had led to the question of how life originated. For a long time, however, speculations of this type lacked sufficient scientific basis.

Some scholars assumed that a kind of very primitive bacteria-like living organism had evolved on other celestial bodies in circumstances unknown to us, and had then reached the earth in the form of resistant spores, together with cosmic dust or through radiation pressure. This hypothesis was first proposed in 1865 by a physician, E. Richter; later discussed by the physicist W. Thomson (Lord Kelvin) and the physiologist H. Helmholtz; and then further elaborated as a "panspermia hypothesis" by the astronomer S. Arrhenius. But it was only of interest as long as it seemed improbable that spontaneous generation might have taken place on this planet.

It was only when the existence of *viruses* was discovered and electron microscopic studies had revealed their structure that the fascinating problem of how life began could be discussed on any sound basis of facts. With regard to several characters these organized bodies stand midway between the animate and the inanimate. They resemble living organisms in three ways: (1) they contain the same essential chemical compounds, that is to say, nucleic acids and protein bodies; (2) they are capable of identical reproduction, i.e., they reproduce and transmit their typical characters to the following generations; (3) they are subject to mutation and therefore to natural selection and to evolution. Unlike living organisms, however, they have no metabolism of their own, and they can only reproduce within some living substance. With bacteriophages, a branch of the viruses, it is known that only their nucleic acid enters the bacterium affected, and multiplies there. This means that at one phase of their development they consist only of a molecule, a nucleic acid, which then alters the bacterial metabolism in such a way that it synthesizes the protein bodies peculiar to phages.

These protein bodies form a covering round the molecule of nucleic acid, and restore the phage to its former phase with a "head" and a process, which later on can attach itself to a bacterial membrane and dissolve it.

The large viruses, the Mycoplasmatales (Microtatobiotes, or Cysticetes [Ruska]), which induce pleuropneumonia in cattle, psittacosis, etc. are more complex, and in this they rather resemble very simple bacteria. Besides nucleic acid and protein of the covering, they contain enzymes, carbohydrates, fats, and phospholipids. Moreover, some species are not confined to living protoplasm and can be grown on culture media.

Altogether, then, viruses possess a number of the characteristic features of living organisms, and they may at least serve as models to illustrate the successive emergence of life from inanimate matter. Some authors (e.g., R. S. Edgar and R. H. Epstein 1965) have even called them "the simplest living things." Whether or not this seems justified depends of course on one's definition of life.

At first, however, the problem of how primitive viruslike prestages of life could have emerged from inanimate matter seemed insoluble. It appeared most unlikely that highly complex substances like protein bodies and nucleic acids—or even their building material, the peptides, and amino acids, nucleotides, sugar, purine and pyrimidine bases—could have been formed. And it was still more difficult to imagine how protein bodies and nucleic acids could have combined to form stable self-replication systems, and then unite with other substances and develop complicated cell organelles such as ribosomes and mitochondria.

The first decisive step toward a solution of these problems was taken by S. L. Miller, who published his astonishing experimental findings in 1953. He produced a mixture of gases similar to what it was assumed the earth's original atmosphere had been at the epoch when life presumably began (see below). In this mixture of hydrogen, methane, ammonia, and water vapor, strong electric currents produced complex organic compounds including various amino acids and urea. Soon after Miller's discovery, other researchers synthesized many other amino acids and substances essential to the most primitive organisms, under a variety of conditions assumed to represent those of a prebiological epoch on earth. They also succeeded in combining amino acids into peptides and polypeptides. Above all, they were also able to produce the elements of nucleic acids—ribose and deoxyribose, purine and pyrimidine bases, and the resultant nucleosides and nucleotides (see C. Ponnamperuma and R. Mack 1965; G. Schramm 1965b). The energy required was supplied by strong electric currents or by hard ionized rays, ultraviolet rays for example, like those reaching the earth's surface before it had a protective layer of ozone, or by high temperatures comparable to those when the earth's mantle was being formed (see J. Keosian 1964; S. W. Fox 1965). G. Schramm (1965b) even succeeded, by bringing together nucleosides and metaphosphate esters, in producing polynucleotides, i.e., nucleic acids, some of them capable of at least imperfect self-replication. It seems possible that in prebiological epochs, natural selection gradually brought about a predominance of such nucleic acids as were best able to reproduce themselves, as, for instance, unbranched chains competing with branched chains.

Nucleic acids have the peculiarity of attaching polypeptides and protein bodies in a certain order, and they can also provide information for the specific synthesis of these substances (Ch. 4 E), many of which can act as enzymes and catalyze further chemical processes. There has been a great deal of experimentation and discussion about how these first prebiological systems capable of forming and replicating themselves came into being, how they survived, and how they could find suitable material in the "primeval nutritive soup" to synthesize and multiply. It was possible to test experimentally various processes which might have taken place, under what was presumed to be the prevailing conditions when the earth's firm mantle was being formed. S. W. Fox and his co-workers (since 1955; see 1965) applied dry amino acids to lava heated for some hours to between 150° and 200°C, and achieved polymerizations with proteinoid qualities. On being dissolved in water, they appeared as microscopic "proteinoid droplets." Electron microscope investigations showed these so-called microspheres to be encased in a double membrane. They remained stable for several weeks. When adenosinetriphosphoric acid (ATP) was added, a substance which when converted releases a great deal of energy, the droplets began to take up protein bodies and to grow. The particular importance of this experiment lies in the fact that ATP is present in all living organisms and that it could also be produced under prebiological conditions. If these proteinoid microspheres took up polynucleotides also, the result could resemble the "coacervate droplets" which A. I. Oparin (1957, 1965) and others had produced experimentally from polypeptides and nucleotides. The capacity of such droplets to adsorb is of significance for the development of living organisms.

Another relevant factor which has possibly contributed to the origin and multiplicity of complex organic compounds is that clays have a considerable capacity to adsorb organic substances (J. D. Bernal 1965). It is also possible that under prebiological conditions chemically stable, indissoluble, melanin-like substances were produced, the complicated surface structure of which may have initiated polymerizations of various kinds (M. S. Blois 1965). In this context it is of interest that H. Staudinger has calculated that twenty amino acids occurring in organisms if polymerized so that 150 form a macromolecule would be capable of producing 20^{150} polypeptide isomers (by comparison, the entire oceans of the world contain only about 4×10^{66} water molecules).

Of course we are only beginning to understand how life probably began. But we now know that a good many chemical prestages and precellular structures might develop from inorganic material under conditions comparable to those that preceded the evolution of true organisms. And precellular forms of this kind already possessed several main features of living organisms. When they became capable of self-replication, they could spread and their diversity could lead to natural selection, to adaptation to varied environments, i.e., to evolution.

However, all such prestages as well as the viruses, which could serve as models, still differ greatly from the most primitive true organisms, because they do not possess a regulated metabolism; and as organelles at least ribosomes are absent. In spite of this important difference, a clear definition of life is becoming more and more difficult.

We have noted that several of these prototypes quite possibly developed simultaneously along more or less parallel lines. But genuine living organisms apparently sprang from a single uniform early group. This is suggested by the fact that chromosomes, ribosomes, and mitochondria, or their equivalents, are structural elements common to all living organisms. Besides, the axiom *omne vivum e vivo* has invariably proved true.

Remains of organisms found in very early Precambrian rocks, which can be dated by the rate of decay of radioactive inclusions, indicate that life originated at least 2.8—and probably 3.5(−4?)—milliard years ago. (What appears to be remains of blue-green algae in flints from Ontario have been dated as 2×10^9 years old; algae remains from Rhodesia at $2.6–2.6 \times 10^9$ years; and traces of carbon in schists from Kansas at 2.5×10^9 years.) This would mean that some kind of primeval life originated fairly soon after the earth's mantle was formed some 4.5 milliard years ago, under very different thermal, atmospheric, and geological conditions from those which obtain at the present time. (The age of rocks on St. Peter and Paul Island in the Atlantic east of Brazil has been calculated on the evidence of the rubidium decay to be 4.5×10^9 years; see St. R. Hart.)

With regard to our scientific and philosophical conceptions, it is also of interest that we may assume the existence of different *organic* compounds, of the same type as those on the earth's surface, in other parts of the universe. Astronomical findings and chemical analyses of meteorites have shown this to be probable. If the same sequence of comparable conditions as on earth were to have occurred on some other celestial body, life could then have evolved, because of the presence of the same chemical elements and compounds and the effect of the same physical and chemical laws. Such equality or similarity of conditions seems to be rather unlikely; but in view of the almost limitless number of celestial bodies, this may possibly have occurred more than once. The number of stars (suns) in the Milky Way is estimated to be some 100 milliard, and the number of galaxies (spiral nebulae) of a similar type as the Milky Way at 80 million. The most distant known nebula, "3 c 295," is some five milliard light-years away. In this infinity—which may quite literally be an infinity of space and time—life may possibly have originated many times already. (If there is any truth in the very hypothetical assumption that the universe "began" by expansion, then the suggested span of 10-15 milliard years would also allow sufficient time for this.) If external conditions differed markedly from those on earth, organisms with different structures would of course have developed (see B. Rensch 1947, 1954, 1960a, Ch. 4).

The increasingly probable assumption that life emerged successively from inorganic matter leads to another consequence of philosophical significance. As organisms have also developed psychological characters, these too must have their origin and their prestages in inorganic matter. This is one of the deliberations which leads toward an identistic conception. We shall be discussing the problem in Chapter 6 L.

F. The Problem of Entropy

According to the second law of thermodynamics, in an irreversible process within a closed physical system, states of increasing probability run off until a state of maximum probability, that is to say, a state of equilibrium, is reached (see H. J. Meister 1960). The greater probability means greater disorder, because heat conduction and equalizing of the temperature (dispersion) cause a loss of free energy. Hence, when work generates heat, this cannot completely be transferred to other forms of energy.

The principle of entropy has often been interpreted too widely, as if the entropy in the world were tending constantly toward a maximum, and in the end all differences of temperature would cease to exist and no mechanical or other forms of energy would be recoverable. The world would then perish from excess of heat. But the principle of entropy is only valid with certain restrictions. It does not apply with temperatures near absolute zero (Nernst's "heat theorem") ; with microphysical processes, like the formation of atoms and molecules from the plasma which predominates in the universe; or when celestial bodies successively originate. The principle of entropy does not exclude the lawful development of higher levels of order, for example when the properties of molecules distributed at random in a liquid combine into crystals.

Like other physical laws, the tendency toward a general increase in disorder by conduction of heat applies also to processes within living organisms (see Ch. 4 B). In many physiological processes, heat is generated and dispersed by friction, and this means that entropy increases. When material is dissolved into body fluids or decomposed into its gaseous components, carbon dioxide for example, the increase of disorder is obvious. On the other hand, the maintenance of order and progression, both ontogenetic and phylogenetic, toward a higher order of complexity are precisely the characteristic peculiarities of all organisms. A relatively simple germ cell successively produces different types of tissue, organs, and systems of organs; and parallel to this morphological differentiation, higher levels of functional order develop. A corresponding increase in systems of order took place in the course of phylogenetic evolution. This contrast to normal physical processes, which are much more dominated by the principle of entropy, has given rise to a great deal of discussion. And some authors have tried to explain this peculiarity of living organisms in a vitalistic or "neovitalistic" manner (e.g., O. Costa de Beauregard 1963).

It must be remembered, however, that the principle of entropy applies only for processes taking place within a closed system. Most life processes do not fall within this category. Organisms are open systems, for their metabolism maintains a constant exchange of materials and energy with the world around them. As L. von Bertalanffy (1949, p. 120) pointed out, the principle of entropy is therefore not applicable here. E. Schrödinger (1944, 1954) was the first to state that organisms are kept at a relatively stable level of entropy by their metabolism, which prevents decay into the inert state of equilibrium. They free themselves of superfluous entropy by giving off the heat generated by metabolism. Besides, *order produces order* in organisms. This fact became fully conceivable only after nucleic acids, and in particular DNA, had been identified as structures capable of replication. Schrödinger's starting point had necessarily been the conception held at that time, that genes were unknown discrete molecules or molecule groups. But he anticipated the development of molecular genetics in some degree when he wrote in 1954 (p. 86): "We believe a gene—or perhaps the whole chromosome fibre—to be an aperiodic solid."

Today we are in a position to explain in more detail how it is possible for an organism "as it were to attract a current of negative entropy." The hereditary process is based upon the polynucleotides and primarily the DNA which is effective in the cell nucleus (or equivalent structures in simple organisms), the plastids, and the mitochondria. It is the DNA which, by identical replication, takes on the structural order of a very long and complex molecule fibre from that of its predecessors. The sequence of the bases in the DNA determines the sequence of the bases of the messenger RNA. These in turn determine the synthesis of specific protein bodies on the ribosomes, by matching their codons with the anticodons of the different transfer RNAs; and these protein bodies, partly acting as enzymes, ensure further specific syntheses within the

cell (see Ch. 4 E). The originating protein and lipoid laminae produced in this way are paracrystalline and mesomorphic structures which
act as matrices capable of steering the formation of specific structures
of molecules and cells (see H. Seifert 1966). This "stream of order" is
continued in the ontogenetic processes of differentiation, by means of
timely gene action and of inductors which promote the formation of various specific tissues, organs, organ systems, and so the whole body structure (Ch. 4 F). *The discovery of this automatic sequence of order emphasizes once again the enormous significance of having uncovered the
molecular structure of the nucleic acids.* And it explains the peculiar
position of the living organism in regard to the principle of entropy.

Mutations of the nucleic acids have the effect of producing *new sequences of order*; and for this reason also phylogenetic development
eludes the negative influence of entropy. The emergence of higher orders
of structures and functions has been determined by the evolutionary
rules or laws already mentioned. *Natural selection in particular has operated to prevent any increase of entropy, because it eliminated any disorder which tended to arise and promoted an increase of more efficient
and often more complex structural and functional orders* (by division of
labor, centralization, and regulatory mechanisms).

G. Summary of the Most Important Facts and Theoretical Conclusions

The factors involved in *the formation of races and species*—mutation,
gene flow, gene combination, influence of population size, selection, annidation, isolation, and hybridization—have already been analyzed quite
intensively. It became clear that speciation took place in different ways,
depending on which factors predominated. Usually, new species evolved
by way of former geographical and ecological races.

One very important discovery has been the surprising realization that
the hereditary substances, the DNA molecules, of all organisms differ
only in the sequence of their four complementary paired bases and in
the length of the molecular fibers. Accordingly, gene mutations primarily
depend on changes in the sequence of these pairs. Chromosome mutations and mutations of the whole chromosome set, resulting from polyploidy of herteroploidy, are also the result only of quantitative changes

(in the arrangement of the genes or the number of chromosomes). A causal explanation for these mutations has already been established in some cases, though mainly in viruses and bacteria.

Adaptation to environmental conditions is brought about through natural selection (and a certain degree of noninherited physiological versatility). The enormous overproduction which leads to the premature destruction of at least 95-99.9 percent or more of the progeny has been a particularly important factor for the efficiency of selection. No less important was the emergence of sexuality, which ensured that sufficient varied combinations of hereditary characters are always offered to natural selection.

Transspecific evolution, leading to the development of new organs and types of construction, and so to new genera, families, orders, classes, etc., depends on the same factors and can be regarded as a continuation of infraspecific evolution. But longer lines of descent also show certain evolutionary rules (or "laws") : strong radiation of forms after the emergence of new types of structure, followed by a gradually reduced rate of phylogenetic differentiation; maintenance of certain trends of evolution, even if they entail disadvantageous, exaggerated characters; parallel evolution of similar forms, and therefore sometimes polyphyletic development of higher taxonomic units; and a tendency toward evolutionary progress (anagenesis). There is no need, however, to suppose that some hitherto unknown evolutionary force is involved; mutation and selection afford quite a sufficient basis, especially in well-documented phyla, to explain the emergence of these rules and also the origin of new organs and new types of construction.

Although natural selection operates on certain characters, it always affects the entire individual, and promotes or restricts the propagation of the whole gene set or destroys it. This explains why selection may also produce disadvantageous or superfluous structures (excessive organs, vestigial structures or functions, "unnecessary" detours in individual development), when these are correlated with some more important advantageous characters. For example, during the Ice Age it was advantageous for mammals to develop a larger body size, which loses relatively less heat, although this entailed certain harmful features such as excessive antlers or tusks (in accordance with positive allometry, that is to say, more rapidly growing than the body as a whole).

Another consequence of selection affecting the whole individual has been that at every stage of phylogenetic development *all* the structures of an animal must be kept in harmony if the species is to survive. This explains the so-called co-adaptation or synorganization—the fact that many organs were often altered simultaneously, in order to adapt themselves to a certain bodily structure.

For a philosophical assessment of evolution it is very important that a causal explanation has already been found for a relatively large number of observed regularities. The hundred *rules* mentioned above should serve to illustrate their variety. It is clear, then, that evolution is a largely determinate process. The rules ultimately represent the effects of the general causal law operating without any interruption; but as the effects of the special causal laws often interact, the rules entail many exceptions and we do not speak of "laws." In general, it thus seems probable that all the processes of evolution are determined.

Already in the prestages of living organisms, the first mutations and the selective processes, caused by the developed variability, necessarily produced a kind of species formation leading to a vigorously ramified phylogenetic tree, because of the steady continuation of such speciation. During this process, the evolutionary rules have manifested themselves successively in the sense of an epigenesis as potential effects of the cosmic laws in the same measure as the requisite conditions were developed. Mendel's laws, for instance, could not arise until sexuality and fertilization processes had developed, nor could Bergmann's rule become effective until warm-blooded animals had appeared. In this sense, then, *we can envisage the entire phylogeny of living organisms as a predestined process running off in space and time, and unfolding itself as the shifting point of present time moves forward.*

It has come to seem increasingly probable that *the beginning of life* on earth can be traced back to a sequence of causal chemical processes leading successively from inanimate matter to the earliest organisms. Under conditions probably prevailing in a prebiological phase, about the time that the earth's mantle was formed, a number of different organic compounds were synthesized from inorganic ones including the highly complex substances essential to life processes, such as amino acids, prestages of proteins, components of nucleic acids, and adenosinetriphosphoric acid. Proteinoid microspheres and coacervate droplets, produced under these prebiological conditions by Fox and Oparin, already displayed some of the features essential to the origin of hypothetical prestages of life. Viruses, too, can be regarded as models which may help us to understand the gradual development of genuine organisms.

The fact that neither during the ontogenetic nor the phylogenetic development of organisms does disorder increase according to the principle of entropy, but that *successively higher systems of order* originate, does not imply any special vital forces or similar vitalist interpretation. Organisms are open systems and can therefore get rid of their superfluous entropy by giving off heat which is produced by metabolism. In individual as well as in phylogenetic development, order originates from order. This "stream of order" begins in ontogenetic development with the information mediated by DNA, and is continued by the production of specific enzymes, setting up further chains of reactions; for instance, the development of certain specific inductors and structures of cells, which in turn determine morphogenetic processes. During phylogenetic development, natural selection has prevented the origin of structural and functional disorder.

Chapter Six

Epistemological and
Ontological Problems

A. The Ideological Significance of Epistemology

In the preceding chapters we have not characterized the psychic phenomena in detail and have mainly opposed them to matter in the sense
of a widely held dualism. Because of their importance for the philosophy
of nature, we must now concern ourselves with psychological processes
and their analysis. This is necessary because animals are obviously capable of sensations, memory, and generalization, and because the evolutionary progress of mental processes has been decisive for the emergence
of man. Furthermore, all our psychological processes are closely connected with physiological processes in the brain and sense organs. And
all knowledge depends on the association of various perceptions and
mental images, on judgments and conclusions; in other words, on mental activity. So we have to ask what factors our knowledge depends upon,
how we arrive at it, how reliable it is, and what its limitations are.

Our task here is to use the means and methods of epistemology (gno-
seology), which is concerned with these problems, to examine the process of acquiring cognition and to assess its results. In doing this we
must endeavor to start with facts that can be regarded as *absolutely
certain*. These are not the material processes but the mental processes
which each one of us is aware of within himself. We must then examine
how perception leads on to mental images, how we come to frame concepts, judgments and conclusions, and what reliance we can place upon
them. We shall need to see how we arrive at our ideas of matter, space,
and time, and at the concept of self; and we also have to inquire into
the processes of feeling, attention, and volition, and determine what part
imagination and intuition play in cognition.

The aim of epistemological discussion is always to establish the most
reliable foundation for a consistent philosophical picture which does jus-

tice to scientific and philosophical statements and leads to some general theory of being, to ontology. Such a philosophical view may also be of considerable heuristic value for future special studies in various branches of knowledge. In order to build up a picture as close to reality as possible (Ch. 6 D IV), we must above all avoid weakly based presuppositions and any kind of dogmatism. We must not start from "self-evident phrases" or "primae veritates" in the Aristotelian sense, or the "revealed and eternal truths" on which some religious systems are based. We must not place too much reliance on widely accepted assertions—for example, that a proton is indivisible, that the self represents an indivisible and primary unity, or that "souls" live on after death. Their epistemological basis will have to be tested with the help of particular findings in the relevant field of knowledge. On the other hand, there is no ground for complete skepticism; a considerable body of knowledge does, in fact, exist. What we shall have to examine skeptically, however, is the degree of certainty we can ascribe to our knowledge.

Our chief aim will be to combine the results of epistemology (largely acquired through introspection) with the results of scientific investigation (normally obtained regardless of philosophical implications), in order to construct a reasonably consistent, comprehensive picture. We shall see that the phylogenetic development of psychic phenomena is a particularly significant problem in this connection.

B. Epistemological Theories

Before we discuss phenomenological analyses and the main problems of epistemology, and try to combine them with the results arrived at by science and in particular biology, it is necessary to make some assessment of the most important epistemological theories. We shall first outline the different theories so that the reader can keep in mind not only the nature of the problems but also the views not shared by the present writer. It will also become clear that many versions contain a certain degree of "truth," but as they correspond to the state of knowledge and ideas of the time of their origin, they are often expressed in a different manner from present-day theories. In addition, a survey of this sort may be welcome to scientists who have only a superficial knowledge of epistemological problems. And this seems to be true for many scientists, not only those of the younger generation.

The view held by those not especially versed in philosophy is mostly a *naive realism*; that is to say, they identify the content of perception with the extramental reality. They attribute color, smell, hardness, etc., to objects, without reflecting that these qualities of sensation are only correlated with physiological processes in the brain (and possibly in the sense organs as well) and are released by certain stimuli deriving from the objects—for example, electromagnetic waves of a certain wavelength. The naive observer fails to realize that our sense data are no more than indications of characters of the extramental world and that the objects themselves have no color, hardness, or smell. The naive realist is unaware of the fundamental principle of epistemology.

As already mentioned, all kinds of *dogmatism* must be avoided in any consideration of epistemological theory. Almost all systems of religious and supranaturalist philosophy (anthroposophy, for instance) are dogmatic, as they are based not upon reliable evidence but upon accepted articles of belief such as the existence of a god or gods, metempsychosis, nirvana, astral bodies, etc. Kant even called all epistemological theories dogmatic which do not examine critically the process of cognition, the "powers of reason" (*Kritik der reinen Vernunft*). This would include a good many philosophical conceptions which assume "self-evident truths," an "indivisible self," or an "immortal soul," or perhaps only some primal "substance." Such theories should not be rejected totally, but they have to be subjected to epistemological scrutiny and modified at least in regard to the reliability and significance of their dogmatic assumptions.

So a certain *skepticism* is essential before any epistemological theory can be propounded or accepted. Every philosophical system has emerged as a result of skepticism applied to naively realistic conceptions. From classical times to the present, philosophers have postulated some skepticism. Descartes, however, was the first to establish it as a basic principle. His starting point, *de omnibus dubitandum est*, which served to free him from the philosophical dogmas and authoritarian religious claims of his day, enabled him to build up a new theory of knowledge which may be said to have inaugurated modern philosophy. (In spite of this principle, some dogmatic ideas which prevailed in the seventeenth century are still to be found in Descartes' work. He held that the existence of God was proved by the fact that man can conceive the idea of a divine being; and he accepted the existence of bodies because of the veracitas dei.)

A skeptical scrutiny of the bases of cognition then led to awareness of the limitations of human experience. The application of this *principle of immanence* led men to realize to what extent philosophical theories may overstep these limitations, if they incorporate transcendent elements which are by definition incapable of proof, or if the inferences which they draw directly from the phenomena are merely transgressive (Ziehen calls them "transgredient") and can often be fully confirmed or at least supported by further analyses and syntheses. (Kant made no distinction between these two ways in which the limitations of immanence may be exceeded.)

Strict application of the principle of immanence led to *positivism*, which in principle bases its deduction on the experienced psychic phenomena alone, and attempts to analyze them with the help of our innate powers of thought and to investigate their relationships and laws. In this way it is hoped to elucidate such problems as matter, space, time, memory, conceptualization and generalization, the will, self, values and logical processes of thought. This was the method inaugurated by Locke, Berkeley, and Hume, and continued by the Encyclopedists d'Alembert, Diderot, and Maupertuis. Kant's idealistic philosophy is founded on the same principle of immanence; but he disregarded it in his a priori conceptions of space and causality, and the assumption of a categorical imperative. It was Comte who first used the term "positivism." Since then, positivism has been further developed in different directions by Spencer, Laas, Mach, Riehl, Avenarius, Ziehen, and others. There are realistic, idealistic, identistic, and relativistic varieties of positivism, as well as phenomenalism. Being based on incontestable phenomenal statements, it naturally appeals to the scientist, who in his field also works exclusively with sure premises and as far as possible with confirmable data.

Any satisfactory philosophical picture must include very different statements arrived at by scientific as well as epistemological analyses involving knowledge in many diverse fields. It is thereby dependent upon the state of research and the accepted theories of the day, as well as upon other cultural traditions. Understandably, therefore, no generally recognized or even predominantly accepted epistemology as yet exists. In our own century, as in the past, theories based on dualism, monism, and identism exist side by side.

Dualist interpretations appear to be the most frequently held. Matter and mind are regarded as something fundamentally different. Descartes spoke of two "substances," of a res extensa and a res cogitans. The antithesis between "body and soul" or "object and subject" expresses the same contrast. The intimate relationship of both principles within the brain, i.e., the coordination of perceptions and mental images with certain physiological brain processes, is usually thought of in terms of mutual causal influence. But this kind of *psychophysical reciprocal effect* presents difficulties. According to the law of the conservation of energy, no energy should either be lost or gained in the course of excitations of the brain. Yet both would have to take place if physiological processes were to have psychological "effects," or if psychological processes were to "release" physical ones. As this cannot be the case, because excitations in the brain have to be regarded as energetically gapless processes, many philosophers, psychologists, and biologists assume that psychological processes only run parallel to physiological ones. Of course, such *psychophysical parallelism* cannot explain why certain specific excitation patterns in the brain (and sense organs) are always coordinated with certain specific phenomena. For example, the stimulus set up by light of a wavelength of 670 nm releases the sensation "red" while that of 525 nm induces the sensation "green." Descartes, Geulincx, and Malebranche ascribed this to divine influence in each case (occasionalism). Leibniz assumed the existence of a "pre-established harmony" in the sense of two synchronized clocks. Ziehen sees this fixed mutual relationship as the effect of "laws of parallelity" which had to be accepted as something primarily "given" like causal or logical laws, and which cannot be further explained—or traced back to something else. (He incorporates them, however, into an identistic conception.)

The close connection between mind and matter led other philosophers to build up *monistic theories of knowledge*. The most questionable form is the *absolute materialism* by which everything that happens is traced back to physical processes, and psychological phenomena become equated with physiological functioning of the brain. In its extreme form this theory was largely confined to the eighteenth century, influenced by the first investigations of cerebral processes. It was held by Holbach, La

Mettrie, and Cabanis. Schopenhauer characterizes this extreme materialism in Part 2 of his *Welt als Wille und Vorstellung* as "the philosophy of the subject that forgets to take account of itself."

Functional materialism, which is more widely held, recognizes psychological phenomena but sees them erroneously as "functions" of the brain, like other physiological functions. This view involves denying the existence of an immortal soul (Vogt, Moleschott, Büchner). This type of materialism has naturally always had a wide appeal for scientists; for their chief aim is to examine all the processes of nature with reference to their causal relationships.

But these materialistic theories ignore the fact that what is absolutely certain is not the material processes but our phenomena, the perceptions and processes of thought. This emphasis on the reality of phenomena was the basis of *spiritualism* as it was mainly held by Berkeley. Spiritualism assumes that only psychic phenomena exist, and that matter means purely the content of our perceptions, an "objectivation" of the mind. As only the phenomena of the self can be regarded as absolutely sure—since other "selves" consist merely in what our own self perceives and images—it is theoretically possible even to argue that the only thing which exists is the consciousness of that self. Yet such *solipsism* is hardly discussed seriously.

A similar or more or less equivalent point of view is expressed by *absolute idealism*. This is also monistic, as it contends that all being is psychic in character. Fichte, Schelling, Hegel, and other more recent philosophers have held various versions of this view. It makes little appeal to many scientists, because astronomical and geological findings have inevitably led them to realize that a material world must have existed before living organisms came into being. And most scientists would also say that this inorganic world is of an entirely nonpsychic nature.

Critical idealism, propounded by Kant, does not deny that things exist outside our consciousness; but it does not attempt to make any statement about the nature of the "thing in itself." All the qualities and spatial and temporal properties of our phenomena are only "appearances" of the world outside ourselves.

As ultimately only psychical phenomena represent the "given" reality, yet material processes obviously take place whether or not there are any living organisms to apprehend them, many philosophers have attempted to find a uniform theory to cover the disparate sets of facts arrived at by the two different methods. An acceptable solution for both philosophers and scientists was an *identistic conception* which also complied with monistic tendencies such as those we have outlined in Chapter 2 D. These identistic theories also exist in various versions. Giordano Bruno, Campanella, Spinoza, and some other philosophers assumed that the ultimate "something" which underlies all experience and all events is neither psychical nor material, but possesses material and psychological properties or "attributes." Other philosophers hold the view that only the psychic phenomena exist (panpsychism), but that they have properties whose effects we apprehend as material processes. According to Ziehen the "ultimate something" to which we have to attribute "consciousness" in its widest sense (though not at all in the sense of *self*-consciousness) is governed by general laws: the causal law, the logical laws and the laws of parallelity by which the phenomena are coordinated with certain cerebral processes. So far as causal relationships are involved, we speak of material processes.

We shall see that the identistic theories, insofar as they are not too speculative, are supported by investigations on the correlation of phenomena and cerebral processes, the study of individual and phylogenetic psychical development, and an analysis of the phenomena themselves, and of matter. An identistic theory also offers a particularly plausible explanation for the relationship of phenomena to being. The next chapters will deal with these problems in greater detail, and will lead to the author's own panpsychistic, realistic, and identistic conception, which will be treated in Chapter 6 L.

This brief survey is intended to do no more than outline the most important epistemological theories. A number of other interpretations have been omitted because they are too speculative (e.g., Heidegger's existentialist ontology), or they simply represent variants or combinations of the theories mentioned (psychism, conscientialism, hylozoism, the empiriocriticism of Avenarius), or they deal with only partial aspects of cognition (sensualism, actualism, relativism, or Verworn's conditionalism).

C. The Psychophysical Substratum

1. Development of the Problem. The survey outlined in the foregoing chapter showed how the judgment on the psychophysical connection between "body and soul" determines any epistemological interpretation. So we must now ask what can be confidently asserted about this connection. We may assume that psychic phenomena and matter are completely different, or else that they are merely different aspects or components of something identical. Whichever conception we choose, we are faced with the problem that "matter" (or its causally determined components) is divided into two categories. Some parts of it, particularly certain regions of the human brain (and possibly other portions of the nervous system and sense organs), are characterized by the fact that the neurophysiological processes involved can be coordinated with conscious phenomena. In other parts of the brain or body, as well as in all external objects, this is not the case. We shall refer to all structures falling within the former category as the *psychophysical substratum.*

What can we state about the morphological limits and the structural and physiological peculiarities of this unique substratum? And how does it come about that physiological processes within the same substratum are sometimes not combined with phenomena, as for example, in deep sleep, or in a state of coma or narcosis, or even when an action has become automatic? Philosophers, scientists, and physicians since classical times have been occupied with this problem, but without reaching any satisfactory conclusion.

Greek records of the sagacious physician and philosopher Alkmeon (6th century B.C.) cite him as teaching that "all sense organs and sensations are connected with the brain; if the brain is shaken and its position changed, they are paralyzed; for then the passages leading to perception are blocked. . . . It is the brain through which we have sensations of hearing, seeing and smelling; these lead on to memory and mental imagery, and when these have settled and become fixed, the result is knowledge" (quoted after W. Kranz 1949). This clear idea which is already similar to our present conceptions was not always shared by later scientists and philosophers. Some thought that psychological activity was also bound up with the action of the heart or the circulation of the blood. And often no clear distinction was drawn between the psychic phenomena and the principle of life which governs the body's activities, or it was totally identified with the vital force (Thales, Democritus, Plato, Aristotle, and others). These ideas survived into medieval times and were held, for example, by Thomas Aquinas (1225-1274), who tried to combine Aristotelian philosophy with Christian ideas. He taught that the soul is the formative entelechy of the body, but that when it leaves the body it is immortal.

Only in the Renaissance, when scientific research received a new impulse, when anatomical studies were carried out and experience was recognized as the most important source of knowledge, was the connection between body and mind again examined on a sounder basis of facts, and the importance of the brain emphasized. This influenced the development of all later philosophical systems. Locke, physician and philosopher, wrote in his *Essay Concerning Human Understanding* (1690) : ". . . it is evident that some motion must be thence continued by our nerves or animal spirits, by some parts of our bodies to the brain, or the seat of sensation, there to produce in our minds the particular ideas we have of them." According to him, a stimulus deriving from the object observed is transformed into an excitation within the eye: ". . . it is evident some singly imperceptible bodies must come from them to the eyes, and thereby convey to the brain some motion." N. Malebranche (1674) pointed out in his *De la Recherche de la vérité* that after amputation of a limb, men may feel pain in the lost part of their body, and he concluded (1938, pp. 120-121) : "Car toutes ces choses montrent visiblement, que l'âme réside immédiatement dans la partie du cerveau à laquelle tous les organes des sens aboutissent. . . . Si le mouvement des fibres de la main est modéré, celui des fibres du cerveau le sera aussi." ("For all these things demonstrate that the soul resides in that part of the brain at which all our sense organs terminate. . . . If movement of the fibers in the hand slackens, so does that of the fibers in the brain.")

Eighteenth-century studies in the physiology of the brain confirmed the view that the brain, and in particular the forebrain, has to be regarded as the psychophysical substratum. The physician and philosopher J. O. de la Mettrie (1748) had observed that the psychic functions are altered during an attack of fever, and he concluded that thinking can be regarded as a physiological function of the brain. D. Hartley (1749) and J. Priestley (1777) held the same view, and so did the nineteenth-century materialists, like Moleschott, Büchner, Vogt, and Cabanis.

Kant disagreed with this interpretation. In his *Träume eines Geistersehers* (1766), written in the fervor of his youth, we read:

"Wo ich empfinde, da bin ich . . . Ich fühle den schmerzhaften Eindruck nicht an einer Gehirnnerve, wenn mich mein Leichdorn peinigt, sondern am Ende meiner Zehen. Keine Erfahrung lehrt mich . . . , mein unteilbares Ich in ein mikroskopisch kleines Plätzchen des Gehirnes zu versperren. . . . Daher würde ich einen strengen Beweis verlangen, um dasjenige ungereimt zu finden, was die Schullehrer sagten: Meine Seele ist im ganzen Körper und ganz in jedem seiner Teile.

("Where I sense, there I *am* . . . I feel the most painful sensation when my corn torments me, not in a cerebral nerve, but at the end of my toes. No experience teaches me . . . to shut up my Ego into a microscopically small place in my brain. . . . Thus I should demand a strong proof to make inconsistent what the schoolmasters say: My soul is in my whole body, and wholly in each part.") (1922a, p. 399)

Unfortunately Kant did not treat the problem of localization in more detail in his later work. He makes a brief reference, however, in his *Anthropologie in pragmatischer Hinsicht abgefasst* (1800, § 17). "Der Sinn der Betastung liegt in den Fingerspitzen und den Nervenwärzchen (papillae) derselben . . ." ("The sense of touch lies in the fingertips and the papillae themselves.") Kant's attitude places the problem of the psychophysical substratum in a new light. As sensations, like all phenomena, are directly given, their spatial differences—for example, pain *in* a toe and *in* a finger, or *in* a tooth—must also be regarded as indubitable reality. Later, when it came to be more generally accepted that only certain parts of the brain represent the psychophysical substratum, an attempt was made to fit the phenomenological findings of the spatial relationship between sensations into a "local sign theory" (see below). In many cases, however, the problem was simply ignored.

II. Localization of the Psychophysical Substratum. In the nineteenth century, when more intensive histological and physiological brain research began, the work of F. J. Gall (1809), M. J. P. Flourens (1824), and others led to the view that mainly the cortex of the forebrain must be regarded as the psychophysical substratum. Pathological experience and experimental results served to strengthen this opinion. It was established that disease or injury in different cortical regions was accompanied by certain nervous and psychical disturbances or inhibitions. Thus destruction of the area striata in the posterior region of the forebrain led to cortical blindness (E. Hitzig 1874). By studying diseases in a certain region of the lateral lobes of the forebrain, P. Broca (1878-1879) was able to localize the motor speech center. J. F. Ferrier (1874, 1875) had already attempted to show that different regions of a monkey's cortex had specific functions. And G. Fritsch and E. Hitzig (1874) had succeeded in marking off several functional areas by local electric stimulation of various cortical regions. A large number of histologically different cortical regions could also be distinguished, regions which often corresponded more or less accurately with the functional regions established on pathological and physiological evidence. (Th. Meynert 1867-1868, 1872; H. Munk 1881, 1890; A. W. Campbell 1905; O. Vogt 1910; K. Brodmann 1909, 1910); C. von Economo and G. N. Koskinas 1925). All these studies led to regionalization of the cortex of the forebrain and a corresponding *localization theory* which maintained that specific mental, sensory, motor, and associative activities were connected with certain histologically distinguishable cortical regions (K. Kleist 1934; A. Hopf 1964; and others).

More recently, strict localization of functions could be confirmed only for certain sensory and motor cortical regions. Electrophysiological and exstirpation experiments have shown that psychical activity often corresponds to processes of excitation in larger parts of the cortex. It has been found, too, that the sensory fibers of some sense organs lead to other cortical regions as well, and that some sense organs are doubly represented in the cortex (C. von Monakow 1910; K. Goldstein 1927; K. S. Lachley 1929; E. Th. von Brücke 1934; J. G. Dusser de Barenne 1937; E. D. Adrian 1947; O. Bumke 1948; and others).

Moreover, some other writers were able to show that processes of excitation in cerebral areas below the cortex might also be accompanied by conscious phenomena. Th. Meynert (1872) held the opinion that processes in the human diencephalon (thalamus opticus) and midbrain (corpora quadrigemina) could be correlated with phenomena. Others considered this was probable or possible, at least in the subcortical regions of the forebrain (Th. Ziehen 1921; C. von Economo and G. N.

Koskinas 1925; K. Kleist 1934; E. Th. von Brücke 1934; D. G. Marquis and E. R. Hilgard 1936; H. Rohracher 1948; etc.).

Some went so far as to believe that physiological processes in the sense organs themselves are accompanied by conscious processes. This view, similar to that of Kant, was expressed by R. Whytt (1768), J. Purkinje (1823), J. Henle (1838), and J. Müller (1840). Later on it was sometimes mentioned as a possible hypothesis, for instance by E. Hering 1905), A. Stöhr (1922), E. Lindemann (1922), F. B. Hofmann (1925), and F. Budde (1933). A. Bier (1944) wrote: "Wenn bestimmte psychische Vorgänge, und zwar die höchsten, die Bewusstseinsvorgänge, zweifellos an das Gehirn gebunden sind, so ist das so zu erklären, dass die Seele zwar im ganzen Körper sitzt, dass aber das Gehirn das Hauptinstrument ist, auf dem sie spielt." ("If certain psychical processes and in particular the highest ones, the conscious processes, are undoubtedly connected with the brain, this is explicable by saying that though the soul is present in the whole body, the brain is the chief instrument upon which it plays.") And O. Bumke (1948, pp. 214-215) expressed a similar view: "Nach meiner Überzeugung liegen bis heute keine Tatsachen vor, die uns zwängen, den von Kornmüller und seinen Vorgängern beobachteten Aktionsströmen andere und innigere Beziehungen zum Seelischen zuzusprechen, als sie etwa die Reizung der Netzhaut durch einen Licht-oder des inneren Ohres durch einen Schallreiz auch schon besitzt." ("In my opinion, no facts have yet come to light which would oblige us to regard the action potentials observed by Kornmüller and his predecessors as standing in a different or more intimate relation to the psychic processes than does stimulation of the retina by light or of the inner ear by sound waves.")

As the main reason for including the sensory cells in the psychophysical substratum, one may mention the discrepancy already referred to between the spatial arrangement of excitations in the sense organs and those in the brain. For example, if our eyes are shut, and someone touches spots on our thumb, second finger, and nape, we feel that the first two spots are close together, but that the third is relatively far away from the others, corresponding to the different relative positions of the sensory cells or nerve endings involved. Yet all the projection fields lie in the gyrus centralis posterior of the cortex, almost equidistant from one another, because the hand is represented there by a much larger area than the adjacent one for the nape.

The same is true of visual excitations. We see the external world more or less as reflected in the retina, though inverted; and we see it in front of us. However, in the corresponding projection fields, situated in the convoluted occipital lobe of the cortex, the pattern of excitations is quite different: the area of acutest vision in the retina, the macula, which measures only about 4 × 2 mm, corresponds to an extensive cortical area, and all the other parts of the retina are represented by a much smaller one (Fig. 4). And the great number of primary excitations of the visual cells is by no means reflected in the visual field, the "cortical retina." What happens is rather that excitations transmitted by several rods are united in the bipolar retinal cells, and many of these in turn in a smaller number of multipolar ganglion cells. The axons of these cells form the optic nerve. St. Poljak (1941) has calculated that the impulses from four million cones and 110-125 million rods of the retina are united in a mere 800,000 to one million fibers of the optic nerve. After these excitations have been switched to other neurons in the lateral geniculate body of the diencephalon, their axons branch out, and the excitations then penetrate into the visual center of the occipital lobe which contains perhaps a hundred times as many neurons as the lateral geniculate body (see K. Chow, J. S. Blum and R. A. Blum 1950). Besides, even in the fovea centralis, where vision is most acute, the pattern of excitations is sharply deflected toward the rim of the fovea. Throughout their further path, especially in the Gratiolet's optic radiation leading to the visual center, the course of the fibers shows many crossings and irregularities. So there can be no question of an exactly corresponding arrangement of the excitation pattern in the retina and in the visual center of the brain. This could indicate that the sensory cells belong to the psychophysical substratum.

I myself formerly shared this view, and I collected a good deal of data in its favor. I intentionally put forward this "aesthetophysical hypothesis" (1952) in an acute form, in order to provoke discussion about these biophilosophically important problems. The response was unfortunately limited. Since then, investigations have made it seem more probable that the psychophysical substratum is confined to certain portions of the brain (see below). One ought not to lose sight of the prob-

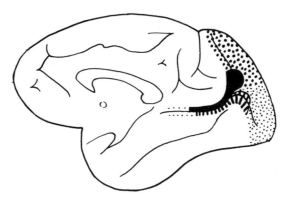

Figure 4. Termini of retinal impulses in the visual center of a monkey. Longitudinal section of the forebrain from the medial side. Projection fields of the upper (dorsal) retinal area are shown black and with larger dots, the lower (ventral) part hatched and finely dotted. The macula (area of acutest vision) dotted, the upper part with larger dots. (After B. Brouwer, G. J. van Heuven and A. Biemond)

lem, however, in view of its connection with the phylogenetic development of the psychophysical substratum (Ch. 6 H). We must perhaps at least recognize that sensations and mental images need not be limited to the forebrain, for it seems to be improbable that a fish, the visual center of which is the tectum opticum of the midbrain, and also a cuttle-fish, with its entirely different brain, should not be capable of seeing.

It was recognized at quite an early stage that the discrepancy between the immediate experience of the spatial relationships of sensations and the different excitation pattern in the cerebral cortex makes for difficulties in localizing the psychophysical processes in the brain. To surmount these difficulties, R. H. Lotze (1852) and A. Bain (1864) put forward a *local sign theory* which was then developed further by F. Hering, W. Wundt, Th. Ziehen, C. Stumpf, and others. In its original form the theory was not particularly convincing. It was assumed that a local sign, which determines the local relations, is assigned to each individual excitation which reaches the cortical projection field. At first these signs were thought of as qualitatively distinct. But this seemed improbable, for it would have necessitated millions of differing qualities for cutaneous tactile sensations and the spatial arrangement of the excitations in the visual field; and the arrangement of these qualities would have to be strictly determined by hereditary factors, in spite of the individual differences in the structure of the cortex in each case. A more satisfying assumption is that of "conductive local signs" (as Ziehen called them); this is based on differing movements in response to impulses from different nerve fibers or on the memory of such movements.

The "empiricistic" version of the local sign theory, which Helmholtz put forward, is still more in line with the present state of knowledge. According to this, when a stimulus is shifted, a sequence of corresponding termini in the excitation pattern of the cerebral region involved is affected. This means that in the course of time a definite movement image, which remains the same in relation to those of neighboring cells, becomes coordinated with every excited cortical cell. For instance, a definite movement image essential to the assessment of angular distance would be coordinated with every cell in the visual center of the cortex. All our perceptions do in fact depend upon the movement of stimuli. But if the spatial relationships of sensations were the outcome of a sequence of sensations and images of movement in this way, then infants would begin by having a completely disorganized image, which would then change very rapidly into one which corresponds to the external world. Since this seems unlikely, we must assume that we are *born* with a precise pattern of distribution of neurons and connections of nerve fibers, through which sensations of movement are conveyed.

According to this "ordinative" local sign theory, the excitation pattern in the cortex does not correspond to that of the retinal image. Its spatial arrangement is a different one, but it disposes of local signs, conditioned by the different susceptibility of the neurons to different sequences of their excitation. This permits perception of the spatial order of the world in an adequate manner. That such establishment of the right order of phenomena is possible is already indicated by the fact that we perceive the inverted retinal image "the right way up," and that we do not notice the blind spot, a blank area near the center of our field of vision, where the optic nerves leave the retina. An "ordinative" version of the local sign theory is in fact plausible.

Recent physiological investigations have supported this "nativistic local sign theory" in a remarkable way. R. W. Ditchburn and D. H. Fender (1955) have proved that a shape or a single dot can be visually apprehended only by very fine trembling movements of the eye, during which the fixation point is continually shifting. When such shifting is made impossible by a small plastic shield, placed over the cornea and moving with the eyeball, the image disappears. D. H. Hubel and T. N. Wiesel (1965; D. H. Hubel 1963) could also prove by electrophysiological investigations of the visual center of the cat, that the neurons there show almost no reaction to diffuse or punctiform light stimuli, but that they react to more complex stimuli such as small strips of light which are apprehended by fine eye movements. Some neurons react only to horizontal shifting, others only to vertical shifting, or to rotation of a strip.

The assumption that the psychophysical substratum is limited to certain parts of the brain is also supported by the law of specific sensory responses. For example, pressure on our eyeball produces a sensation of light (pressure phosphene), and so does electrical stimulation of the retina. The primary processes in the sense cells, however, are very different from the photochemical processes set up by light stimuli. Yet the excitations transmitted to the visual center are similar in both cases. In this context it is also of interest that W. Penfield and his co-workers (1950, 1958) found that electrical stimulation of the temporal region of the cortex of patients under a local anesthetic produced vivid and detailed visual and auditory images. The great importance of the hippocampus region of the forebrain in the development of memory has not yet been sufficiently explained (see P. Glees and H. B. Griffith 1952; W. Penfield and B. Milner 1958). As this region is a part of the limbic system, it may perhaps only contribute to the fixation of memory by reinforcement of the excitations.

The only finding which appears to contradict the idea of limiting the psychophysical substratum to certain parts of the brain is that we perceive the spatial relationships of sensations at the points of stimulation, in our fingers or toes, and not along the sulcus centralis of the forebrain, and that we localize a picture in the eye, and not in the occipital lobe of our cortex. There can be no doubt that this a reality. Though appearing strong at first, this argument loses a good deal by the fact that according to the ordinative local sign theory there is no need for the actual spatial relations of the nerve ends in the brain to correspond to the spatial properties of the accompaning sensation.

III. Separate Regions for Sensations and Mental Images? Though most writers on the subject agree that the psychophysical substratum is limited to certain portions of the brain, it has not yet been established whether the brain, or the cortex in the case of man and mammals, contains particular regions for sensation and mental images or whether both groups of phenomena belong to the same psychophysical substratum.

In favor of the view held by Meynert, Munk, Ziehen, and others that separate spheres exist, H. Wilbrandt and A. Sänger (1904, pp. 338-390) put forward the following arguments. (1) In cases of cortical blindness, the patient may still have sensations of color or brightness although he has lost the ability to apprehend, remember, or imagine shapes (see also A. Gelb and K. Goldstein 1920; W. Benary 1922). (2) Visual reproduction may take place without any impulse being transmitted from the sensory cells. (3) Sensations presuppose the occurrence of a physiological process, which rapidly ceases after the excitation has faded, whereas we must suppose permanent structures (engrams) to be a necessary feature of mental images. (4) Mental images can be called up at will; sensations cannot. (5) Mental images lack sensory vividness and are shadowy. (6) The field of perception is projected into space, while that of the mental image is not. Ziehen (1924, pp. 301-302) noted in addition that in cases of feeblemindedness, whether congenital or acquired, sensations are normal whereas memorial images can be seriously disturbed.

All these arguments, however, are largely inconclusive. The apprehension of shapes and the corresponding memorial images and concepts, and the whole structure of mental imagery, all presuppose synaptic connections among a large network of nerve fibers. If illness interrupts these connections at their weakest point, the synapses, complex mental images or connections between mental images become impossible. Those fibers which are still left here and there cannot any longer connect up with other parts of the brain, in particular with the motor speech areas, through which the patient would be able to give any information. Yet he may have visual sensations, and be able to follow moving objects with his eyes. But he has no clear idea what it is that he is seeing. The reason that sensations are normally much more vivid than mental images, that they can also be projected into space, is because the stimuli received from the environment are very much more intense than impulses which originate in the brain and set up mental images. If such brain impulses become unusually intense, hallucinations may arise, which are then as vivid as normal sensations or even more vivid. The fact that mental images can be produced at will, and sensations cannot, might better be used as an argument in favor of the assumption that the two spheres are identical, for the weaker cerebral excitations can initiate a sequence of mental images but cannot prevail against much stronger excitations coming from the sense organs.

A strong argument in favor of the identity of the spheres of sensations and those of mental images is that sensations and their memorial images agree. The latter are only much weaker and more shadowy and often incomplete or "ragged"; but this is because the excitations which arouse mental images are weaker than the excitations set up by external stimuli, and because only a part of the network of nerve fibers excited during the sensation is activated. The "ragged" and partial nature of this activation is already typical of the positive after-image which follows on a particularly intense sensation. The reason why memorial images in dreams are more complete is probably because they are not continually being disturbed or suppressed by sensory excitations.

IV. The Memory Trace. Up till now, the morphological and physiological nature of the memory trace, the *engram*, has been almost unknown. Recent studies by H. Hydén (H. Hydén and E. Eghazi 1962; H. Hydén and P. W. Lange 1966) have shown that neurons which are active over long periods produce large quantities of ribonucleic acid (RNA). This finding is one of the bases on which the Swedish researcher founded a preliminary engram hypothesis. Its main arguments are as follows. Since varied genetic information can be conveyed by ribonucleic acids, these may also form the cytological basis of the various memory traces. Specific excitation patterns transmitted through the afferent nerves to the cerebral neurons may bring about sequence alterations in individual polynucleotide chains. As each RNA molecule is capable of many millions of possible alterations, this could explain the more or less infinite diversity of memorial impressions, particularly as the molecules are fairly stable. Experiments have revealed that the cerebral neurons of rats which had learned something (to change from the preferred to the other hand, or to climb over wire netting to get their food) displayed a different arrangement of bases in the RNA of brain neurons from those of the control animals. Hydén furthermore assumed that an engram becomes activated again, and a memorial impression is aroused, whenever the same specific excitation pattern occurs in the cell or cell group to which the RNA structure had been adjusted by the sensory excitation.

It remains to be seen how far this hypothesis can be substantiated. One could argue against it that the polynucleotide molecules differ only in the sequence of the same four (or five) bases; the claim that the images coordinated with the engrams are so extraordinarily diverse is hard to reconcile with this fact. It is possible, however, that the function of the RNA is only to mediate the synthesis of proteins in the brain neurons and that the engrams are represented by protein bodies which are much more multifarious. The following experiments which we have conducted show that this assumption is at least discussible. We injected radioactively tagged amino acid (3H-histidin) into fishes, fixed in a special device, and stimulated one eye by one narrow vertical strip of light or by two such strips, and then investigated cross sections of the tectum opticum of the midbrain by autoradiographical methods. We were able to state that one or two regions of the tectum showed one or two bands of increased protein synthesis marked by the incorporation of the radioactive histidin (B. Rensch and H. Rahmann 1966; B. Rensch et al. 1968). These bands indicate at least the possible extension of the engrams produced.

It is not yet known whether the engram is formed within the cellular body of the neurons or in its fibers. The latter seems the more probable interpretation. Electrophysiological investigations have shown that almost all sensory excitations reaching the cortical neurons spread over a considerable network of neuron groups, or even over several regions of the brain. So it may be that memorial traces originate by facilitation and histological fixation of such pathways. This view, which has been expressed in recent times by A. Weber (1955), J. Z. Young (1962), and G. S. Grosser and J. M. Harrison (1960), largely corresponds to A. Goldscheider's conception, which was already developed in 1906

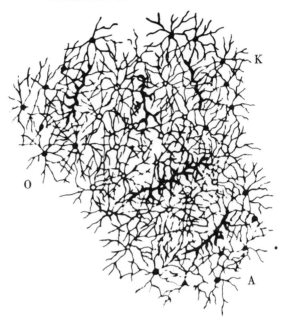

Figure 5. "Classical" diagram, already sketched in 1906 by A. Gold-scheider, representing "engrams" as a pattern of associated fibers. O = an optic center; K = a kinesthetic center of the eyes; A = an acoustic center. Interconnecting traces resulting from the frequent passage of impulses are marked as thicker lines.

(Fig. 5). W. Penfield assumed (1954) that memorial traces in man are represented by a pattern of neurons extending from the temporal cortex to a subcortical region.

If we assume that engrams are reticular structures and that aware-ness corresponds to the passage of excitations, it is conceivable that memorial impressions are only revived when an impulse passes through the same network. If some and not all parts of the engram network are involved, only incomplete, sketchy, or "tagged" recollection re-sults, such as is typical of most mental images. Partial revival of this sort would also provide a physiological basis for the comprehension of similarity with the original recollection. It also points to a possible explanation of conceptualization and generalization (Ch. 6 F II). On the other hand, it is possible that engrams involve macromolecular connections *within* individual cells. Another view put forward is that recognition may take place because proteins at the synapses direct the flow of ions by transformation of their structure or by acting as selective enzymes. So far, however, none of the engram hypotheses could be satisfactorily substantiated (see surveys by G. C. Quarton, T. Melneschuk and F. O. Schmitt 1967; F. O. Schmitt 1967; E. R. John 1967; and others).

Whatever its morphological basis may be, the engram is a structure composed of chemical compounds, probably largely protein bodies or polynucleotides, which have not yet any inherent psychophysical properties. In the case of short-term memory, when the image fades fairly soon, cellular metabolism relatively rapidly resolves these structures, perhaps only at the synapses, and the closed excitation circuits become interrupted. With long-term memory, the structures are more stable, and the breakdown occurs less readily or only when the neurons begin to decay because of aging. However, the compounds are all made up of the same atoms and the same elementary particles as constitute non-living material. So we cannot define the psychophysical substratum satisfactorily in terms of specific atoms.

V. Processes of Excitation and Consciousness. As the material bases of psychic phenomena do not reside in their molecular or atomic properties, we may assume that they are to be found in the system of integration of their elements or in the physiological processes which these cause. The main characteristic of nervous processes is the origin and *course of excitations.* Can we therefore take this to be the physiological correlate of phenomena? At first sight this appears to be unlikely, because there are many nervous processes, reflexes, automatic actions, and motor impulses, which are not accompanied by psychic processes. Our phenomena are only correlated when the excitations enter into a continuous connection which we term the *stream of consciousness.* We almost never experience sensations or images in isolation; they are always embedded in a stream of consciousness.

It is however possible, in spite of this fact, to assume that *all* nervous excitations are correlated with conscious processes, whether general or specific ones. It may be that we do not experience many of these, simply because they do not enter our stream of consciousness, which corresponds to our spiritual being. It is significant that most "sensory" excitations, for instance those caused by the contact of our skin with our clothes or those which come from the periphery of the retina, are cut off by special mechanisms, mainly by the formatio reticularis, and by inhibiting mechanisms in the cortex. This ensures that attention can be confined to more essential excitations (see Ch. 6 F IV). Nor are we aware of efferent motor impulses because they leave the central nervous system and disrupt the connection with the stream of consciousness. All we have is secondary information about motor excitations, sent back from the sense organs after a material reaction has taken place. When actions such as walking upstairs become automatic, the sensory impulses at first accompanied by consciousness are also gradually cut off, and cease to reach the central connection of consciousness.

It may at first seem rather improbable that unassociated components of consciousness can exist. But this becomes easier to accept if we reflect that there is no reason to assume any dualism between mind and matter. On the contrary, we apprehend the existence of matter solely by reduction of certain properties of our psychic phenomena, and in this process of reduction we do not abstract from awareness (in its widest sense). Such an identistic, psychistic theory, held by many philosophers (see Ch. 6 L) and also shared by the author, can be supported in different ways: on a study of the probable course of phylogenetic psychogenesis, on ontogenetical psychogenesis, on an analysis of the psychophysical substratum and analysis of the phenomena themselves, and of matter. These will be discussed in sections E, F, H and I of this chapter.

Such a psychistic hypothesis could possibly also lead to the assumption that the sense organs could be included in the psychophysical substratum, despite the objections discussed above. This would at least make it easier to understand how the very different sensory qualities and modalities—sight, smell, taste, hearing, touch, etc.—have developed. For the sensory excitations are released by very different external stimuli and different initial physicochemical processes; the brain, however, receives only patterns of potential oscillation of varied frequency and amplitude. This translation of the various primary sensory excitations into a uniform language of action potentials, whose only differences are quantitative, enables the brain to combine all the sensory information, weigh it up, and make a resulting response. It seems more probable, however, that these quantified excitation patterns, reacting upon the engram structures, which we may think of as specific proteins, initiate various processes to which qualitative differences of the phenomena correspond. If this is so, then it is the effect of the excitations upon the engrams, and not the excitations themselves, which would constitute the real psychophysical processes. The sensations would correspond to the formation of engrams by means of specific excitations.

We must at least briefly mention that we know equally little about the physiological basis of *feeling tones*. These exist merely as properties of sensations and mental images, but they can influence other trains of mental images and produce "moods." If a microelectrode is inserted into the septal region of the hypothalamus of an animal, and this region is then stimulated by an electric current, the animal will behave as though it had pleasurable sensations. J. Olds and P. Milner (1954) connected the electrodes with an electric current so that the animals (rats), when they pressed a lever, became stimulated in the septal region. A positive feeling tone could be assumed to have resulted, for the rats continued to press the lever, even up to 1,920 times an hour. But if the electrode was inserted into a different center in the hypothalamus, the animals reacted as if they felt pain. Similar results have been obtained with other mammals and also with humans, and some drugs introduced into the hypothalamus through fine canullae had the same effect. Acetylcholine, for instance, injected into the septal hypothalamus area induced feelings of pleasure (see R. G. Heath 1964). However, these results must not be taken to mean that the physiological processes in these parts of the brain immediately correspond to positive or negative feeling tone, for the hypothalamus is connected by fiber tracts with many other regions in the forebrain and midbrain.

Some observations suggest that facilitation of the passage of impulses along the nerve tracts also contributes to positive feeling tones. The positive effect of acetylcholine would then be explained by the fact that this substance has a facilitating effect on all the cholinergic neurons at the synapses. Positive feeling tones are released by various other substances which promote the passage of the excitations and remove or weaken inhibitions. These include alcohol, lysergic acid preparations, and other compounds. The consummative actions of instincts in man and very probably in animals also (eating, sexual processes, etc.) are accompanied by feelings of pleasure; and these processes too are characterized by strong excitations, facilitation of relevant fiber tracts, and removal of inhibitions. This would imply that feeling tones are not linked with any specific structure in vertebrates; for we may assume that instinctive activity of this kind produces positive feeling tones in invertebrates as well (insects, snails, cuttlefish, etc.).

It is of interest in this context that aesthetic feelings of pleasure are produced if the apprehension (the "complexibility") is easier, as produced in the visual field by symmetry, rhythm, balance of right and left parts, and organic shape, and in the auditory by consequence, rhythm, and recurrence of a leading motif. Negative feeling tones on the other hand are generally associated with irregular shape, dissonance, etc. So far, however, all attempts to discover the physiological bases of feeling tones are still entirely hypothetical.

We know still less about the physiological basis of our *logical faculties*. We have not yet any indication why some people—mathematicians for instance—have a particular gift for logical inference, while others, even if they make great efforts, show only restricted logical abilities. Apparently this has nothing to do with the number and arrangement of engrams. It might perhaps be a matter of the engrams being less precisely confined.

Finally, it may be mentioned that the degree of *intensity in normal consciousness* is correlated with oxygen consumption and the intensity of metabolism of brain cells, and with the intensity of excitations. The intensity of consciousness decreases when there is a lack of oxygen (when we faint, for example) or when many afferent impulses are switched off (as in a drowsy state). On the other hand, if the brain is powerfully stimulated, by a lysergic acid preparation for example, perception and mental images may become very intense.

We do not yet sufficiently know how the mechanism works which switches off and on when we fall asleep or wake. We know that the formatio reticularis is involved, a network of functionally nonspecialized neurons reaching from the spinal cord to the diencephalon and spreading diffusely from the hypothalamus to the subcortical regions of the forebrain. This formation can inhibit sensory impulses before they reach the cortex, and so induce sleep; and impulses from the sensory organs or the cortex can activate it, and wake the sleeper up. Falling asleep and waking can also be induced by electric stimulation of two antagonistic hypothalamus systems. Destruction of these centers may lead to a permanent state of sleep or sleeplessness. It seems that the "stream of consciousness," which is typical of the waking state, depends upon the cortex receiving a constant succession of impulses from the formatio reticularis.

VI. Summary. In surveying what we have been discussing in this chapter, we are obliged to state that despite the efforts in various fields of science, we still know relatively little about the morphological and functional bases of the psychophysical substratum. We cannot state precisely that it is limited to certain parts of the brain. This seems probable; but the possibility that the sense organs too are included cannot definitely be ruled out. Nor can we say exactly which portions of the brain are involved; we only know that we can apparently exclude parts of the brain stem, the cerebellum, and motorial tracts. It is not clear whether the spheres of perception and mental images are separate or identical; the latter seems more likely, however, since sensations and mental images are largely identical, and mental images as vivid as sensations occur in the hallucinatory state.

The structure of the engram is still unknown. We have several indications that it is a network consisting of established tracks of nerve fibers. It is not clear whether engrams are formed by specific RNA or specific protein bodies, or by other substances. Finally, we do not know if psychic phenomena correspond directly to processes of excitation or to the formation and revival of engram substances caused by the excitations. It seems probable that phenomena can only be "experienced" if they enter the stream of consciousness which represents the basis of the self-consciousness possessed by the subject who is experiencing and assessing the sensations and mental images.

D. Phenomenal Reality

I. Development of the Problems. When scientific investigations uncover certain verifiable facts and causal relationships, and rules or laws can be established, we assume that the results are "true." These investigations deal with the so-called objective world, even if they are analyzing the activity of the psychophysical substratum. Most scientists do not normally examine the mental processes which lead to such knowledge, and they have certain reservations about their significance, as one can assess these ever-changing perceptions and mental images only in the light of his own consciousness. Yet it is absolutely essential to determine the reliability and limits of scientific knowledge; and the only way to do so is by the examination of one's own mental processes.

As already mentioned (in sections A and B of this chapter), very different epistemological theories exist. If we are to assess these theories, we must examine the problems without any presuppositions and take as our starting point something which cannot be called in question. But the only fact we can know with absolute certainty is that we are experiencing perceptions, mental images, and thought processes.

Plato had already expressed this clearly. Discussing the nature of cognition in the *Theaetetos*, he makes Socrates say: "My perception is true for me; for it is the perception of my present being." Aristotle writes in a similar way in the *Three Books on the Soul*: "The perception of what can be perceived is always true . . . but it is possible to cognize falsely, and this belongs to nobody who has not reasoning power." St. Augustine makes the same statement in his *Soliloquy with Eternal Reason* (R) (about A.D. 387); and he already develops the problem further: "R: Do you know that you think? A. I know. R: Is it then true that you think? A: It is true." This leads on to the statement that things only acquire their characteristics through being present in our minds. "R: There are then no stones in the unseen bowels of the earth, nor anywhere where there are not those who will perceive them . . . nor can yonder wood be wood, except on its surface." Descartes, in one of his early works, *Regulae ad directionem ingenii* (published posthumously in 1701), inquired still more closely into the nature, certainty, and limits of knowledge. His principle was to doubt everything at first, and he found himself absolutely certain of nothing but his own processes of thought. In his later *Principia philosophiae* (Bk. I, 7, 1644) he expressed this in his much-quoted "Cogito, ergo sum."

Epistemological theory began to emerge in this way as a distinct and specific philosophical discipline. Its real founders, however, were Locke, Berkeley, Hume, and Kant, who conducted more detailed and critical analyses of perception and thought processes. Locke, who inquired into the origin, certainty, and extension of knowledge and belief, also emphasized in his *Essay Concerning Human Understanding* (1690) that one can make reliable statements only about his own perceptions and mental images: "What perception is, everyone will know better by reflecting on what he does himself, what he sees, hears, feels, etc., or thinks, than by any discourse of mine" (1877, p. 253). This led him to state that all phenomena are based on perceptions: "Perception then being the first step and degree towards knowledge . . . the inlet of all knowledge in our minds" (pp. 261-262). Kant (1787) examined all the problems of human cognition in a very critical and careful manner. He assumed that some cognition exists a priori, before all experience. By accepting the transcendental nature of phenomena and examining the categorical functions indispensable to cognition, this great philosopher laid the foundations of all subsequent epistemological discussions.

More recently, the findings of psychology and science have extended the area of epistemological inquiry. The research undertaken by Fechner, Wundt, Ebbinghaus, Mach, Ziehen. Carnap, Schlick, Einstein, Schrödinger, Heisenberg, and others has been instrumental in establishing a basis for our present-day philosophical knowledge. In particular, Ziehen (1898, 1913, 1934, 1939), originally a psychiatrist but possessing an encyclopedic knowledge in all branches of science and philosophy, has provided us with a masterly exposition and a clear analysis of the problems, and has offered new solutions. His works will therefore frequently be referred to in subsequent passages.

II. Facts Which May Be Accepted as "Given". What then are the facts of consciousness that can be claimed as undeniably real and the sole *absolutely* reliable basis for any philosophical interpretation of the universe? As already mentioned, adult man experiences in his waking state a *stream of consciousness* within which he can only artificially delineate various components. These are mainly sensations of different kinds, conveyed by sensory excitations, and mental images. As sensations normally occur in conjunction with mental images, and not in isolation, it is more apposite to describe what we experience as perceptions. Only rarely, when half awake or in some equally drowsy state, may we experience a sound, perhaps, as an unrelated sensation. Or we may have very intense experiences, so intense that they are practically devoid of associated mental images, as may occur with acute pain or profound sensuous delight—perhaps when carried away by music or by a pungent taste, or the satisfaction of a drive (the sexual drive, for example). Sensations are never completely unrelated, however, because all "statements" about them are only possible in the framework of our consciousness. In dreams, on the other hand, when perceptions are entirely excluded, we can experience trains of pure imagery.

Other categories of phenomena within our consciousness include feelings and processes of volition. Feelings, as we shall see, are simply properties of sensations and ideas. But if they are sufficiently intense they may almost predominate, and they may also affect other groups of sensations and ideas. Processes of volition, being also phenomenally distinct, are ultimately only sequences of mental images directed by a dominant image.

It is usual to describe all these components of consciousness as *phenomena* in the Kantian sense. But the term is not a completely happy one, for the word means "appearances." As sensations and mental images not only "appear" to have certain properties but are immediate reality, Ziehen used the more neutral term "gignomena" (= something which comes into being) to denote them. Unfortunately this term has not been adopted in international philosophical parlance. So we shall use the term "phenomena," but always in the sense of "data," something directly "given" whose reality is indisputable.

It is a characteristic of all phenomena that they cannot be defined. We cannot explain "blue" or "red" to a color-blind man, nor can we help him to form the corresponding mental images. All we can give is a negative definition, saying that it is neither black, grey, nor white. In the *Theaetetos* Plato makes Socrates say that the "primary elements" of all that we perceive "are such that no explanation can be given of them. Each of them just by itself can only be named"; and later, "it [the element] can only be named, for a name is all there is that belongs to it." Locke (1690) was no less explicit about "simple ideas." "These, like other simple ideas, cannot be described, nor their names defined; the way of knowing them is, as of the simple ideas of the sense, only by experience."

Another characteristic of the stream of consciousness is that we experience it as a temporal sequence of phenomena. Each perception clearly belongs to a "present," and most mental images have a peculiar character of being related to a "past" ("Vergangenheitstönung"); reflection and planning often have the character of being related to a "future" ("Zukunftstönung").

Another characteristic of our consciousness is *subjectivity*, the connection with our own self, in the sense of *self*-consciousness. Kant describes this relationship as the "synthetic unity of apperception" (*Kritik der reinen Vernunft*; 2nd ed. 1787, I, 2, § 16, 17). As we shall see, however, the concept of the self is something which has gradually developed both ontogenetically and phylogenetically (Ch. 6 F II, H III). So, although the subjectivity of consciousness is typical of adult man, it is not a basic property of phenomena, but a peculiarity of a secondarily developed complex of mental images.

The sequence of perceptions and ideas within the stream of consciousness is not an arbitrary one. It is governed partly by association of ideas —in extreme cases by morbid flight of imagination—and partly by dominant ideas which direct the choice of a certain train of association of ideas, in the sense of an act of volition. In such cases, ideas are often connected to one another in the form of judgments and conclusions. The *judgments* involve peculiar functions; we recognize sameness, difference, and similarity (categorical functions), and are capable of analysis and synthesis. All these functions, which Ziehen aptly terms *differentiation-functions*, are also basic phenomenal properties which we must accept as "given" data.

Conclusions are made possible because we are able to think logically. And this too is something primarily "given." We shall see, however, that thought itself has developed phylogenetically by successive adaptation to the logical laws that govern the universe.

The fact that our thinking is based upon innate differentiation functions and logical capacities led Kant to his a priori theory. He held that what our experience teaches us is simply to recognize connections between phenomena. We recognize that these connections are causally or logically determined, independently of and before any experience. This "transcendental" character is "given" a priori. If viewed in the light of phylogenetic development, however, this a priori character, which later philosophers interpreted in either a more psychological or a more logical sense, loses much of its metaphysical particularity. Psychophylogenesis proves to be a gradual process, directed by natural selection, whereby the cerebral and the correlated psychic functions adapted themselves to the causal and logical laws that govern the universe. Man, however, only becomes aware of these two systems of regularities inductively, through experience, in other words a posteriori.

III. The Nature of Cognition. If we want to develop reliable ideas of matter, space, time, lawfulness, the self, the freedom of the will, etc.— that is, of concepts which are vital to problems of biophilosophy—we must first establish in what way these ideas are based upon the "given" data, upon indubitable reality. At an early date philosophers began to distinguish two groups of facts: on the one hand, the phenomena themselves; and on the other, the logical combination of such phenomena by means of thought processes and in particular of conclusions. Thomas Aquinas made a distinction between "cognitio sensitiva" and "cognitio intellectiva"; Roger Bacon spoke of empirical and inferential cognition; William of Occam of "intuitive" and "conceptual" cognition; Spinoza of "imaginatio" and "ratio," and also of intuitive cognition. Locke called these basic types "sensation" and "reflection"; and Kant wrote of "perception" (Anschauung) and "thinking." In more recent times, logic has usually been considered as a separate discipline on account of its particular nature and of its scope; but it belongs to epistemology.

The aim of all cognition is "objectivity" and "universal validity"— which is often termed "truth" (though the precise meaning of the term is not always clear; see the next section). The "will to know" is a prerequisite of all cognition.

Cognition is limited by man's powers of comprehension and is therefore never absolute. Many philosophers, however, have ignored this *principle of immanence* and have based their theories—or have been obliged to do so—on certain maxims or religious dogmas, in other words, on transcendent ideas not derived from human experience and often incompatible with it. The biologically based philosophy expounded in the present work does not involve the acceptance of any transcendent "subjects," "objects," or processes. In this respect it adheres to the *positivistic principle* of restricting itself to the analysis of phenomena and their mutual interrelations, and the syntheses which can be derived from these. (Positivism, which is based on the psychic phenomena, has sometimes been equated with "materialism"; but this can only be the result of either total ignorance or deliberate distortion of philosophical concepts.)

We exceed the bounds of immediate experience in different degrees, by conclusions, abstractions, generalizations, and the establishment of laws. Ziehen (1913) has aptly termed this activity *transgredience* as opposed to transcendence; and it should be noted that it does nothing to invalidate the principle of immanence. All we are doing is to use the innate functions of differentiation and logical thinking.

While we generalize and establish laws, however, we must of course critically examine the principles of human reasoning, the psychological processes accompanying cognition, as well as our logical operations. In other words, we must carry out a *critique of cognition*; for it is always possible to be deceived by psychic phenomena or wrong conclusions (see the following section).

IV. The Problem of Scientific Truth. All scientific cognition aspires to "objectivity" and "universal validity." These two terms indicate the most essential properties of what in the epistemological sense we may call "truth." This is a concept which has engaged the attention of almost every philosopher; the many definitions which have been put forth agree in essential points, but they differ in a number of respects. It is not within the scope of the present book to discuss the various versions and their historical development (the reader may refer to the writings of E. Husserl 1900; R. Eisler and R. Müller-Freienfels 1922; E. Tugendhat 1964; and others). It is, however, necessary to state briefly what I understand by "truth," and to distinguish this concept from others which are sometimes used in a similar sense.

All truth demands factual "correctness" and formal, i.e., logical "correctness" in the process of thinking. Factual correctness means being concordant with relationships based upon experience. The content of ideas or judgments must be as consistent as possible with what is supposed to exist outside the mind, or with the content of other phenomena which are being assessed. But as we can never assess what exists "objectively," we cannot claim that the content is fully consistent. There will always be only an *approximation* to "objective" reality.

Another important criterion of scientific truths is their consistency, the fact that nothing is contradictory and that they fit in with existing laws. In other words, truths must be as universally valid as possible. Normally, scientific truths are successively developed and are first established only as hypotheses and theories (Ch. 2 B). Hypotheses, especially contrasting ones, often contain only partial truths. So truth has to prove itself.

Correctness can be equated with truth only in cases where both factual and formal logical correctness are assured. Many philosophical systems illustrate the fact that despite correct conclusions and logically incontestable methods, they may be "factually" false, because they are based on inadequate or wrong premises.

The concept of *reality* only partly coincides with that of truth. We denote all our psychic phenomena as real; and in general we say the same of the knowledge which we acquire transgrediently through *direct* associative and logical treatment. For example, if we look two or three times in quick succession at an object and recognize it as the same one (see p. 255 for the problem of sameness), we say that it "really" exists. But we may be wrong about the true nature even of phenomena. If we look at an illusory pattern, Zöllner's for example, showing a number of parallel lines with alternate oblique hatching, we see these as convergent and divergent. Here we have a case of phenomenal reality but not factual truth, for the lines exist extramentally as parallels. Truth, then, is always dependent upon our thought processes. It represents the theoretical validity of the content of our thinking.

When we come to consider causality we shall see that it is doubtful if a priori "transcendental" truths exist. And the transcendent religious "truths" of various dogmas are only human postulates.

During phylogeny behavior appropriate to "truth," in the sense of what exists outside the mind, has obviously been developed because it was disadvantageous to behave otherwise. With hereditary reactions, this happened by natural selection of individual variants (Ch. 5 B, D I). But nonhereditary behavior as well, based on learning and generalization and in particular planned actions (involving "foresight") of higher animals, has always been guided by selection, for reactions which were not appropriate to the real circumstances of the situation resulted in failure and sometimes in loss of life.

Much the same may be said of man's actions. In spite of many set-backs and the fact that men have sometimes clung for centuries to false ideas, cultural development has been a progressive process of adaptation to external "truth." So we would do well to heed Schopenhauer's warning (*Die Welt als Wille und Vorstellung*, Part 2, Ch. 6): "Daher kann nicht zu oft wiederholt werden, dass jeder Irrtum wo man ihn auch antreffe, als ein Feind der Menschheit zu verfolgen und auszurotten ist und dass es keine privilegierte oder gar sanktionierte Irrtümer geben kann. Der Denker soll sie angreifen, wenn auch die Menschheit, gleich einem Kranken, dessen Geschwür der Arzt berührt, laut dabei aufschrie." ("So it cannot too often be repeated that every error, wherever we meet it, is to be pursued and rooted out as an enemy of mankind; there can be no privileged or sanctioned errors. The thinker should attack it even though he makes humanity shriek like a patient when the physician touches his ulcer." The twentieth century has seen very rapid progress in approximation to external properties and relations which have been recognized as true. In this sense the conduct of life has become much more scientifically based. Tugendhat (1964) has described this process as "radicalization of the interest in truth."

E. Analysis of Sensations

1. Properties of Sensations. Within the stream of consciousness sensations can be distinguished from mental images because sensations are much more sensuously vivid and always relate to the actual present. They correspond to excitations initiated by stimuli outside or inside the body of the subject (exteroreceptive or interoreceptive stimuli).

Every sensation has certain basic properties: quality ("red," "sweet," "rough," "cold," "thirsty," etc.); intensity; spatiality (location, shape, and direction of single components to one another); and temporality (reference only to the actual present). Many sensations are also characterized by a feeling tone (pleasurable or unpleasant). These basic properties can be experienced but not defined; at best (and especially the more complex among them) they can be characterized to a limited extent. The fact that individual sensations correspond to different causally determined physiological processes (for instance, qualities of color to excitation patterns set up by light waves of different lengths) has nothing to do with the possibility of defining the qualities of the relevant sensations.

1. *Qualities* group themselves into more or less distinct *modalities* which are coordinated with the different sense organs. Thus in man the qualities of visual, auditive, and temperature sensations are so different that they do not melt into one another. On the other hand, we experience a uniform complex quality when we eat a fruit: the "taste" of a pear, for example, is made up of the qualities of taste (sour and sweet), of smell (scent of the fruit), and of touching its flesh. We do not yet know if the uniformity of this complex quality corresponds to a single physiological process in the brain (the result of a mingling together of the excitations from different sense organs), or if it is only caused by the simultaneity of the different excitations. The first of these interpretations would fit in best with an identistic conception, which is also supported by several other considerations.

Some qualities are arranged in a definite manner within the relevant modality. Most color qualities blend into the next in a color cycle or, if all mixed colors are included, into a "color body" in the form of a double cone (see W. Trendelenburg et al. 1961). Qualities of pure tone, warmth, and cold can be arranged in a monodimensional series. Qualities of smell, on the other hand, though they fall into distinguishable groups (flowery, fruity, putrid, pungent, etc.) cannot be arranged in any series or satisfactory order.

2. All qualities can occur with different *intensity*. The degree of intensity is dependent upon that of the relevant causal process, i.e., the excitation and the releasing stimulus; but there is no total parallelism. The degrees of intensity can be arranged in a series, but the scale is limited by the upper and lower thresholds of sensitivity. Intensity of sensation, being a property of a psychic phenomenon, cannot be measured, although one can of course coordinate the intensity with the measurable stimulus or excitation. According to Weber's law (F. H. Weber 1834), in the medium range of intensities the relative "differentiation threshold" remains constant, whereas it drops with lower grades of intensity and rises with higher ones. With a weak concentration of a color in the range of pink, for example, a much slighter increase of color is sufficient for an increased intensity to be apprehended than with a

denser red color. Besides, Aristotle had already expressed the same basic principle as that contained in Weber's law when he wrote in his *Metaphysics* (X, i) : "In the case of anything of larger size, any addition or subtraction might more easily escape notice than in the case of something smaller." Th. Fechner's logarithmic formulation of Weber's law seems more precise, but it is contestable, because intensities of sensations cannot be measured, as E. Zeller, F. A. Müller, and others have already pointed out (see also Ziehen 1934, pp. 111 ff.).

3. *Local properties* of sensations can best be recognized in the relationships between the individual components of a sensation, the different parts of a shape for instance, or the interrelations between several sensations, for example when several parts of the body are touched simultaneously. These local properties are most pronounced with sensations of sight, touch, pressure, tension, pain, and temperature—those sensations where the relevant excitations are released by receptors which are arranged in surfaces. They are less pronounced in the case of taste, and still less with smell; but these are supplemented by attendant sensations of touch, as the tongue is touched or air passes through the nostrils.

Spatial properties relating to sensations of shape and position of several sensations indicate the direction and also the distance of the individual components. The perception of direction predominates in binaural hearing. This is accompanied by a sensation of position in the region of the inner ear.

The local properties of all sensations are serial data and merge to form a single homogeneous, limitless, three-dimensional space. All positions, directions, and distances are normally related to our own body. But because of the homogeneity of spatial sensations, it it possible to set the zero point for the three dimensions anywhere within an apparently existing absolute space, though this is only concluded from the correspondence of the local properties of all sensations. (Direction and distance from such a point vanish, while the local properties as such will remain; see Ch. 6 I.) With sensations of movement, successive sensations merge, and intensity is experienced as speed.

4. The *temporal properties* of all sensations are also homogeneous and can be arranged serially, but only in a monodimensional sequence, which differs radically from the spatial dimensions in having an ever-shifting point of present time and in being irreversible. As the temporal

property of sensation is identical with the present, time thus becomes a "reality-coefficient" (Ziehen).

As all sensations possess temporal properties of the same kind, an idea of absolute time has developed, similar to that of absolute space. Absolute time, however, is arrived at purely by a process of logical deduction; it is not primarily given. The problems involved here will be discussed in section I of this chapter.

We experience the relations of temporal properties as simultaneities and sequences. The present can only be experienced as a point of time, but we also experience a quality which indicates that a sensation is already known to have been formerly perceived (H. Höffding's "Bekanntheitsqualität"). This gives temporal properties their tone of present or past. And as we can establish connections in our minds with something anticipated in the future, we may also experience a tone of futurity. These three tones, like all the properties of sensations, constitute the primary "data" of our phenomena, which cannot be explained to any further degree.

5. Many sensations are characterized by *positive or negative feeling tones*. The latter often protect us from the onset or threat of something harmful. Strong negative feeling tones are almost always present if the qualities of sensation reach a very high degree of intensity, indicating that the relevant sense organ or the body may be in danger: sensation of a glaring light, an over-loud sound (noise), extremes of heat and cold, extremely sour or salted food, very putrid or pungent smells. Negative feeling tones also accompany perception of possible danger. Positive feeling tones are mainly experienced with sweet-tasting food and harmless scents, or when instincts vital to the survival of the individual or the species are satisfied (hunger or thirst, sexual activity, maternal reactions, etc.). As already mentioned in section C V of this chapter, positive feeling tones are also present in most cases when sensations are fully comprehensible (complexible): rhythm in music; rhyme or alliteration; symmetry in the visual realm; rhythmic repetition of identical or similar elements (the pleasure of recurrence, as in music and poetry); steady curves (circles, spirals, etc.). On the other hand, negative feeling tones normally accompany incomplete rhythm, insufficient symmetry or congruence (principle of disappointed expectation), and other

sensations the apprehension of which is hampered (see Ziehen 1923, 1925).

The intensity of feeling tones is parallel in some degree to the intensity of the sensations involved. As feeling tones are linked to widely differing qualities of sensations, it might seem that many types exist besides negative and positive ones. However, most so-called feelings are complexes of ideas involving positive or negative feeling tone.

Feeling tones are unlike other properties of sensations in that they may irradiate from one group of sensations to other sensations or trains of ideas, and may evoke a certain "mood" such as melancholy, grief, joy, or euphoria.

II. Phylogenesis of Sensations. When we discuss the phylogenesis of psychic phenomena (Ch. 6 H III) we shall see that it is very probable that invertebrates which have sense organs and a nervous system have sensations. Even unicellular organisms may possibly have isolated sensations, or prestages of sensations. All these animals react to diverse stimuli, some of them to chemical stimuli, gravity, or touch, and many to light. This raises the question whether the differences of the qualities and modalities of sensation appeared independently of one another during the course of evolution, or whether they gradually evolved from one another. This problem is of vital significance in the context of biophilosophy. For as far as we humans are concerned, it is the *qualitative* differences of the psychic phenomena which distinguish them fundamentally from the "material" physiological processes which in the last instance are largely determined by *quantitative* differences (Ch. 6 K V, L).

Results in comparative anatomy, comparative embryology, and phylogeny lead to the assumption that the different qualities and modalities of sensations have been developed gradually, together with the corresponding sense organs, by mutation and natural selection. And only such sensory reactions have been developed as referred to biologically significant stimuli: stimuli suitable for spatial orientation, recognizing food, enemies, and members of the same species, vital light rays, sound waves, and tactile influences. In addition to making this selection from among the numerous external stimuli, the sense organs further restricted the reactions to those grades which were important for their life (upper and lower thresholds), and to an advantageous scale of differential thresholds. In this way sense organs with very different structures and functions developed, and at the same time the different phenomenal qualities and the intervals between the modalities arose.

Is it possible that qualities evolved out of one another by a process akin to phylogenetic branching? On the basis of an identistic conception this seems very probable. In the same manner as atoms with very different properties arise as a result of only *quantitative* differences, combinations of protons, neutrons and electrons and greatly differing molecules arise by only quantitatively differing combinations of atoms; in just the same manner, then, protopsychical properties uniting in various ways may have produced new qualities (see B. Rensch 1961a, pp. 321-322). In this context we must take into consideration that the ultimate units known to us, the elementary particles, have basic properties which cannot be traced back to other sources: energy or mass, momentum, spin, charge, and position in space and time (Ch. 6 I, L). Moreover, throughout the course of phylogenetic development, whether in the extramental or the phenomenal sphere, the processes in question have not been summations but integrations leading to the emergence of new *system properties*, due to general system laws.

In the course of evolution it has always been an advantage for an animal to make a sharp distinction among qualities of sensations transmitted through the different sense organs, and also the qualities of the corresponding mental images produced by them. Visual sensations, for example, which enable an animal to perceive objects at a distance (to recognize its enemies or its prey) have quite a different significance for its behavior from sensations of touch or taste. As the reacting structures, the sense organs and the central nervous system, form part of the psychophysical substratum, we may assume that natural selection has been concerned with the differences and favored those variants which could distinguish qualities and modalities more sharply. But this kind of "psychological" selection operated on the "material" structures and functions of the sense and nerve cells. In the majority of animals, therefore, we find those particular modalities most markedly differentiated which have a very different significance with regard to the reactions they set up. Sensations of sight do not mingle with those of touch or taste. On the other hand, differentiation is less marked among modalities which induce similar reactions. In humans, and no doubt in many animals too, the "taste" of food is a complex quality in which qualities of taste, smell, and touch belonging to three modalities are blended.

It is *imaginable*, then, that lower animals which react to differing stimuli in the same or a similar way may have correspondingly undifferentiated sensations. O. Mangold (1935) has shown that earthworms respond in the same way to substances which humans distinguish as sweet, salty, sour, or bitter. If different combinations of these substances are presented, they produce an additional effect, and nothing indicates that several qualities are distinguished. However, the evidence for plurimodal sensation which could be obtained from such findings cannot be ranked very high, because earthworms are only capable of a very few relevant reactions, and conclusions by analogy are rather vague in the case of lower animals.

It is perhaps worth noting here that auditory sensations in vertebrates may have developed from sensations of vibration, an assumption already made by E. Mach. And the more general assumption that all modalities and qualities evolved from original plurimodal sensations has been accepted by several scientists and philosophers from Democritus onward to H. Spencer, W. Wundt, F. Leydig, O. and R. Hertwig, W. H. Nagel, Th. Eimer, and L. Plate. It remains a hypothesis, however, which cannot be proved. But the important point is that there is no objection in principle to a gradual phylogenetic differentiation of sensations, and that this conception accords very well with an identistic interpretation of the universe.

Since their origin, sensations have probably shown different *degrees of intensity* in accordance with differences in the relevant stimuli. The upper and lower limits of intensity, developed during phylogeny, were determined in the first place by the biological significance of the intensity of the releasing stimuli, and by the physiological and morphological possibilities of developing a lower and upper threshold of excitation of the sensory cells in question.

As sensory cells developed and multiplied in the course of evolution, they became coordinated with corresponding neurons or neuron-groups within the central nervous system. Apparently at the same time, the individual components of sensations of shapes developed their appropriate spatial "local signs." As the phylogenetic increase of such local signs proceeded step by step, it becomes conceivable that in higher animals with many more neurons, it was no longer necessary for the arrangement of the central excitations to correspond to that of the stimulus pattern on the sensory surface, for instance, on the retina. As already mentioned in section C II of this chapter the excitation pattern in the convoluted visual center of higher mammals and man no longer represents a replica of the retinal excitation pattern, whereas in fishes and amphibia, electrophysiological studies have shown that the retinal pattern still largely corresponds to that in the tectum opticum of the midbrain (see R. M. Gaze and M. Jacobson 1963; M. Jacobson and R. M. Gaze 1964; R. M. Gaze 1958; and others).

The reason why temporal signs did not develop is that sensations always relate to the present time. Only perceptions which are associated with mental images have past and future tone. The gradual fading of a mental image with the lapse of time may in such cases be an indication of past tone.

It is highly probable that positive and negative *feeling tones* developed at an extremely early phylogenetic stage as they served to guarantee the course of vital instincts and to shield the organism from harm. We may therefore assume that positive natural selection always promoted the development and maintenance of the advantageous properties of sensations.

F. Analysis of Mental Imagery

1. Properties of Mental Images. Mental images, like sensations, can only be artificially isolated in the stream of consciousness. But they can be more clearly assessed; for in some conditions, such as drowsing with eyes closed or especially in dreams, we can have a relatively prolonged sequence of mental images totally unconnected with sensations. In normal daily life, on the other hand, mental imagery is often overlaid by sensations, and certain mental images are almost always immediately associated with sensations within the stream of consciousness, so that we experience "perceptions" and not pure sensations.

Mental images are distinguished from sensations by being much less sensuously vivid and less complete. They often have a tone of past or future, and only partially of the present, though of course they can only be experienced in the actual present. One typical feature peculiar to them is the link with corresponding earlier sensations which is the basis of recognition. This "intentional" connection manifests itself as a *quality of familiarity* (H. Höffding's "Bekanntheitsqualität": 1911), which cannot be traced back to any other quality. It corresponds to Kant's "synthesis of recognition in the concept" (*Kritik der reinen Vernunft*). As Ch. Bonnet had already suggested (1770-1771), this quality of familiarity may be due to the greater ease with which an existing engram is revived, compared with the building-up of an entirely new one.

The physiological basis of sensations is obviously quite different from that of mental images. Sensations are related to excitations (or rather, if one accepts an identistic view, they correspond directly either to them or to their effect on the structures of neurons in the brain) produced by an external or internal stimulus, and they either alter an existing engram or build up a new one. Mental images, on the other hand, become conscious when excitations which arise in the brain "call up" existing engrams.

There are borderline cases, however, where sensations and mental images are not clearly distinguishable. If the excitations of the brain are greatly enhanced, often as a result of a pathological condition, this can produce hallucinations. Although these are not initiated by external stimuli, they have the sensuous vividness of sensation, as well as its reference to the actual present. These hallucinatory mental images may even be so strong as to overlay normal sensations of sight and hearing, and the patient actually sees phantoms or hears sounds. Hume had already recognized this intermediate stage between sensations and ideas. In 1739 (Part I) he wrote: "Thus in sleep, in a fever, in madness, or in any violent emotions of soul, our ideas may approach to our impressions: as on the other hand it sometimes happens, that our impressions are so faint and low, that we cannot distinguish them from our ideas."

This borderline condition, when the subject is unable to distinguish between sensations and ideas, could be experimentally proved. When O. Külpe (1902) presented his students in conditions of darkness with short, faint light stimuli, they frequently could not distinguish these from purely subjective appearances of light. Perry (1920, cited by D. G.

Ellson 1941) obtained still clearer results. He placed his subjects in a darkened room, and then projected the image of a colored object onto an opaque glass screen. After it had been switched off, he directed his subjects to reproduce the picture as a mental image, so that they saw it, so to speak, "on" the screen. Meanwhile, he projected the real but very faint image onto the screen from behind, at irregular intervals between the reproductions of the mental image. He then found that his subjects often failed to distinguish between the "objective" image and the mental image.

We have already noted that this general degree of concordance between sensations and mental images supports the view that both groups of phenomena belong to the same region of the psychophysical substratum (Ch. 6 C III). Apart from the distinguishing characteristics we have mentioned, therefore, all other properties of mental images correspond to those of sensations. Mental images also possess *qualities*. When dreaming or simply when our eyes are closed, we may "see" colored or noncolored things of all kinds. Normally we also experience the quality of familiarity. Sometimes, however, we have the impression of something new and never before experienced, and in such cases this typical criterion is lacking. Mental images, too, can have different intensity (usually termed "energy"). If mental images have a certain shape, in particular like visual images, they have *spatial* properties; and they also have *temporal* properties, in that they are experienced either as present or as past, especially if the quality of familiarity is very pronounced. Finally, mental images are often connected with positive or negative *feeling tones*. We sometimes delight in agreeable memories or are harassed by disagreeable images, anxious dreams, etc.

Because sensations and ideas have so much in common, it was realized at an early stage that the latter are determined by the memorial traces (engrams, residua) left by the processes which produce sensations. When these traces are reactivated and mental images are revived, their components may be combined in different ways. This is what happens in certain abstractions and generalizations, or when logical principles are applied (the ability to analyze and synthesize). We have already mentioned that Alkmeon (5th century B.C.) is reported to have taught: "It is the brain which enables us to have perceptions of hearing, sight and smell; these lead on to memory and ideas; but when these are once fixed and have come to rest, then knowledge results " (see W. Kranz 1949). If we take "fixed" to refer to the formation of engrams and associative links, and "come to rest" to mean that they remain unconscious until revived, the conception of this great Greek physician and philosopher may be said to be already in line with modern thought on the subject. Similar views were held later on by Aristotle, Diogenes Laërtius, Leonardo da Vinci, Campanella, Hobbes, Gassendi, Locke, Hume, Leibniz, Condillac, Holbach, Helvetius, La Mettrie, Cabanis, Moleschott, Feuerbach, and others, and by more modern philosophers such as Wundt, Ebbinghaus, and Ziehen. Locke's well-known statement "nihil est in intellectu quod non prius feurit in sensu" has been supplemented by Leibniz, who added "nisi ipse intellectus" (1702; see 1906a, p. 54). He apparently wanted to draw attention to the capacities of logic and volitional thinking, to which he referred as the "particular materials of thought, particular objects of reason," in a letter he wrote to Queen Sophie of Prussia (1702; contained in the second volume of his *Major Works* under the title "Sur ce qui passe les sens et la mattière"). Leibniz spoke as well of the imaginative faculty which can also be used in mathematical conclusions (1906a, pp. 412-413). But he also derived the materials of thought, the ideas, from perceptions. This sensualistic conception has been fully confirmed by later results in psychology and physiology of the brain.

Kant's doctrine, that there is also a priori knowledge independent of experience, is incompatible with such a sensualistic interpretation. It contradicts the inductive basis of our knowledge (see Ch. 2 A). In my opinion Kant's a priori theory has been sufficiently refuted already (Ch. 6 K I and Ziehen 1913 § 51). However, Kant was right in pointing to the spontaneity of thinking, the "capacity to produce ideas" (*Kritik der reinen Vernunft*). The physiological basis of such active production apparently lies in the steady activity of the brain, which we can recognize and analyze by recording the electrical action potentials which act like a motor, keeping up the flow of thought and making possible new associations of ideas as well as the processes of analysis, synthesis, and conceptualization.

This peculiarity of mental images, that they are linked by association, is also very characteristic. We must assume this peculiarity in all invertebrates and vertebrates which are capable of associative learning, and therefore probably have memory, in other words, mental images. When, for example, an animal is trained to discriminate among optical patterns, and the choice of one of them is always rewarded, an association will arise between pattern and food or expectation of food. Such association in animals, as well as in man, may be established *by contiguity*, by the experienced sequence of perceptions or mental images—by an animal for example opening a lid with a pattern and then finding food. Or the association may be established *by similarity*, when perceptions or mental images have some identical components (an animal may recognize a cross once learned, even though one of its arms is missing). Aristotle had already distinguished these two types of association, but added "association by contract," which can, however, be explained by contiguity. Association by similarity always happens when a sensation becomes connected with an existing similar mental image. This is often the case when a "perception" arises. We then often experience the "quality of familiarity" already mentioned. Presumably the same holds good in higher animals.

II. Abstraction and the Formation of Concepts. Ideas develop by stages. A single sensation may leave a trace making a short- or long-term recollection possible, according to the intensity, feeling tone, and type of other ideas associated with the original sensation. This kind of *primary mental image* has usually already lost much of its experienced spatial and temporal connection. If the same or similar perceptions recur, normally all the spatial and temporal characters fall away, and a *secondary idea develops*. On further repetition and abstraction of unimportant details, these may become *concepts*, such as "brimstone butterfly," "cherry blossom," "gold." The next step is taken when separate identical properties of several similar perceptions and mental images are isolated, so that more abstract concepts arise such as "butterfly," "insect," "yellow," "flying." By picking out particular common features belonging to more widely differing phenomena, we then arrive at very general concepts like "color," "activity," "cause," "analysis"; and in science and mathematics, we build up universal symbols such as RNA, Hz, %, $\sqrt{\ }$. In this sense all the words we use—or more precisely, our perception of words seen or heard, or the corresponding mental images of their sound—represent *symbols* for objects, activities, properties, causal and logical relationships, etc. And it is this "thinking in symbols" which makes a considerable degree of abstraction possible.

We are able to form concepts because we can recognize identities and similarities, and are capable of analysis and synthesis. We must accept these *differentiation functions* as primal phenomenal data which cannot be traced back to other phenomenal elements (see Ch. 2 A). Their physiological basis is not yet known, although an analysis seems not to be impossible. We may guess that the apprehension of identity and similarity depend on the fact that the relevant engrams are localized in reticular arrangements of neurons; excitation of the complete engram corresponds to identity, and partial excitation to similarity (Ch. 6 A IV). We are capable of analysis and synthesis because we can direct our attention to different parts or components of the reticular engram by inhibiting the excitation of other parts of components. Synthesis, too, is normally caused by a dominant idea—or rather the excitations corresponding to it—which guides the train of association (Ch. 6 G V).

Another important point in the "psychical" and physiological activity involved in synthesis is that the thought process has been adapted to the causal and logical laws of the universe. This process of adaptation has probably taken place during the whole course of the evolution of man's ancestors, beginning with the most primitive animal groups capable of memory, because natural selection always excluded or at least reduced any brain structures leading to processes and activity contrary to the causal and logical laws. As vertebrates and higher invertebrates can learn and generalize as though they were capable of averbal concepts, we must admit that they also possess the differentiation functions, at least the ability to compare and analyze. At any rate, their brains function in a way that corresponds to these capacities.

Experiments with animals show that averbal concepts may be formed in a purely *passive* manner. For instance, if we train a fish always to select the smaller of two black circles, and then gradually alter the size of the circles, we find that usually the fish will then spontaneously also select the smaller of two rectangles, triangles, or other shapes—in other words, it behaves as though it had developed a concept of "larger" or "smaller." When long series of similarly altered pairs of pattern have been offered, some birds and higher mammals have developed much more abstract averbal concepts. A chimpanzee trained in our Zoological

Institute, who had learned to use a piece of iron resembling a chisel, with a short blade 1.5 cm in width, to unscrew countersunk screws of suitable size, spontaneously took a red-handled screwdriver it had never seen before, with a thin 8.5 cm pin and a blade only 4 mm in width, and used it to unscrew a very small round-topped screw (B. Rensch and J. Döhl 1967). Ch. B. Ferster (1964) even succeeded in training two chimpanzees to count up to seven differing shapes (e.g., triangles) which were projected onto a glass screen, and to represent the number by switching off and on three lamps ranged in a row, using the binary code (e.g., 2 = dark-light-dark, 5 = light-dark-light).

Man's "genuine" concepts differ from animal concepts in a decisive manner by being normally combined with words. When reading, that is, when seeing word pictures, and when thinking in sounds of words, we can arrive at abstract ideas much more easily than if we only thought in pictorial images as animals do. And by forming words, we can produce grammatical structures—substantives, adjectives, predicates, conjunctions, etc.—and can also express causal and logical relationships. Above all, we are able to produce concepts *actively*, that is, we can select concordant components belonging to different types of phenomena and combine them at will.

In regard to epistemology, the *concept of the self* occupies a unique position. Many philosophers and psychologists have discussed the problem of the self, and they have often taken very differing views. For a brief discussion it will be best to consider the biological and physiological bases for the development of the concept in childhood. During the first few weeks of life, the infant is governed primarily by instincts and reflexes. Before long, however, engrams become established, and we must assume that the phenomena which the young child experiences become more and more interconnected. At first he makes no distinction between the perception of his own body and of the world around him. But as soon as the child begins to grope about and to see more accurately, a change sets in. He begins to distinguish between perceptions which relate to his own body and to those of the world outside. The former involve reciprocal sensations: sensations in his hand as he touches his own body and sensation in the part touched. Moreover, perceptions relating to his own body are often accompanied by a much stronger feeling tone, pleasure at satisfying thirst, and especially sensations of pain. In this way he gradually learns to mark off his own body as distinct from the world around him, or, in more strictly epistemological terms, he makes a distinction between groups of phenomena relating to his own body and others related to his environment. A great-niece of my own demonstrated this most strikingly during her first weeks of life. She began tugging at her unusually thick shock of hair; this obviously caused her considerable pain, and she began wailing. Yet she repeated the action several times during the next days. By then, however, experience of pain and touch had taught her to include her hair in the emergent concept of her own body, and so of her self.

The following represent the principal factors which lead successively to the formation of the concept of the self (see also Ziehen 1913).

1. As soon as phenomena begin to be associated with one another, continuous consciousness is experienced in the waking state.

2. Reciprocal sensations occur only when one is touching one's own body.

3. Strong feeling tones are particularly associated with sensations in one's own body; pain especially is accompanied by a powerful negative feeling tone.

4. Active sensations relating to position, that is kinesthetic sensations, and active sensations of movement are experienced only in connection with one's own body.

5. The sight of one's own body helps in distinguishing it from the world around.

6. When feeling tones spread, they give rise to moods and emotions such as comfort, joy, fear, rage, and grief, and these strengthen the formation of a primary concept of the self.

7. The experience of deliberate attention and acts of volition increasingly contributes to the same result.

8. A second stage in the concept of the self is reached as recollections begin to accumulate, recollections of experiences, of what one's limbs can do and what they cannot, of contacts with others, and of connecting one's own name with the developing concept of self.

9. Only two complexes of psychic phenomena were so far distin-

guished. But when speech begins to develop, one progresses to a distinction between "thoughts" and extramental "things" including one's own body, although it occupies a unique position among the "things."

10. Only as the individual grows up, as he discovers his own capacities and gifts, develops his own likes and dislikes, follows ideals, maxims, moral and ethical norms, does the concept of the self become transformed into one of his own "personality."

It is clear, then, that man's concept of the self evolves step by step in a predetermined manner. This peculiar self is then characterized by the specific continuous flow of consciousness in the waking state, and the associative connection of many perceptions with this concept. After interruption during sleep, memory restores the continuity of consciousness.

So there are no grounds for assuming that the concept of the self is based on "intuition" or some particular primal "feeling," or that the self represents an incomprehensible unity or the "immaterial carrier" of all psychic phenomena. Hume's contention that the self is simply a "bundle or collection of perceptions" was correct at least in principle. Although philosophers and psychologists differ on several points, most agree that the self-complex can be recognized by a characteristic steadiness of the stream of consciousness. This corresponds more or less to the view taken by Herbart, Kant, Schopenhauer, Ribot, Ebbinghaus, Wundt, Riehl, Husserl, Rehmke, Schuppe, Lipps, and others.

One very important point is that many observations indicate that animals already possess at least a limited concept of the "self." We may also take it that this was true of the animal ancestors of man. (We shall be giving some examples from among the monkeys and apes in section H II of this chapter.) It is safe to assume that all higher social animals develop at least some prestages of the concept of the self, because they arrange themselves in an order of precedence, either by threat or combat; and those which stand highest continue to be respected and to dominate those below them. (It is difficult to say how far this holds for young animals in relation to their parents, for their behavior is affected by hereditary instincts as well as by learning processes.)

Of course, animals cannot develop such precise and abstract concepts of hierarchy as man, who denotes and fixes his concepts by words. Animal concepts of this kind are at best a pattern of association connected with the image of those who stand either higher or lower. Precedence also provides the basis for averbal concepts of "rights." For instance, a monkey shows a primitive concept of ownership when it defends a toy against its fellows. However, the innate instinct to defend the food is certainly involved as well in this action. When a vertebrate defends its territory or breeding ground from others, there is a similar cooperation of instincts and acquired knowledge of boundaries.

It is important that apes and monkeys also learn to behave as if they had formed averbal *concepts of value*, at least in regard to material objects. They can be trained to recognize the different values of iron rings of various colors if each color means a different amount of food (Ch. 6 H III). Monkeys living in freedom learn to distinguish fruits which taste better from others of lesser "value."

All these more abstract averbal concepts and prestages of concepts have developed among the higher animals because of the advantages in competition among individuals of a population or among different species, and because they contribute to the survival of the species.

III. Feelings, Moods, Affects. Philosophers and psychologists have developed greatly differing views on feelings. They have been called a "particular subjective side of the inner life," "the primal element of the soul," "the immediate assessment of the self in relation to its experiences," "the initial impetus of an endeavor," or "the ingredients of consciousness" (R. Eisler and R. Müller-Freienfels 1922, p. 226; F. Krüger 1937; and others). C. Stumpf (1907) assumed feelings to represent a particular category of sensations. Again we must dispense with any attempt to deal with the vast literature on the subject and confine ourselves to a brief criticism of some of the more usual interpretations and an outline of the conception which is most compatible with biological and psychological findings, mainly based on Ziehen's penetrating analysis (1913, 1915, 1924).

Feeling tones are never isolated phenomena but *properties* of sensations and mental images. Feelings only show relative independence insofar as they can irradiate to other sensations and mental images, and impart a positive or negative tone to them. The clearest cases occur when the spread of feeling produces a lasting *mood* such as joy or sadness, or an *affect* such as anger, rage, or fear. As in these cases a whole complex of ideas is given the same feeling tone, we receive the impression that a great many feelings exist. In fact, however, all we have is a complex of ideas with positive or negative feeling tone, which are then sometimes termed "feelings of form," "aesthetic feelings," "logical," "ethical," and "religious" feelings, or in a more comprehensive manner "feelings of value," "feelings of content," etc. It is, however, quite sufficient and more comprehensible in regard to the physiological basis to assume only positive or negative feeling tones, i.e., feelings of pleasure or distaste, of varying intensity.

The same feeling tone recurs more or less regularly with certain sensations. A positive feeling tone accompanies sweet taste, the scent of flowers, some gentle mechanical stimuli such as stroking, harmony of sounds, etc.; negative feeling tone accompanies pain, harsh and bitter taste, putrid odors, intense pressure, dissonances, etc. These primary *sensory feeling tones* are often carried over to corresponding mental images, where they form intellectual or *ideative feeling tones* of lesser intensity. But the irradiation of feeling tones can quite easily alter in mood; negative feelings of ideas can shift over to positive ones, or the reverse. W. Wundt (1908) added two further pairs of feeling tones to those of pleasure and unpleasantness: excitement and appeasement, and tension and relaxation. But these represent merely perception of differences in intensity of excitation, connected with perception of the temporal course of the alterations.

As already mentioned (Ch. 6 F I), feeling tones may protect the subject from harmful stimuli and may favor aspiring to more advantageous ones, especially the satisfaction of instincts promoting individual or specific survival. Hence, feeling tones of sensation probably developed very early in phylogeny. Moods and affects probably also developed in higher invertebrates like insects and cuttlefish, for after repeated strong stimulation their behavior corresponds to the emotive behavior of vertebrates in a similar situation. Besides, these "moods" which release vital instincts, the mood of reproduction for example, very much resemble the moods in higher animals and man. However, in the context of instinctive behavior the term "mood" means a drive in which all actions which lead toward satisfaction have a positive feeling tone.

Feeling tones are no less important in the normal course of associations. They lend a positive or negative tone to many concepts, especially those of value, and they often direct the association along certain lines. This is of special importance for the understanding of volitional or logical thinking (Ch. 6 F IV, G, K VI). As feeling tones are subjective, it follows that *we cannot ascribe any objective character to values.*"

Unfortunately we know very little about the physiological basis of feeling tones. We have already noted (Ch. 6 C V) that electric stimulation of the rear portion of the septal cortex region may induce a feeling of pleasure, while stimulation of the hippocampus region leads to displeasing feelings or pain. Various drugs too can induce corresponding feeling tones. The positive feelings of aesthetic sensations may perhaps come about by a facilitation of excitations. All relevant investigations will have an important bearing on epistemology, if we accept a panpsychistic, identistic conception, because then the physiological process in the psychophysical substratum can be equated with the psychic phenomenon.

IV. Attention and Volitional Processes. Our ideas do not normally succeed one another in a random course prompted simply by the existing trains of association. Such a "flight of ideas" only takes place when we are very tired, under the influence of alcohol or certain drugs such as lysergic acid preparations, when we are dreaming, or in some cases of brain disease. On the contrary, we pursue definite trains of ideas, that is, we direct our attention and guide our thoughts along certain lines by volitional processes. Attention and processes of volition thus represent specific phenomena.

So the question arises, are these specific phenomena primal and irreducible data, or are they only complex phenomena? Many philosophers and psychologists are convinced of the former. But detailed analyses of these phenomena show that we are probably dealing here with constellations of phenomenal factors, in which the degree of intensity of mental images and perceptions and their feeling tones are decisive.

Attention, characterized by accentuation and more conspicuous experience of individual phenomena, can be caused in different ways. *Sensorial attention,* by which a single perception is emphasized, already comes about *passively* when the perception is exceptionally intensive or accompanied by a very pronounced feeling tone. But we can also direct the sensorial attention actively towards certain components of perception and even toward some quite unobtrusive element of perception. This happens, for example, when we are trying to find an insect with protective coloring. At such times attention is directed by a process of volition. But since attention can also be excited passively, it would be an error to interpret attention generally as a function of the will, as some philosophers have done.

Another type is represented by "ideative" or "intellectual" attention. Here, certain components of a group of ideas are passively or actively singled out. This is a highly important process, which is essential for the apprehension of identity, similarity, or difference, and also in analytic and synthetic processes of thought, where the differentiation functions are involved. However, actively directing the attention to certain trains of thought already constitutes a process of volition.

Ideative attention, too, can be directed by the intensity and the feeling tone of certain mental images, and also by the type of associative connections. So we may consider both passive sensorial and ideative attention as *determinate* processes.

The phenomenal peculiarity of passive sensorial attention arises because sensations are usually accompanied by accommodation movements in the appropriate sense organs, for instance in focusing the eyes, and by slight sensations of tension which then melt into the general phenomena. This phenomenal peculiarity is much less marked in passive ideative attention, or it may be totally absent. But if ideative attention is actively directed toward some train of ideas, as when we seek to recall a name, then it is an act of volition.

Processes of volition also have a phenomenal peculiarity. Many philosophers and psychologists therefore regard the will as a "specific, original, psychical process which cannot be ultimately derived from any other process" (R. Eisler and R. Müller-Freienfels 1922). What is peculiar about it is that the train of ideas appears to be actively directed, and that the attention is involved. A type of "pure will" determined a priori, which Kant postulated (*Kritik der reinen Vernunft* and *Grundlegung der Metaphysick der Sitten*), does not however occur as an isolated phenomenon; in each case directed trains of ideas are involved.

Contrary to the conception of extreme voluntarists such as A. Schopenhauer who regarded the will as a primal phenomenon of all being (though Schopenhauer himself saw it rather more in terms of a universal drive), the processes of volition have proved to be complex phenomena (see Ziehen 1913, 1915, 1924 and especially 1927b). *In processes of volition, a train of ideas is directed by a dominant idea; in deliberate attention there is an immediate association with a dominant idea.* The directive dominant ideas may be quite precise, but usually they are of a more general nature, as in the case of a sum or puzzle where an unknown solution is aimed at and guides the direction taken by the train of ideas. Here all approaches toward the solution acquire a positive feeling tone, whereas all associated trains of thought which lead away from the particular aim acquire a negative feeling tone. Even when our volitional thought is directed toward some disagreeable aim, the intensity of the dominant idea and the feeling tones are the decisive factors. This is true for instance when an aim fraught with discomfort or connected with danger is associated with the idea of duty.

Now the directing dominant idea, like all other mental images which participate in a volitional process, is ultimately determined by present and former perceptions and by earlier trains of ideas; in physiological terms, it is governed by engrams and in general by inherited structures and functions of the brain. This means that *processes of volition are also determinate*, and the existence of a "free will" is in principle improbable. Because of its vital implications for epistemology and practical life, this important problem will be discussed in special chapters (6 G and 7 A).

Despite this determinate character (which can often be analyzed) and the fact that volitional thinking only involves mental images and sensations, it undoubtedly has a specific epistemological peculiarity which is probably due to the following factors: The attention is always involved, and is often accompanied by sensations of tension. As we usually think in words, the tongue in particular is subject to faint excitations. Indeed, with persons who tend to part the lips a little while reflecting intensely, one can frequently see the tongue moving slightly in the tempo of speech. Mental images of motorial processes also often play a part in volitional thinking, and contribute to a "feeling of activity." Moreover, the complex of ideas involved is often connected with the concept of the self. This is most noticeable when we intend to do something which is unpleasant, when we "ought" to do something, and the dominant ideas are those of duty or morals. All these factors fuse, and they lend the process of volition its distinctive phenomenal character.

This fusion is apparently due to the inertia of the corresponding physiological processes in the brain and in the sense organs, for example, when we see a revolving propeller as a circular disk, or a quick succession of static pictures as a moving film, or when tasting our food we experience one complex quality composed of different modalities. If we base our biophilosophical considerations on a panpsychistic and identistic conception, which appears to me to be the most acceptable of all, and identify psychic phenomena with physiological processes in the psychophysical substratum, then the "constellation theory" of volition already discussed also provides a causal explanation for processes of volition.

If all the continuous and manifold impulses from countless sense organs and free nerve fibers were to lead on to conscious phenomena, then physiological inertia in brain processes would produce phenomenal chaos. We know that this does not occur and that the formatio reticularis of the brain stem acts as a filter, so that only some of the impulses reach the cortex (see H. H. Jasper 1954; P. Glees 1957, Ch. XIV; K. Poeck 1959; and T. Weiss 1965). This system of neurons allows concentration of certain processes of the brain to take place, as well as an accentuation of special excitations, and thus contributes to the development of sensorial attention.

This selection of afferent excitations and corresponding psychic phenomena contributes to the origin of the *narrowness of consciousness* which is further caused by suppressor zones in the cortex and by electrophysiological balance between nervous messages (when action potentials "fall into step"). Besides, Aristotle was already familiar with the "*narrowness of consciousness*." In his book *On Sense and Sensible Objects* (see 1847, vol. 2, p. 227) he writes that it is doubtful "whether it is possible to perceive two things in one and the same indivisible time or not. . . . For men do not see things impinging on their eyes, if they happen to be concentrating on some thought, or in a state of fear, or listening to a loud noise." One might imagine "that we see and hear simultaneously, because the intervening time is not noticeable. However, this seems not to be true. . . ." (p. 230)

Such structures, which prevent the brain from being flooded with impulses, have probably developed with the progressive evolution of higher animals by natural selection. Without some such filter mechanism, the life processes of higher vertebrates would not function. With lower animals, the different force of individual impulses and the accompanying feeling tones seem to be enough to direct the central nervous processes in a meaningful manner (positive feeling tone if food or a mate is perceived, and negative tone if some enemy or other danger threatens).

V. Judgments, Conclusions, Thought Processes. The actions of perception, directing the attention, and volitional thinking, which we have been considering, already make up a large part of what we usually call the processes of thought. A number of philosophers and psychologists, such as Schopenhauer, Paulsen, Höffding, Jodl, James, Wundt, and others would even trace back all thinking more or less to voluntarily directed trains of ideas. But are there specific modes of thought or some other types beyond volitional thinking or a kind of "pure" thought? And how can we characterize such types?

Kant, as we have said, stressed the spontaneity of thought, which we have traced back to continuous physiological activity of the brain during the waking state. It seemed unlikely, however, that "pure thought" also exists in the sense that "things" or "relationships" could be recognized a priori. But the phylogenetic adaptation of our thinking to the universal logical laws has led to specific, largely volitional, processes of thought which we term judgments and conclusions.

Our differentiation functions enable us to apprehend identities, differences, and similarities (this capacity partly corresponds to Kant's "unity of apperception"). We express these varying degrees of coincidence of components or relations of our phenomena in the form of positive or negative *judgments*. In addition, these almost always involve analytic and synthetic functions. Many very different types of judgment are possible: designatory, descriptive, analytical, synthetic, problematic, hypothetical, apodictic, assertorial, causal and evaluatory (Ziehen 1920). Feeling tones may profoundly influence judgments. The awareness of validity and truth, which is peculiar to many judgments, has already been treated (Ch. 6 D IV). Relationships stated in a judgment are usually expressed by a predicate: the apple *is* sweet. Animals are not capable of such clearly delineated predicate concepts as they do not associate them with words. Yet higher animals may act as though they were framing judgments (for further treatment, see Ch. 6 H III).

If a new judgment is derived from one or more previous ones, it constitutes a *conclusion*. This occurs mainly when statements which are expressed in the judgments coincide: Thus, stag beetles have six legs; ground beetles, leaf beetles, and water beetles have six legs; therefore: beetles have six legs. This is a case of inductive conclusion from individual cases to a general statement. Higher animals also are capable of this type of conclusion, which may even lead to more universal concepts. On the other hand, if we infer a particular instance from a general statement, we have to do with a deductive conclusion (see also Ch. 2 A). The behavior of some higher animals, especially anthropoid apes, shows that within limits they too can act as *though* they were drawing deductive conclusions. This is particularly evident when the conclusions which they reached inductively have led on to a generalized knowledge of certain causal relationships, and this then forms the basis for a planned course of action in a new situation (see examples in Ch. 6 H III). Man's deductive conclusions are particularly important insofar as they concern purely logical relationships, as in mathematics. Within the framework of conclusions one can distinguish various types: simple and compound, correct and false, hypothetical, categorical, classificatory, conclusions by analogy, etc.

Every instance of judging and concluding, however, can be traced back ultimately to associations of ideas and the use of differentiation functions. *As in all volitional thinking, the process involved in judgments and conclusions is absolutely determined by the constellation of current ideas and perceptions at the time in question and their accompanying feeling tones.* With regard to our biophilosophical picture it is also important that *therefore only two basic types of phenomena exist: sensations and ideas, and that primary mental images are residua of sensations.*

As to its particular direction and motivation, thinking may therefore be determined in very different ways: (1) sensorially, led by dominant perceptions; (2) instinctively, directed by innate drives which in consequence of their strong feeling tones may overrule associations based on perceptions and ideas; (3) ideatively (intellectually) by dominant ideas; (4) by judgments; (5) by conclusions; and (6) by moods and emotions.

During ontogenesis man begins by forming associations on the basis of reflex actions and instincts. The earliest memorial impressions lead on to uncoordinated and soon to coordinated trains of mental images. Later on, by abstraction and generalization, averbal and then verbal concepts and finally judgments follow. Gradually phenomena from the "objective" world around become separated. The child now begins more and more to draw conclusions. His thought is still largely animistic at first, but later it gradually grows more causal and logical in character (see J. Piaget 1927). This development, then, is similar to that which we may assume took place in man's animal ancestors.

But there are certain other particular thought processes that do not fall within the categories just mentioned: imagination and intuition. Insofar as *imagination* is a matter of deliberately establishing or trying out new associations, it may be interpreted in the same way as other processes of volition. Besides this *active* and therefore creative type, which is of great importance for the sciences and arts, there is also a *passive* kind of imagination, which is not immediately directed by a dominant idea. In such cases the ideas are simply linked by associative connections. We experience this type in dreams or in "dreamy" meditation, when we are "lost in thought," or in some morbid flight of ideas. But sometimes passive imagination is given a certain direction by a prevailing feeling tone, mood, or emotion, or by some innate drive like fighting or the sexual drive, maternal instincts, etc. The physiological basis of passive imagination is the continuous activity of the brain, especially of the cortex of the forebrain, during the waking state.

The behavior of higher animals indicates that they too seem to be capable of active and passive imagination. This is best illustrated by the experimental playing of mammals and birds, insofar as this has nothing to do (or only indirectly) with their natural mode of life, and is very well exemplified by the play of monkeys, when they pile up children's bricks, or when a mongoose rocks to and fro in an upturned wastepaper basket, or when young carnivores turn somersaults (see M. Meyer-Holzapfel 1956; B. Rensch and G. Dücker 1959b). W. Köhler (1921, p. 61) detected active imagination in his chimpanzees. They played a game of holding out bread through a wire fence to some hens; when the hens tried to peck at it, they snatched it away or poked at them with a stick. Besides, any planned action on the part of an ape entails some active imagination (Ch. 6 H III). In theory, any animal we think capable of association of mental images may display imagination. As animals are much less able than man to sustain a volitional train of mental images, they are probably more given to undirected trains of association.

Intuition is defined by most philosophers as an immediate grasp of an essential new idea without any consciousness of the associative links required in normal reflection, that is to say, without being aware of any process of reasoning. This manner of acquiring knowledge, which Plato, Spinoza, Schopenhauer, Fechner, and others have recognized as a particularly important one, is apparently bound up with the "narrowness of consciousness." In the act of thinking, chains of fibers in the neurons of the brain are excited which are linked by association. But many of these excitations lack any accompanying mental images, until they lead into the complex to which the stream of consciousness corresponds. When this happens, it is often only after a certain time that a new and appropriate connection of ideas may suddenly spring into being.

To sum up, we may state that *all thought processes depend upon the associative constellation, the intensity, and the feeling tones of the relevant perceptions and ideas*; this has already been shown by Th. Ribot, A. Binet and others, and especially by Ziehen (1913, 1915, 1927b). This means that *our thinking is necessitated, and it is only because the factors involved are so complex that its processes appear to be autonomous and "free"* (Ch. 6 G). If we accept a panpsychistic and identistic conception which equates the physiological processes in the psychophysical substratum with the phenomena, this would imply that thought processes too can be causally analyzed, though only to the very limited extent to which at present we can assess the causal relationships involved.

G. The Problem of Free Will

1. The Problem. All the facts we have discussed, and the implications drawn from them, lead to the conception that everything which occurs in the universe is controlled by eternal basic laws; and these find expression, according to the stage of development reached by the various celestial bodies, in a great number of separate laws. What exists outside the mind is assumed to be protophenomenal. The psychophysical substratum which emerged during the phylogeny of living organisms has evolved in the same manner within the framework of universal laws. The physiological processes in the brain, which I identify with the psychic phenomena (instead of considering them as running "parallel" to these), are governed by causal laws.

One could then hardly conceive that the universal laws of the determined world could have been broken by the thought processes of man, and that man's will could be free. Man has only been in existence for some 0.3 percent of the age of our planet. Even if we ascribed free will to the higher animals as well, this would still cover only a small percentage of the supposed age of the earth, and a minimal fraction of the "eternity" of the universe. It is possible to assume that some primal explosion brought the world into being some 15 billion years ago; but this opinion is hard to reconcile with epistemological considerations, and it becomes increasingly improbable from an astrophysical point of view. If "free will" really existed it would have emerged in the head of man, thereby disrupting the causal law which governs the processes of the brain.

Obviously, one may doubt such an assumption and ask whether the will does not simply appear free because the brain processes in man are so complex that it is impossible to analyze all the causal relations involved. If one were to base such considerations on a dualistic conception (and I have already given reasons why I regard this view as untenable), it might be easier to admit that a "soul" which is largely independent of the body could also be independent of the universal laws. But then one would still have to ask where during the course of phylogeny such souls could have originated, how they could arise during embryological development, and where their lawless, "free" activity could have come from. And one would have to remember that the concept of "activity" already involves causal processes.

As we shall see, however, a good deal of evidence suggests that thought and action which appear to be "free" are determined by hereditary factors, education, former experiences, physical condition of the body, and actual perceptions. This evidence is further supported when we analyze the behavior of animals of different phylogenetic levels, for their actions are more predictable than those of man.

One is naturally reluctant to deny that the will is free. The statement seems to place the whole of man's freedom in jeopardy, and to endanger the significance of ethical maxims and religious convictions. But we shall see that this is not so, at least in practice, for the pursuit of freedom, ethical maxims, and religious ideas themselves constitute determinants of our thought. On the other hand, I have already stressed that human maxims and religious convictions cannot form the cognitive basis of any philosophy (Ch. 6 B). However, such maxims can be legitimate consequences of a biophilosophical conception exempt from any prejudice whatever.

II. Historical Development of Interpretations. Philosophers in all ages have studied the problem of free will, but without arriving at a definite solution. Even today, determinists and indeterminists confront one another, although most indeterminists concede that volitional thought is causally conditioned to a considerable extent. Various attempts to find a compromise must unfortunately be regarded as unsuccessful, because even limited freedom would infringe the law of conservation of energy.

We shall not deal in any detail with the historical development of the relevant views; we refer the reader seeking a fuller treatment to the writings of L. Müffelmann (1902), M. Offner (1904), W. Windelband (1904), J. Rehmke (1911), G. F. Lipps (1912), Th. Ziehen (1927b), H. Groos (1939) and others, and to relevant passages in almost any work of the classical philosophers. In our context it may be sufficient to outline some characteristic conceptions.

Thomas Hobbes was the first among more "modern" philosophers to regard all human thought and action as causally determined. In his work *De corpore* (1655) he wrote: "Even wanting or not wanting freedom is not a more pronounced trait in Man than in any other creature. Wherever an urge arises, a sufficient cause for it was already there. . . . Neither Man nor beast can claim a freedom which would imply freedom from necessity."

Spinoza is no less unequivocal; in his *Ethica* (Part V, Proposition I) he wrote: "Prout cogitationes rerumque ideae, ordinantur et concatenantur in Mente in Corporis affectiones, seu rerum imagines, ad amussim ordinantur et concatenantur in Corpore." Demonstration: "Ordo et connexio idearum idem est ac ordo et connexio rerum. . . ." ("Just as thoughts and ideas of things are arranged and connected in the mind, so in the body its modifications or the modifications of things are arranged and connected according to their order." Proof: "The order and connection of ideas is the same as the order and connection of things.") "Voluntas non potest vocari causa libera, sed tantum necessaria." ("The will can only be called a necessary cause, not a free one.") Spinoza did indeed write of man as free and governed in his actions by "ratio"; but what he obviously meant was not freedom from causally determined processes, but only a mode of thinking governed by the laws of reason, for his identistic epistemology is founded on the conviction that ideas are associated in the same way as the physiological activity upon which they are based.

Leibniz was not quite so consistent; though he wrote "omnes actiones sunt determinatae," he also recognized a certain degree of freedom, for he declared that motives only "drive" but do not "compel" our actions. On the other hand, in his *Théodicée* (§45) he states quite unequivocally: "Even supposing one takes a certain decision out of caprice, to demonstrate one's freedom, the pleasure or advantage one thinks to find in this conceit is one of the causes tending toward it."

Kant at first also regarded human thinking as completely determinate. In his essay entitled *Idee zu einer allgemeinen Geschichte in weltbürgerlicher Absicht* (*Idea for a Universal History in Cosmopolitan*

Intention) (1784), he stated: "Was man sich auch in metaphysischer Absicht für einen Begriff von der *Freiheit des Willens* machen mag, so sind doch die *Erscheinungen* desselben, die menschlichen Handlungen, ebensowohl als jede andere Naturbegebenheit, nach allgemeinen Naturgesetzen bestimmt." ("Whatever one's metaphysical concept of the *freedom of the will* may be, its *manifestations*, human actions, are like every other occurrence in nature in being determined by the universal laws of nature.") He said much the same thing in his *Kritik der praktischen Vernunft* (*Critique of Practical Reason*); but he applied this only to the world of appearances—by so doing, however, he related it precisely to the reality of phenomena! He feared that man's ethical freedom which he had postulated would be jeopardized, so he insisted on the will being also a "thing in itself" of "intelligible character," endowed a priori with freedom. But we have already noted (Ch. 6 F IV) that it is impossible to accept the allegation that volitional processes can be ascribed to anything other than phenomena.

Schopenhauer follows much the same lines of thought as Kant. In his chief work, *Die Welt als Wille und Vorstellung* (1818), he holds a completely determinist view: "Das Gesetz der Motivation ist ebenso streng wie das der physischen Kausalität, führt also einen ebenso unwiderstehlichen Zwang mit sich!" ("The law of motivation is as strict as that of physical causality and is therefore just as irresistibly coercive.") On the other hand, he assumed that the "will" was a "thing in itself," not subject to the causal law; he was thinking, however, of a "universal world-will," the force behind every event.

Later philosophers have been divided in their opinions. Those of the Enlightenment—La Mettrie, Holbach, Helvetius, Laplace, and Voltaire—were all outright determinists. In the first chapter of his *Système de la Nature* (1770) Holbach wrote: "All our ideas, our will, our actions, are the inevitable results of the being and properties with which nature has endowed us, and of the circumstances which we necessarily experience and which affect us." And in Voltaire's *Philosophie générale* (1785) we find the following passages:

"Il n'y a rien sans cause. . . . Toutes les fois que je veux, ce ne peut être qu'en vertu de mon jugement bon ou mauvais; ce jugement est nécessaire, donc ma volanté l'est aussi. . . . Nous pouvons réprimer nos passions, . . . mais alors nous ne sommes pas plus libres en réprimant nos désirs qu'en nous laissant entraîner à nos penchants; car dans l'un et l'autre cas, nous suivons irrésistiblement notre dernière idée; et cette dernière idée est nécessaire; donc je fais nécessairement ce qu'elle me dicte."

("Nothing exists without a cause. . . . Every time I exercise my will, I can do so only by virtue of my judgment, whether good or bad; and as this judgment is necessary, so therefore is my will-power also. . . . We can restrain our passions, . . . but even then we are no more free in restraining our desires than if we yield to our inclinations, for in either case we are irresistibly giving way to our latest idea; and as this is necessary, it follows that I am bound to do what it dictates.")

Materialists of the nineteenth century such as Vogt, Büchner, Moleschott, and others took a similar view: "Es denkt in uns und wir werden uns des Gedachten bewusst . . . es will in uns und wir werden uns des Gewollten bewusst!" ("Thought takes place in us, and we become conscious of what has been thought; will is exercised in us and we become conscious of what has been willed.") (J. C. Fischer 1871)

Among more recent exponents of determinism we may mention A. Riehl and W. Wundt. Wundt did indeed speak of "free, voluntary decisions," but all he meant by this is that the complex basis of motivation cannot be fully explored. The clearest refutation of free will comes from Ziehen (1927b), who has shown that arbitrary thinking and moving (acting), volitional reproduction of ideas, and volitional attention are completely determinate. We notice this most clearly after a long period of hesitation; it is, then, the relative strength or feeling tone of a motive that brings about a decision. In the third edition of his *Leitfaden der physiologischen Psychologie* (1896, p. 184) Ziehen already drew the conclusion from his investigations: "Unser Denken ist nie willkürlich, es ist wie alles Geschehen nezessitiert." ("Our thought is never arbitrary; like everything which happens, it is governed by necessity.") And in the twelfth edition (1924, p. 498) he formulates this still more precisely: "Wir können nicht denken, wie wir wollen, sondern wir müssen denken, wie die gerade vorhandenen Assoziationen bestimmen." ("We cannot think as we will; we must think as the then existing associations determine.") We have already mentioned his constellation theory, by which he explains volitional processes (Ch. 6 F IV; see also H. Rohracher 1960). H. Groos (1939) also stated quite clearly:

"Der Mensch ist also ein hochorganisierter Durchgangspunkt im allgemeinen Geschehen." (p. 139) "Aber der Wurm wie der differenzierte Mensch tut nichts weiter, als seiner Natur, d.h. der Gesamtheit der für ihn geltenden Gesetze zu leben, seine Natur, die zwar ihm gehört . . . , die er aber sich nicht selbst geschaffen hat, auszuleben, zu erfüllen, zu vollenden." (p. 141)

("Man is a highly organized spot through which the course of universal events flows. . . . But the worm, no less than highly specialized man, does nothing else than live simply according to his own nature, i.e., the sum of the laws that relate to it, living to the full, to fulfill and to complete his nature which he did not create.")

Though determination of volitional processes has been generally accepted to a large degree, a restricted measure of free will has often been recognized because educational maxims, ethical responsibility, legal necessity, and religious convictions appear to demand it. According to E. von Hartmann (1879), in addition to the laws determining volitional action, there exists "eine sittliche Grundgesinnung, die als übernatürliche Funktion zu der Abstraktion des natürlichen Menschen hinzukommt" ("a basic moral attitude, a supranatural function added to the abstraction we call natural man"). O. Külpe (1895, 1923) claimed that determinism must be limited on ethical and juridical grounds. N. Hartmann (1925, 1949) assumed that besides the determination of the processes of thought and motivation imposed by natural laws, some "personal determinant" must be operative as a "primal, irrational and irreducible element." He declared: "Die Freiheit der sittlichen Person ist erstens ethisch notwendig und zweitens ontologisch möglich" ("The freedom of a moral individual is ethically necessary and ontologically possible"); but he admitted that "ein strenger Beweis der Willensfreiheit lässt sich nicht führen. Er müsste gerade die ontologische Notwendigkeit erweisen." ("No convincing proof of free will can be put forward. It would have in particular to demonstrate ontological necessity.") M. Planck (1923, see 1933b) at first assumed that "strict causality" governs all thought processes, in the sense that "selbst der Geist eines jeden unserer allergrössten Meister, der Geist eines Kant, eines Goethe, eines Beethoven, sogar in den

Augenblicken seiner höchsten Gedankenflüge und seiner tiefsten inner-lichesten Seelenregungen, dem Zwang der Kausalität unterworfen war, ein Werkzeug in der Hand des allmächtigen Weltgesetzes." ("Even the mind of each one of our greatest masters—Kant, Goethe, Beethoven and even in the moments of sublimest feeling, was subject to the coercion of causality, and remained an instrument of the all-powerful universal law.") This view is compatible with parallelistic as with identistic conceptions. But Planck, following Kant, felt obliged to in-troduce a limited free will, because as subjects involved in the process of cognition we are not capable of a causal judgment. Man's practical needs, Planck argued, make it essential for him to have fixed princi-ples: "instead of the causal law the moral law, ethical duty, the cate-gorical imperative." Yet Kant was unable to give satisfactory proof of such an absolute ethical principle existing a priori (see Ch. 7 B).

But if our aim is an unbiased interpretation on an exact epistemo-logical and scientific basis, we must not introduce unsupported maxims which are incapable of proof, simply because of their practical claims in regard to our way of life. Of course, it is impossible to produce any cogent proof of determination in acts of volition. Our human brain-processes are so complex, and the engrams involved so inter-woven, that an exhaustive analysis is almost never possible. But we shall see that the weight of evidence is far greater *against* freedom of the will than *for* it. This evidence comes from findings in the fields of epistemology, human and animal psychology, physiology, genetics, and phylogeny. Besides, the ethical and juridical objections to free will are by no means as valid as is generally believed (see Ch. 7 and B. Rensch 1963).

III. Causal Determination of the Processes in Animal Brains. If we are seeking to answer the question of determination or free will by means of phylogenetical considerations, we must bear in mind that there are certain inevitable limitations. In Chapter 6 D we stressed that we can infer psychological processes in animals only by analogy. Besides, it is difficult to declare which kind of animal behavior can be regarded as volitional and which cannot. Yet these borderline cases are naturally of particular importance in studying how the will has evolved and is determined.

We have a rather more dependable basis if we assume that the psychological processes are inextricably linked with the physiological processes of the brain in the sense of psychophysical parallelism, or that the two kinds of processes are identical, so that the process in the psychophysical substratum and the psychic phenomenon would be the same. In my opinion the latter conception offers the more satisfactory interpretation of all data.

The behavior of warm-blooded animals, especially monkeys, carnivores, ungulates, parrots, and crows, often shows such a resemblance to human acts of volition that it is hardly possible to doubt that the two are based on similar physiological processes in the brain. When a chimpanzee carries through an action entailing foresight (as illustrated by Figure 6), or when he piles up two or three boxes in order to reach a fruit hanging high above them, we are surely entitled to call this an act of volition. Or if we call a dog several times and he does not come because he expects to be beaten, we may describe this behavior as volitional; that is to say, as behavior in which the anticipation of pain, accompanied by a negative feeling tone, is determining the trains of association. Less highly organized animals such as fishes and even cuttlefish, and some insects, also sometimes act in such a way that it is at least possible to assume volitional impulses.

In most of these cases we do not know enough about the underlying central nervous processes to decide whether we can speak of volitional actions, and whether they are completely determinate or not. In particular, we can rarely judge what traces of former behavior, what engrams, have contributed to the reaction in question. So we shall tackle the problem best by studying those cases where it is possible to estimate the factors relevant to motivation. In experiments with animals this can be achieved to a certain extent by selecting individuals which are genetically and physiologically alike—of the same sex and age, and brought up and fed in the same way—and by making the experimental conditions as unambiguous as possible, calculating average values from a fairly large number of cases, and comparing the results with those obtained from an equal number of control animals.

The following arbitrarily selected examples are intended to show how the difficulty of judging what may be called volitional behavior increases when the central processes become more complex, and also how experiments with animals of different phylogenetic levels give us grounds to assume a general causal determination of all cerebral activity, including the processes of volition.

1. We shall begin by examining the behavior of a planarian, one of the flatworms. If we put this worm in an aquarium with a few stones on the bottom, the animal will crawl about for a while, and then disappear under a stone and come to rest in the dark. All planarians of the same species behave in the same way. If half the aquarium is darkened, they will gather on the dark side. So we may assume that hereditary negative phototaxis is determining this behavior, that the processes which induced restless crawling-about only come to an end when the light stimulus is absent. There is no occasion for thinking that any kind of volitional process is taking place.

2. The behavior of a frog might perhaps seem to offer a rather more persuasive example of a primitive act of will. If we gently stroke a certain region on its right flank a few times for a fraction of a second with the end of a piece of fine wire, the frog will wipe that region with its right hind foot. If we stroke the lower part of its back, it will wipe there with its heels. These reactions are normally useful to it, for it would be wiping off what was stimulating the skin. But in this experiment the action is meaningless and obviously determinate, for the wire is not wiped off. The movements may look purposive, and give the impression of being volitional; but we may definitely say that the whole is a reflex action and is physiologically determined. Even when its brain or its entire head is removed, the frog reacts by the same movements. The impulse from the stimulated region is carried by afferent nerves to the spinal cord, where neurons respond by setting up the motorial action for the wiping movement. The whole reaction is caused by a hereditarily determined reflex arc and runs off automatically.

3. In the summer months, as soon as the morning is warm enough, the cabbage butterfly begins to flutter about. Following a somewhat wavering course across a dry meadow, it will accidentally come to a spot where the scabious is in flower. It alights on one of the blossoms and begins to suck the nectar. Has the butterfly sought out these flowers intentionally? Was its action governed by a mental image of the aim? Investigations of instinctive behavior in insects lead to the conclusion

that this is not the case. What happens is that the butterfly's physio-logical condition, which one might call a "mood of hunger," simply releases a certain "appetitive behavior," in this case flying about in the open meadow. As it did so, its eyes received a certain optical stimulus by the color of the flowers (mauve, for human eyes), and this released nervous excitations in its brain, causing it to fly in that direction. A certain scent then acted as a second releaser, and it settled and stretched out its proboscis. In other words, these two release situations evoked in turn the appropriate hereditary nervous response, resulting in the reflex activity of settling, stretching out the proboscis, and sucking.

Many instinctive processes run off in a corresponding manner. For example, a fox will normally begin to roam about as dusk falls, and some time or other he will chance upon a releaser—perhaps the scent of a young hare—to which he responds by seizing it and giving it a fatal bite. The appetitive action of running about is clearly hereditary in character; for foxes in zoos, where they are amply fed during the day, also begin to roam their cages as the light fades. So this appetitive running-about is not volitional. The animal's experiences can of course help to determine his special route; and certain nervous constellations determined by will may play their part as a kind of directive aim; he may know a farm where he has carried off a hen in the past.

4. All purely instinctive behavior is clearly determinate, since nerv-ous response is inherited, and all animals of the same species react in the same way. But when actions are based on experience, behavior is much more difficult to trace back to its cause. If experimental condi-tions are well controlled, however, it is often possible to establish the causal determination or at least to show the causal connections to be probable.

For example, we can train a hen to distinguish between two patterns. This is best done in a soundproof room, where the hen is placed in a closed box lit from within. When an opaque trapdoor in the center of the box is raised, the bird sees two food containers in the farther divi-sion, now accessible to it, each covered with a square white cardboard lid. One lid shows a black cross and the other a black triangle. As the hen has previously learned to get at its food by lifting the lids with its beak, it begins by pushing away one or the other of the lids at random. However, it finds food only under the cross and none under the triangle.

The two patterns are changed over at irregular intervals. In this way the hen gradually learns always to throw off the lid with the cross. After it has mastered this so well that in 99-100 percent of choices it throws off the correct lid only, then *both* containers have a food reward placed in them. But the hen continues in a significant percentage of cases to throw off the lid with the cross, in spite of the fact that when it occasionally makes a mistake, it also finds food under the triangle. The hen's behavior is thus *predictable*. We must assume that memorial traces have been laid down, associating "cross pattern" with "food reward." During the final tests the bird's behavior was not based on "free will." We can, however, say that it "wants" to get food and so it makes a choice and throws off the lid that leads to a reward. But the motive for this volitional action and the associative process that precedes it appear to be completely determinate.

In general, then, well-controlled experiments show that the behavior of even higher animals is largely predictable. Though we cannot always analyze all the complicated physiological brain processes involved, these can be assessed insofar as they concern the observable reactions during the experiment. The animal's decisions as it makes its choice prove to be determinate, not "free."

5. Experiments with anthropoid apes have provided many opportunities for studying complex behavior which is quite obviously volitional but where the coercive associative processes are so clearly delineated that it is relatively easy to assess them. It may be sufficient to illustrate this by a characteristic example. We trained a young female chimpanzee at our Institute to choose between two pathways in a maze, only one of which led to the goal (B. Rensch and J. Döhl 1968). The animal was offered a perspex-covered board with a sunken pattern of pathways in white, only one of which led without a break to the edge of the board. A small metal disk, placed on a raised block in the middle of the far side of the board, could be moved by the chimpanzee along the white pathways with the help of a magnet. The animal had first to draw the disk down off the block, either to right or left, and thus either into a pathway leading to the goal or into one that did not, as it ended somewhere before reaching it. The goal was always the edge of the board, where the disk could be taken out and exchanged for a food reward in a kind of slot machine. There were also other paths that had

no connection with the starting point but also had an exit at the edge of the board. The pattern of pathways was altered each time, and during some thousands of trials the solution was made progressively more difficult. At the beginning of each attempt, the chimpanzee surveyed the maze for some time and looked particularly toward the two or three possible exits at the edge of the board. Almost always, or at least in a significant percentage of cases, she then drew down the disk with the magnet on the correct side, choosing the quickest pathway and avoiding any of the branching dead ends (see Fig. 6).

Experiments of this kind show the animal behaving as though its actions were volitional. The "will" is quite clearly a factor here. The ape wants to reach the goal, and must compare several possible pathways in order to choose the right one. Obviously the goal at the edge of the board is a dominant idea governing the chimpanzee's cerebral activity, when she runs through the possible actions "experimentally" in her mind before coming to a "decision" and initiating motorial nerve processes in the hand that moves the disk. As she shifted the disk along with the magnet, the chimpanzee sometimes paused for a short time and eyed the dead ends that branched off. We may assume that the ideas of the chimpanzee were just those that a man would have had: pulling the disk into one of the side tracks was linked with a negative feeling tone, and drawing it the right way with a positive tone. The whole chain of "volitional" activity was determined by the positive dominant idea of reaching the edge of the board and then getting the reward, and also by the negative perceptions and ideas when a choice had to be made at the points where the dead ends branched off.

So far I have intentionally dwelt upon the final stage of the experiments, to illustrate what is an amazingly complex volitional process for an animal. The fact that the chimpanzee was not acting by trial and error but in an astonishingly purposive manner had been assured by the slow progression of complication of the maze. We began with a half-size maze containing two very simple paths, one leading directly to the goal and the other with a break in the white path (see Fig. 6 above). As we altered the pattern of the maze each time, the chimpanzee was obliged to grasp the meaning of the task and was not merely recollecting certain pathways. In the first few simple trials the animal "reflected" (i.e., the play of association lasted) only for a few seconds. It was only

Figure 6. Maze experiments with a chimpanzee. Above: one of the earliest arrangements. Below: one of the latest ones. The chimpanzee has just placed the wooden-handled magnet on the strip of perspex covering the metal disk; she then pulled it down on the correct side, and drew it to the exit at the side of the board without guiding it into any of the blind alleys.

after hundreds of trials, as a more and more complicated maze was presented, that the period of "planning" grew longer. In the end it covered one or two minutes. But we could predict the result of acting at least in 80 percent of the cases.

This type of experiment may serve as a model of simple volitional actions in man. The physiological processes in our brains are basically of a similar kind. The difference is that they are for the most part much more complex, because our cortex is a little more functionally specialized and contains a greater number of neurons, and above all, because our associative processes are usually matched by verbal images, so that we mainly think in abstract concepts and are capable of logical deduction. However, there are many cases in which we are able to realize that our thought processes and actions are determinate. To assess this we must first inquire into the nature of the determinant factors involved.

IV. Hereditary Determinants in Man's Processes of Volition. Man's thought is in many respects determined by his hereditary constitution. Generally, the particular structures and functions of our brain and sense organs impose certain limits on our trains of thought and consequently on our actions, and direct them along certain lines. The influence of some basic nervous reactions on our behavior can already illustrate this. Rage, for example, produces an exaggerated train of ideas with negative feeling tone, often leading to corresponding action. Coupled with this, as the sympathetic nervous system is strongly excited, more adrenalin is secreted by the suprarenal glands, and this in turn causes a further increase of excitation. This spiral-like increase of emotion, predictable in excitable people, usually outruns the reason for the reaction and demonstrates the cogency of the process.

The involuntary nature of thinking and behavior is just as conspicuous when hereditary instinctive reactions are released (see also B. Rensch 1959, 1965a, Ch. 5 c). For example, when our fingers are very sticky, the urge to wash them soon overcomes any other intention. Reproductive instincts, released by the sexual hormones, are equally powerful, and sometimes they compel the subject to act in a way which runs counter to considerations of reason. Surely few of the many illegitimate children born in every country were deliberately begotten. The juridical concept of rape teaches us that the sexual instinct can lead to acts of violence often quite out of keeping with the perpetrator's usual mentality. The imposing behavior of young men which is also conditioned by hormones, and is often surprisingly like the corresponding behavior of higher animals, is just as involuntary as woman's "submissive" response to it, or as the maternal behavior, dependent upon the level of prolactin, that follows the birth of a child. Although man's fighting drive, likewise dependent on sexual hormones, is often vestigial or suppressed by considerations of reason, it may still lead to specific manifestations, though it more often takes the form of harmless brawls. The drive to gain precedence, having its parallel in many higher social animals, can be a decisive factor in thought and action. Of course, the impulses released by these human instincts are all weakened by other trains of ideas, diverted into more reasonable channels, or totally suppressed. But there can be no doubt that they remain essential determinant factors in our thinking. With compulsive criminals they can entirely govern all trains of ideas.

The limitations in man's so-called free will are also hereditarily determined by the stage of development which the individual's brain has attained. During the first few months of life, the infant is a creature of reflexes and instincts. Even in the first two years, ideation is still so rudimentary that we can assess and predict many of the child's reactions. We therefore do not regard him as a responsible being. Something similar is true of aged persons. The reduction of associative links, and the degeneration of neurons, lead to progressive forgetfulness, particularly of the most recent past, increasing rigidity of thought and behavior, and sometimes a growing disregard for ethical and aesthetic considerations.

Man's thinking, however, is not governed solely by these more general phenomena which we have mentioned, but also depends upon the individual's hereditary abilities and weaknesses. This is very clear in cases of insanity, where the brain fails to function properly. Investigations of related persons and identical twins have shown some forms of feeblemindedness and a tendency to schizophrenia to be hereditary (see Ziehen 1911; A. Juda 1934, 1935; F. Dubitscher 1937; O. von Verschuer 1959; C. Stern 1960). Hereditary mental disorders derive from physiological cerebral disturbances (sometimes recognizable by histological alterations) which interfere with mental imagery. Schizophrenia, which usually becomes more clearly apparent when the sufferer is in his teens or twenties, is variable as to extent and course. The trains of thought are often governed by depressive moods and abnormal affective reactions, and the sufferer displays less and less will power. In some cases progressive imbecility may set in. Hereditary feeblemindedness may range from debility and imbecility to complete idiocy. But in all cases mental activity is impaired; concentration and volitional thought and acting becomes feebler; and the train of ideas is more monotonous.

Several other psychopathic conditions are much less easy to isolate, and in some cases the hereditary basis is not sufficiently proved. Their symptoms may be a lack of will power, unstableness, shallow affective reactions, depressivity, pathological mendacity, inability to control sexual drives sufficiently, etc. These all represent deviations from normal behavior, in which these symptoms help to determine thoughts and actions. Homosexuality is thought to be largely hereditary, and it often leads to actions at variance with moral considerations.

Volitional thought and actions are also much influenced by the particular hereditary type of constitution as manifested in various physical features. These types have been systematized in different ways according to whether they are distinguished in terms of morphology, hormonal constitution, or mental characters. Besides, most people cannot clearly be placed in any one of the proposed categories. The relatively pure types, however, all display reasonably clear psychical characteristics, and these are more faintly observable in the mixed types as well (see E. Kretschner 1921 or 1961; W. H. Sheldon and W. B. Tucker 1940; H. Remplein 1954; R. Knussman 1961; K. Saller in Martin and Saller 1966; and others). The sanguine type of person is more often associated with a positive feeling tone, while the melancholic or choleric type is given to gloomy reflection, and the phlegmatic type is more equable and less involved. The schizothymic type (Kretschmer) is much more capable of concentrated thought and acts of volition than the more sociable and more easily distracted cyclothymic or the sluggish, heavy-blooded, "viscous" type. According to Sheldon, persons of endomorphic type are weak, impressionable, sociable, and somewhat easy-going; the rather more vigorous mesomorphic type is active and forceful, and the gentler ectomorphic type is sensitive, unstable, and reserved. The causal relations between these constitutional types, which are distinguished by differing hormone production, can be fairly clearly assessed. A deficiency in thyroxine production leads to more phlegmatic thinking; too much results in a nervous and irritable state. Deficiency in sexual hormones usually involves frigidity, while excessive production can lead to a lack of sexual restraint. On the whole then, a man's hereditary constitution largely determines the type and tempo of his trains of ideas and his actions.

Aside from constitution, physique, and physiological peculiarities, man shows many hereditary mental capacities or deficiencies. It is a well-known fact that many great painters, among them Bellini, Filippo Lippi, Titian, Van Eyck, Brueghel, and Holbein, sprang from families where the parents or siblings were gifted in a similar way. The same is true of musicians: Bach, Haydn, Stamitz, Mozart, Beethoven, J. Strauss, and many others. Galton (1869) already attempted to show the heredity of special mental endowments. By analyzing 1,162 school reports W. Peters (1915) was able to show that in more than 50 percent of the cases, performances above or below the average corresponded with the achievements of parents and grandparents. M. Skodak and H. M. Skeels (1949), studying one hundred children adopted early in life, found that their index of intelligence was usually the same as that of their natural parents and not of their adoptive ones. In a more exact manner, H. H. Newman, F. N. Freeman and K. J. Holzinger (1937) proved by testing 50 identical and 52 nonidentical twins, and comparing their IQ's, that mental endowments are hereditary. After comparing 100 identical and 50 nonidentical twins, O. von Verschuer (1954, p. 159) wrote: ". . . im Hinblick auf die geistig-seelische Entwicklung der vorherrschende Eindruck die ausserordentliche *Konstanz der Entwicklung*, d.h. das Beharren der eineiigen Zwillinge in dem schon in der Kindheit sich zeigenden Grad von Ähnlichkeit." ("In regard to psychical development, the paramount impression is of extraordinary *constancy of development*; the identical twins retain the same degree of similarity from childhood onwards.") "Es ist erstaunlich, wie die individuelle Eigenart sich oft durch schwere Lebensschicksale hindurch bewahrt." ("It is astonishing how the individual characteristics can prevail in spite of all the blows of fate.")

A comparison between identical and nonidentical (fraternal) twins has also shown that a tendency to crime often has a hereditary basis. A table drawn up by C. Stern (1960, p. 604), and based on material from Germany, Holland, Finland, the United States, and Japan, shows that out of a total of 143 pairs of identical twins, in 68 percent of the cases both twins had committed some culpable offense, whereas the figure for 142 nonidentical twins was only 28 percent, although these last, of course, being siblings also shared a considerable percentage of hereditary factors. That the tendency to crime is often based on hereditary disposition could also be proved by investigation of the parents, who often displayed predominant characteristics such as brutality and pathological mendacity (E. Reiss 1922). Besides, it has been established as probable that 99 percent of a total of 195 criminals were distinguished by "innate affective shallowness, abnormal suggestibility, and exaggerated impulsiveness in the sense of a hyperthymic condition" (F. Stumpfl 1935; see also H. Heinze 1942 and E. Frey 1951).

To sum up: *The volitional thinking and action of man is largely determined by hereditary factors, and in particular by the hereditary structure of the hormone glands, the nervous system, and the special structure of the brain.* The wide difference in mental ability corresponds to marked differences in the absolute and relative size of functional regions of the human cortex. The sensory fields in particular show hereditary differences. The absolute size of the visual area (the area striata), for instance, ranges between 2130 and 4416 qumm, and the relative size between 2.6 and 3.6 percent of the cortex (K. Brodmann 1925). In some cases of pronounced musical talent, the auditory region proved to be highly developed (S. Auerbach 1906). The size of the regions of association in the frontal parts of the forebrain, on the other hand, is also influenced by learning processes (A. Hopf 1964).

V. Nonhereditary Determinants of Volitional Thinking. In most cases it is not possible to analyze a volitional train of ideas so as to uncover every factor involved. But we can often identify some of the chief components. This enables us to more or less predict the course of ideas, and in this way to argue that the course was determined. Besides the hereditary factors, *all* the memorial traces which are connected by association with the ideas involved in volitional thought come into the question. They may embrace experiences in the more recent past or in youth; every memory that one has stored up in the course of education, or from newspaper, radio and television reports, literature, entertainment, etc.; and all the other kinds of experiences.

We can often control the more important relevant content of our thought, and here the train of ideas can be relatively well assessed. This is exemplified in the attempt by a schoolboy to solve a mathematical problem. If we are certain how much he already knows, and if he is concentrating on his task—that is to say if the solution governs his associations as a fairly general but intensive dominant idea—then we can causally analyze the train of his idea accurately step by step, because they proceed logically and digressions are suppressed. But if someone is meditating over a more general problem, even careful introspection will not reveal what ideas played a part as he directed his thought along the lines of a solution. Especially if one ponders over a difficult problem, the interplay of ideas may be multifarious. But at least it is often clear in such cases that a certain complex of ideas finally led to the decision. And we can often trace how the train of thinking has been causally and logically directed by the factors which we either know or do not know.

Our train of thought is also frequently influenced to a considerable extent by pronounced *feeling tones*, particularly by moods and emotions, and furthermore by fatigue. When we are gay or happy, the flow of our associations is easier, and we can solve a problem more readily than if we are vexed or tired, because then our thoughts digress into nonrelevant side issues. On the other hand, a bad conscience or jealousy, for example, keeps our thought returning to the same complex of ideas accompanied by a negative feeling tone. This perseveration clearly illustrates the coercive course of volitional thinking. These perseverant trains of ideas may also occur in conditions of "repression" (frustration), anger, rage, and anxiety, because the ideas are then accompanied by a pronounced negative feeling tone and dominate the flow of thought. This again illustrates how determinate the processes of thought in fact are. Rage is a particularly good example, as we have already seen. The sympathetic nervous system is powerfully excited, so more adrenaline is secreted; this in turn excites the sympathetic nervous system still more, and the victim is caught up in a spiral of exaggerated thought and action, often *against* his will, and certainly far exceeding his original intention. On the other hand, a determinate perseverant process of thought may also result from some idea connected with a strong positive feeling tone, for instance, when we hear an important and agreeable piece of news, for our thoughts then "revolve" for some time around it.

In all these cases the causality of the specific course of thinking can be proved to a certain degree by an analysis of the processes within the psychophysical substratum, processes which are intimately bound up with the phenomena and in my view are identical with them (see Ch. 6 K V). The increase in nervous excitement and the hormone level are both mensurable. And we can understand how other conditions affecting the brain processes alter the thought processes in specific ways. Alcohol first reduces inhibitions, while further intoxication dulls, weakens, and finally almost extinguishes the will power. Stimulant drugs such as lysergic acid preparations cause hallucinatory images which cannot be eliminated by the will. Fatigue and overwork, too, coercively lead to characteristic changes in the thought processes: the consistent and deliberate control over the trains of thought grows weaker, and finally gives place to random associative links. The same can sometimes be observed during the female ovulation cycle, and at the menopause.

Perseverant trains of ideas which repeatedly become dominant may also be induced by *suggestion and hypnosis*. On the other hand, complexes of ideas associated with negative feeling tones, and particularly pathological imaginings, can often be banished through psychoanalytic treatment or by other "psychosomatic" influences: electric shocks, sleep therapy, or relaxation therapy. Despite their name, "*psychosomatic* effects" do not imply a dualistic interpretation. Therapy of this kind achieves a physiological change in the patient's nervous functions, or the psychotherapist's words initiate certain trains of impulses in the patient's brain which correspond to the trains of thought. If one held that the "soul" represents something independent from the body which can be separated from it, it would be hard to understand how, with regard to the "pre-established harmony" between body and soul, such abnormal physiological alterations produced by the treatment could affect psychological processes in this specific manner. Quite apart from this, there is the point that any such effects of soul on body or vice versa are totally inconsistent with the law of the conservation of energy (see pp. 157, 252). On the other hand, if we accent an identistic interpretation, which is much more in line with biological and epistemological facts, the question of "psychosomatic" effects does not represent any problem at all.

Finally, the so-called freedom of the will is restricted not only by the determined course of ideas but also by *actual perceptions* and their accompanying feeling tones. Particularly vivid perceptions such as seeing a close friend after a long absence, or hearing a bird sing, or watching some exciting event, coercively interrupt the current train of ideas and direct it into different sequences of associations.

VI. A General Assessment of Freedom of Will. The past two sections have shown that many diverse factors, hereditary and nonhereditary, govern our thoughts and actions. But the coercive character of our thinking is recognizable only in relatively few cases. We can state it when certain chief components predominate; when, for example, a mathematical problem has to be solved, or some strong suggestive or even hypnotic influences are at work, or some pathological conditions or dominant perceptions or ideas cause perseveration of certain groups of ideas. In such cases the associative course and the main content of the thought are more or less predictable. In this respect man's brain appears to work in a similar way to the brains of higher animals (Ch. 6 H). But in most human thought processes the interplay of determinant factors is so complex that an adequate analysis is not possible. So we cannot state with absolute certainty that all thought is completely determinate and that there is no freedom of the will.

I hope, however, that the phylogenetic considerations have shown that probably many volitional processes in man only appear to be free because of the extraordinary complexity of the associative links involved. This is why the *constellation theory* outlined in section F IV of this chapter seems to me to offer the most satisfactory explanation. And this is the only basis on which it has proved possible to establish *laws of motivation* (see those set out by N. Ach 1935) or to make an apposite prognosis of delinquency, provided one knows enough of the relevant hereditary background, constitution, education, family and social circumstances, previous record of crime, etc. On a broad basis such as this E. and Sh. Glueck (1951) succeeded in predicting future delinquency in children of six or seven years of age. R. E. Thompson (1952), study-ing a large group of school children ten years later, was able to show that the Gluecks' prediction was fully or reasonably accurate in 91 percent of the cases. Further literature on this subject is cited in F. Bauer's very informative work (1957).

An objection frequently put forward to the conception of the com-pulsory character of volitional thought states that *experiences of volition represent a special category of phenomena* characterized by a "feeling of freedom." It is quite true that such a specific "feeling" is present, but the factors that determine it are fully consistent with a deterministic in-terpretation. At first this appears unlikely, because it seems that thoughts are ultimately "free," and that one can prove this by giving one's thoughts a new turn, doing something unusual, or changing a decision if one chooses. However, the objection is inapposite; the point at issue is not whether one can alter the direction of one's thinking, but whether such alteration is determined by specific motives, for instance by the intention to prove the freedom of the will. We have already mentioned how Leibniz disproved this objection (see p. 213). It is always the stronger motives that direct our trains of thought. Thus, if we resist some attraction, we do so because some contrary ideas are stronger; for example, yielding to the attraction may harm us, or being suscepti-ble to it may be considered a sign of weakness. When we refrain from taking any measures to avoid some peril or unpleasant experience, it is usually because the sense of duty, responsibility, justice, or pity is stronger. Such ideas of value are essential to social beings, and it is right that they should become imprinted in our minds as dominant and de-terminative complexes of ideas.

As already noted, a "feeling of freedom" arises because volitional thought and action are frequently or continuously related to the com-plex of ideas concerning the self (Ch. 6 F II), and because we experi-ence a certain activity which manifests itself physiologically by changes in the electroencephalogram (increase in β-waves) and in the circula-tion of the blood, as well as by tonic excitations of certain groups of muscles, the tongue for example (because we generally think in words). In cases of volitional attention and intended actions, an additional feeling of tension in the sense organs involved may strengthen the specificity of the experienced phenomena. The "feeling of freedom" brought about by all these components is particularly pronounced when the "play of motives" lasts a certain time, and various possibilities particularly closely related to the self are balanced one against the other. But if we study instances of this sort in our own experiences, we realize that ultimately the stronger motive was the one that determined the decision. On the other hand, there is no "feeling of freedom" at all if a decision is quickly or easily made, or if the solution of a task is quite simple. However, as we are not normally conscious of the factors involved in an act of will, and may be quite unaware of many of the physiological processes that accompany it, we readily image that the will is free (see Ziehen 1927b).

In my opinion the constellation theory affords a satisfactory explanation of why the will is not free and how the idea of free will arises. Such a deterministic conception avoids a difficulty already touched upon: if the will were free, man's evolution and his complex cerebral activity would contravene the law of causality which has governed the course of events for milliards of years. For every "spontaneous" thought (in a strict sense) which is also matched by processes of excitation in the cerebral neurons would represent the "fresh beginning" of a causal sequence, and would thus conflict with the conservation of energy as well as with the uninterrupted action of the causal law.

To deny that the will is free naturally involves grave consequences for our ethical, religious, and juristic ideas. We shall revert to these problems in Chapter 7, and may indeed find that a solution is easier than might at first seem possible.

H. Psychophylogenesis and Psychontogenesis

1. Development of Conceptions of Consciousness in Animals. One way of throwing light on the problem of the psychophysical substratum is to study the development of psychical processes from both a phylogenetic and an ontogenetic point of view. No living organisms existed on our planet until between three and four milliard years ago, and consequently "psychic" processes (in the customary sense of the term) were still entirely absent. It was only when animals had emerged and higher forms were developing that mental processes first came into being and then grew more complex. The development of the individual living organisms displays a corresponding sequence of "material" followed by "psychical" processes. The fertilized ovum and early embryo of a higher animal or man are normally regarded as lacking any mental qualities. The "psychic" processes evolve only successively as the individual's brain is formed in the course of ontogeny, possibly beginning in humans some time before birth.

How then does this "psychical" element, which is so different from the "matter" of which the body is formed, come into being? According to functional materialism, mental processes are simply a "property" or "function" of the brain (see, in particular, Holbach, La Mettrie, Vogt, Moleschott, Büchner, and others). Cabanis went so far as to state that the brain produces ideas in the same way as the stomach digests food or the liver secretes bile. But this ignores the fundamental difference between psychical and physiological processes. The latter are causal processes, whereas the psychic processes accompanying them happen "instantaneously," without any temporal gap between the material and the psychic event. The psychic processes only run parallel or are identical with the physiological processes (as the author assumes). Moreover, the psychic processes represent undeniably reality; whereas our ideas of "matter" can only be developed by analyzing the contents of our perceptions. We cannot therefore be sure that "matter" is in principle "nonpsychic." As we shall see, an identistic interpretation may in fact offer a satisfactory solution to the problem of psychogenesis.

If we want to acquire a firm and reliable basis for the investigation of the relevant questions, we must begin with our own psychic phenomena. We can assume that others experience the same or similar phenomena as we do, because they can describe them. If and how far animals experience phenomena and of what kind these may be, we can only infer from their anatomical structures and their behavior.

These questions, which are essential to any philosophical theory, have engaged the attention of philosophers and scientists in every age. The observation of domestic animals in daily life led at a very early stage to ideas rather similar to those which we may now base upon scientific investigations. "Animals," however, in those days usually meant mammals and birds, the most highly developed groups. As pertinent remarks were often linked to philosophical aspects of the problem, it may be useful to cite some of these authors.

The ancient Egyptians, Hindus (see the Upanishads), and Buddhists, living in closer contact with animals, came to the conclusion that animals have souls and that after death human souls could pass to animals. The Chinese philosopher Lieh Tse (c. 400 B.C.) assumed that "The thought of animals is of the same nature as that of man" (after H. Hackmann 1927). In Ecclesiastes (iii. 19-21) the Preacher Solomon asks the question later posed also by Leibniz: "Who knoweth whether the spirit of man goeth upward, and the spirit of the beast goeth downwards to the earth?"

The doctrine of the transmigration of souls was also taken up by several Greek philosophers, especially by the Pythagoreans and Empedocles (c. 490-430 B.C.). Other Greek thinkers based their ideas about the psychic processes of animals on biological investigations. Alkmeon (6th century B.C.) taught: "Man differs from all other creatures in that he alone understands, while others merely perceive, but do not understand." In different works, particularly in his *Metaphysics* and *De Anima*, Aristotle ascribed perception and memory to animals, and declared that even lower animals could feel pain, joy, and desire. But he did not consider animals to be capable of either inference or reflection.

St. Augustine (A.D. 397) expressed his views in his *Confessions*: "For even beasts and birds have memory; otherwise they would not be able to find their way back to their dens or nests, or do any of the other things they are used to doing; indeed, without memory they could not become used to doing anything." Descartes (1649), on the other hand, influenced by the causal theory of reflexes then in its early stages, stated that animals have no souls. Leibniz soon opposed this, declaring that they have sensations and memory, but no power of reflection. He even went so far as to assume that the souls of animals live on after death in the same way as man's does.

In his *Essay Concerning Human Understanding* (Book II, Ch. IX, 12) Locke gives his opinion at rather greater length: "Perception, I believe, is in some degree in all sorts of animals; though in some possibly the avenues provided by nature for the reception of sensations are so few, and the perception they are received with so obscure and dull, that it comes extremely short of the quickness and variety of sensation which are in other animals; but yet it is sufficient for, and wisely adapted to, the state and condition of that sort of animals which are thus made."

In Chapter X, 10 he says: "This faculty of laying up and retaining the ideas that are brought into the mind, several other animals seem to have to a great degree as well as man."

A century later Condillac expressed a similar view in his *Traité des animaux* (1766): "Quoi qu'il en soit, il me semble que M. de B. a lui-même démontré que les bêtes comparent, jugent, qu'elles ont des idées et de la mémoire." ("In any case it seems to me that M. de B. himself has established that animals compare, judge, and that they have ideas and memory.") Voltaire, writing on the subject of animals, says: "Ils ont les mêmes organes avec lesquels vous avez des sensations; et si ces organes ne leur servent plus pour la fin, Dieu, en leur donant ces organes, aura fait un ouvrage inutile." ("They have the same organs from which you derive sensations; and if these organs do not perform the same function in them, then when God gave them these organs, he was carrying out a useless task.") In the chapter on "The evolution of the soul in animals" in his *Psyche* (1846), Carus emphasized that it is difficult to say what kind of soul an animal possesses, but that higher animals "reflect and choose" and have "distinct memory," and we may assume them capable of sorrow, joy, anxiety, courage, and anger, but not of reason or imagination.

Finally we may quote Schopenhauer, who wrote in the 1st Book (6) of the first part of his chief work, *Die Welt als Wille und Vorstellung*:

[Animals] "erkennen Objekte und diese Erkenntnis bestimmt als Motiv ihre Bewegungen.—Der Verstand ist in allen Tieren und allen Menschen der nämliche, hat überall dieselbe einfache Form: Erkenntnis der Kausalität, Übergang von Wirkung auf Ursach und Ursach auf Wirkung und nichts ausserdem. Aber die Grade seiner Schärfe und die Ausdehnung seiner Erkenntnissphäre sind höchst verschieden, mannigfaltig und vielfach abgestuft . . ."

([Animals] "recognize objects, and this knowledge determines their movements as motive.—Understanding is the same in all animals and in all men; it has everywhere the same simple form: knowledge of causality, transition from effect to cause and from cause to effect, and nothing more. But the degree of its acuteness and the extension of the sphere of cognition are highly different, multifarious and with innumerable gradations . . .")

Dogs, elephants, monkeys, and foxes are all animals of "great sagacity." "An diesen allerklügsten Tieren können wir ziemlich genau abmessen, wieviel der Verstand ohne Beihülfre der Vernunft, d.h. der abstrakten Erkenntnis in Begriffen vermag" (*Die Welt als Wille und Vorstellung*, Part I, 1859). ("From these most sagacious animals we can pretty accurately determine how much understanding can achieve without the help of reason, i.e. abstract knowledge embodied in concepts.")

These quotations show that almost all scientists and philosophers ascribe conscious processes to animals. They assume them to have sensations, positive and negative feelings, and emotions; and higher animals at least are said to be capable of memory, i.e., of primary mental images and acts of choice. According to Locke, Carus, and Schopenhauer, the most highly developed animals have powers of reflection or "reasoning" and they can comprehend causal relationships. But the general view is that animals have no powers of judgment. It is worth noting that no distinction is made in this connection between the "mind" and the "soul." This distinction was introduced by philosophers under the influence of Christian ideas, and lacks any clear psychological basis. We find the Preacher Solomon still asking and Leibniz confidently asserting that the "soul" in animals as in man survives after death, because the psychic element was regarded as indivisible.

II. Criteria of Conscious Processes. All the conclusions about conscious phenomena referred to in the foregoing section were mainly used on animal behavior. The criteria were not subjected to any more detailed examination. Now in the present century, however, animal psychology has emerged as a discipline working with exact methods. Its findings have been largely complemented and confirmed by physiological study of the senses, nerves, and brain. So we are able to inquire which criteria are at our disposal today, and to ascribe psychic phenomena to animals. We shall first consider "consciousness" purely in its widest sense, including the uncoordinated sensations which lower animals may possibly have. In all the following considerations consciousness is not to be understood in the sense of *self*-consciousness; for that would imply a "self" capable of reflection.

Surely no one would really deny that higher animals, especially mammals and birds, experience phenomena in the form of perceptions, mental images, and feelings. But if we wish to trace the phylogenesis of conscious components further back, we must test those criteria which have a more general application and may refer to invertebrates as well. Two groups of characteristics which some authors have put forward can be excluded at once: purposeful functioning (R. H. Francé, A. Pauly, G. Wolff, A. Wagner; for criticism see Ziehen 1921) and an absence of causal explanation (H. Driesch, F. Dahl). All living organisms including plants possess purposeful structures and functions, and it can be shown that these have developed under the influence of natural selection (Ch. 5 B, C, D). And we have no right to assume that some psychic factor is operative in life processes because they are not yet, or not yet sufficiently, causally analyzed, for causal explanations are continuously progressing (Ch. 4).

So it is morphology and physiology alone, including the study of animal behavior, that can provide us with the most important criteria of conscious phenomena. And we have here to do solely with *conclusions by analogy* which are reliable mainly in the case of higher animals, insofar as the structure and functions of their brains and sense organs are similar to man's and their behavior shows resemblance to those actions of man which involve conscious phenomena. We shall have to examine, however, whether the conclusion by analogy is also valid in cases where the central nervous system is quite different in construction from the human brain, or where, as in lower animals, several nervous centers exist or there may be only neurons with basically similar structures and functions. In the following section we shall see that it is still possible to proceed by analogy here too; but the lower the animal we are considering, the less reliance can we place on such analogies. The principal difficulty lies in the fact that there are no neurons or nerve tracts which are specific for conscious phenomena. This fact has already been stressed in the discussion of the psychophysical substratum.

Most authors who have dealt with the problem agree that we can at least assume conscious phenomena in cases where an animal is capable of memory, or rather associative memory, and choice. But even these criteria are not completely valid. Robots such as those constructed by W. G. Walter (1953) and others are able to learn and to make a choice between two stimuli. Computers, too, are furnished with a storage structure which functions like a memory, and complicated constructions of this kind can make a suitable "choice" from among various possibilities.

Up to a point they can even "generalize," that is, comprehend likenesses and apply their "experience" to new situations (see H. George 1961; K. Steinbuch 1965; and others).

Another criterion that could be put forward is the ability to frame a judgment or to act with foresight. But computers can also simulate this ability to a certain extent, in particular those which are capable of a simple form of chess and can improve their game in the light of their "experience." Besides, these criteria would apply only to conscious processes in higher animals whose psychic phenomena are not in doubt.

As is well known, the failure to establish absolutely reliable criteria has led to behaviorism, which restricts itself to describing and analyzing modes of behavior and tries to avoid the use of psychological terminology. But the latter is not always possible, because the physiological processes corresponding to the complicated psychological processes have in many cases not yet (or not yet sufficiently) been cleared up. Students of behavior and sense and brain physiology are therefore often still obliged to employ psychological terms. It is usual to say that bees "see" and "smell," that they "learn" and possess a "memory," that dogs show "fear" or "joy," and apes can act "with foresight." Yet at least we owe it to behaviorism that these terms are now used with rather more circumspection and care, and that we no longer meet with misleading anthropomorphisms which were fairly common in scientific literature at the beginning of the present century. But if research into behavior had adhered strictly to behaviorist principles, in the case of higher animals many important investigations would never have been carried out, or at least no explanations would have been attempted, for the questioning derives from human psychology (for instance, research into generalization, actions involving foresight, etc.). Moreover, the agnostic attitude would inhibit the study of psychogenesis from the very start. But this is a vital biophilosophical problem which is absolutely legitimate, for even a strict behaviorist would not deny that higher animals see, hear, and feel; in other words, that they experience psychic phenomena.

One important point is that phylogenetic evolution in the animal kingdom has been shown to be a *continuous, gradually progressing process.* Man and animals all belong to an uninterrupted branched stream of life, consisting of unbroken chains of immature and mature reproductive cells (the successive germ tracts) or other totipotent cells, leading from one generation to the next (Ch. 3 A IV). This suggests that psychical processes too have gradually grown more specialized and complex, and have developed progressively in the same way as the psychophysical substratum to whose functions they correspond. If we accept a dualist interpretation and regard "mind" and "matter" as entirely separate, it seems most unlikely that at some stage in the three milliard years or more of phylogenetic development, a psychic element should suddenly have come into being as something fundamentally new, and subject to quite different laws from those affecting matter. This would mean postulating an entirely novel system of psychical laws for which there had been no foundation hitherto. These laws would have arisen at some time in the "eternity" during which our solar system has been undergoing its material development, which had so far been solely governed by the universal causal and logical laws. Besides, it seems difficult to imagine why conscious phenomena should appear, point-like in the universe, in the brains of living organisms alone.

A parallelistic conception avoids such difficulties. Psychical phenomena would then evolve parallel to continuous material phylogenesis. It would be possible to assume protopsychological properties already in the earliest stages of living organisms, running parallel to their material counterparts. Such psychic elements would always have existed; but only after the organisms had reached an appropriate higher stage and in particular had acquired the necessary nervous equipment would they become integrated to sensations. However, a parallelistic view still meets with some difficulties (Ch. 6 K V). But if we take an identistic and panpsychistic view which identifies "mind" and "matter" and only recognizes causally and logically determined relationships within an "ultimate something," then the emergence and further development of psychic phenomena fit without contradiction into a biophilosophical picture of universal evolution.

III. Special Psychophylogenesis. The statement made in the foregoing section, that there are no reliable criteria for the existence of conscious phenomena in animals, and that in all animal groups we are restricted to analogies of very varying cogency, needs a certain modification. Certainly, robots and computers can simulate sensory reactions, memory, choice, comprehension of likeness, etc., but the only genuine analogies which exist in inanimate nature not influenced by man are those which concern sensory reactions (e.g., phototropic growth of crystals). Memory is often compared with certain hereditary phenomena; yet the two are fundamentally different, for the latter are not acquired and consolidated by practice. So memory, generalization, and choice may after all rank as vital evidence of possible conscious phenomena. Hence it would be unwise to ignore these criteria simply because they are not absolutely and invariably reliable. We must remember, too, that these modes of behavior in animals, as in man, are based upon functions of the neurons and nerve systems and, in contrast to robots and computers, upon biochemical processes which do not exist in inanimate nature. Like all symptoms of life, conscious phenomena are properties of a system, and their degree of complexity depends upon that of the organism's nervous structure. The atoms of which this structure is made up are the same as those present in inanimate nature.

When we now try to consider the phylogenetic development of psychic phenomena, to outline the probable *psychophylogenesis*, we do best to start from the phenomena of man, because as already mentioned these have indubitable reality. We may then trace the course of phylogenesis back to successively simpler organisms. We cannot of course make a study of real lines of descent. Only in vertebrates can we make some limited inferences by comparing the size, shape, and structure of cranial casts. For the rest, we are restricted to studying the behavior of recent animal groups at different phylogenetical levels. We shall have to consider how far we may assume psychic phenomena for these groups, and what the nature of such phenomena may be.

Everybody can testify with absolute certainty to the existence of perceptions, ideas, feelings, acts of will, and thought processes in himself alone; but we can be reasonably sure, from their testimony, that our fellow men have the same experiences. The anatomical, histological, and cytological structure of the brain of the anthropoid apes, our nearest relation in the animal kingdom, is very much like ours, and the sense organs are also anatomically and functionally very similar in their ability to comprehend the shape of stimuli patterns and their upper and lower thresholds. So the behavior of apes naturally displays many human features.

Chimpanzees, for example, not only have a great capacity for learning, but also act in such a way that one must assume that they can abstract and generalize, and act with intention and foresight; in other words, they can run through various possible lines of action in their minds, make a "decision," and only then initiate motor impulses in the appropriate muscles. These are processes which we might term "animal thought." But when we are inferring psychological processes of any kind in an animal, it is more correct to say that chimpanzees act *as if* they were abstracting and generalizing, *as if* they had formed averbal concepts (i.e., concepts not connected with words), and *as if* they were proceeding with intention and foresight.

Many performances of this sort have been the subject of special studies (see W. Köhler 1921; R. M. Yerkes and A. W. Yerkes 1929; R. M. Yerkes 1943; B. Rensch 1965b, 1967; and others). In the present context it may be sufficient to refer to the example discussed in section G III of this chapter (p. 221, Fig. 6). The chimpanzee was confronted each time with a maze showing a constantly differing pattern of complicated pathways. She had to choose the one leading to the exit, and to move a metal disk along the chosen pathway with the help of a magnet without pulling it into any of the dead ends on the way. As she did so, we could measure the time of the "play of motives," which often lasted a minute or more. The chimpanzee behaved as if she were weighing up various possible lines of complicated actions in her mind, before her hand received the impulses to set the disk in motion with the magnet. This largely corresponds to what in the case of a human we would call "reflection" (B. Rensch and J. Döhl 1968).

Several other investigations and experiments revealed just such actions involving foresight (see W. Köhler 1921; M. P. Crawford 1937; C. Hayes 1952; J. Goodall 1963; and others). Various experiences involving a comprehension of causal relationships (but not of course amounting to logical "understanding") proved that anthropoid apes are able to discover spontaneously the right solution in a new situation. For instance, they succeeded in opening boxes fastened or locked in different and unfamiliar ways (B. Rensch and G. Dücker 1966; B. Rensch and J. Döhl 1967), or they used a stick to knock down a fruit or piled up empty boxes and climbed on them in order to reach it (W. Köhler 1921; H. C. Bingham 1929; R. M. and A. W. Yerkes 1929). Even capuchin monkeys proved able to use sticks and boxes in this way (H. Klüver 1933). In performing these actions monkeys are also clearly behaving as if they were drawing *averbal conclusions*. The only difference between these and simple conclusions by humans is that the former are not thought out in words. Language, with its concepts such as "on that account," "therefore," and "because," makes it much easier to frame this type of conclusion.

In many experiments apes and monkeys showed that they could act as if capable of a number of different abstract *averbal concepts*. Even *concepts of value* could be developed experimentally. They are of special interest, for hitherto they had been largely regarded as man's prerogative. J. B. Wolfe (1936) and J. T. Cowles (1937) proved that chimpanzees can learn to distinguish three different values of counters (poker chips). A slot machine returned them one grape for a white counter, two for a blue one, and nothing for a metal one. They soon learned to prefer the blue ones, and often hoarded them up before changing them for grapes. T. Kapune (1966), a former student of mine at our Institute, succeeded in training even a rhesus monkey to distinguish five kinds of differently colored iron rings of the same size. In constantly changing combinations of 3, 4, or 5 colors this monkey learned to choose out the most "valuable," i.e., the most highly rewarded of twelve such rings.

We may also assume that monkeys have a more or less distinct averbal *concept of the self*. This probably develops in much the same way as it does with a young child. In the first few months of life the child learns to differentiate between his own body and the outside world by proprioceptive sensory perceptions, and by seeing and touching it. This is made easier by the fact that perceptions of one's own body are often accompanied by more intense feeling tones (pain or pleasure) than perceptions of the environment (Ch. 6 F II). The complex of mental images formed in this way is then successively supplemented as the child begins to remember his own abilities or experiences, and to recognize his own name and his place within the family.

All this applies to monkeys too, in almost the same way, and they also behave as if they had some primitive concept of the self. They certainly show that they know their rank in the social scale of their horde. Observations of wolves, deer, hens, and pigeons have proved that these animals too know their rank in their society. A concept of the self can also be inferred from the experiments of M. P. Crawford (1937) who taught two chimpanzees to pull a box containing a reward to the side of the cage by two ropes. As the box was too heavy for a single animal to move alone, they had to learn to act together. When one of them alone was vainly trying to pull the box along, he then sometimes signed to his fellow or pushed at him, as a signal to come and help him. This behavior suggests that the ape apparently was aware of his own "self" as inadequate for the task.

It may be also mentioned that monkeys sometimes act as if they had *primitive aesthetic feelings*. Experiments with cardboard squares bearing black-and-white designs show that they like to pick out first the ones where the pattern is symmetrical and rhythmic (B. Rensch 1957). When chimpanzees or capuchin monkeys make spontaneous and untaught drawings and painting, they show a tendency to centralization, they often suit their design to the size and shape of the paper; and they sometimes paint with a certain feeling for balance between the two halves of the picture (B. Rensch 1961b, 1965b; D. Morris 1962). The origin of primitive aesthetic feelings of monkeys may be physiologically explicable; patterns involving some repetition of equal components are easier to comprehend, more "complexible" and therefore accompanied by positive feeling tones, as is the case in man (Ziehen 1923, 1925).

Thus we see that the behavior of monkeys is in many respects so similar to that of humans that we may confidently reason by analogy that the psychological processes in such cases are also similar. Their facial expressions, too, are so human, for instance when chimpanzees quite unmistakably laugh and smile, that they look more like caricatures of ourselves than like animals.

The differences in behavior between anthropoid apes and present-day man are mainly determined by certain structures in the forebrain. The frontal part of the cortex and also the basal neocortex are more highly developed in *Homo sapiens*. Impulses of these cortical regions probably prolong the course of excitations in the brain, and so longer trains of thought can be sustained. Furthermore, man's motor speech center allows him to form words, to symbolize concepts by means of words, to think in words, and also to use symbols for causal and logical relationships. Language has enabled him to communicate his experiences to others and especially to the younger generation, and in this way he has built up a tradition which has led to the development of a material and spiritual culture (see B. Rensch 1965a). As already noted, the brain structure and the corresponding behavior at the *Homo erectus* (= *Pithecanthropus*) stage, and at the still more primitive *Australopithecus* stage (living in small hordes, with very slow improvement of the earliest tools), led on to extinct species of anthropoid apes. Hence the phylogenetic derivation of the intellectual life of the hominids from that of the most highly developed animals presents no difficulty.

Tracing the line of phylogenetic levels of animals farther back, "down" to the lower forms, we can report more briefly. Less highly developed mammals than monkeys are also capable of considerable cerebral accomplishments, and one may presume that they experience similar phenomena to those we have ascribed to monkeys. With regard to the structure of their sense organs, we may assume that many diurnal species have similar perceptions. It has also been shown that carnivores are able to generalize to some degree; that is, they act as if they had framed abstract concepts. As dogs recognize their masters by sight, as well as by sound or smell, we may assume that they are capable of forming a complex of mental images corresponding to a generalized averbal concept. An Indian civet cat in our Institute learned to make a choice between two patterns corresponding to an abstract concept of "unlike and like." We first trained the animal to choose a pattern of two circular "spots" (dots) of different sizes and not the alternative pattern of two spots of equal size. Without further training, it was then shown pairs of patterns which were similar, but also successively more and more different in shape, number, and arrangement. However, in almost every case it spontaneously chose the pattern made up of two or more dissimilar units and not the other where the units were alike. The civet cat acted as if it had formed an averbal concept of "unlike" and chose, for instance, a pattern consisting of two lines of different lengths and not the other one resembling "99" (B. Rensch and G. Dücker 1959a). J. M. Warren (1966), employing a different and somewhat simpler method, succeeded in training a domestic cat to pick out only the "odd" units offered together with two "even" ones, although the types of units were constantly changed. Mammals can also experience positive and negative feelings, joy and pain, as the behavior of any dog can teach us.

In *birds*, the cortex is only slightly developed. As far as we can judge, their accomplishments in regard to memory, abstraction, and generalization rest upon cooperation between the midbrain and portions of the forebrain which correspond more or less to the mammalian sub-cortical regions. In spite of this, the best cerebral achievements of some birds are no less than those of rats and smaller carnivores. Crows, parrots, and beos (hill maynahs) can imitate words and phrases quite well; and they can employ some of them appropriately, for instance, when they are begging for food or when they welcome someone they know. Parrots, crows, and related birds can also be trained to behave as if capable of numerical concepts. After a long period of training, a raven and a grey parrot even succeeded in "counting off" between one and seven pieces of plasticine varying in size on a "notice-board." Then the birds proceeded to five food dishes ranged behind the "notice-board," where they chose the one with the appropriate number of plasticine strips on the lid, though the size, shape, and arrangement of the strips did not at all correspond to those on the board. They pushed away the lid and found a reward in the container (O. Koehler 1941, 1943, 1952; H. Braun 1952). Jackdaws and carrion crows may even be credited with primitive "aesthetic" feelings. Like monkeys, they spontaneously pick out symmetrical and rhythmic patterns in preference to irregular ones (B. Rensch 1958b; M. Tigges 1963).

All this goes to show that complex learning processes, generalization, and other cerebral achievements, and the ideas we may assume to correspond to them, do not depend on the development of a cortex of the forebrain. This is also proved by the behavior of fishes, whose forebrain is only slightly developed and whose projection field for optically released excitations and the corresponding visual center of associations, the tectum opticum of the midbrain, has a much simpler histological structure. Fishes too can learn to grasp the significance of up to five pairs of patterns, and they can retain a pair of patterns for some months and can recognize them even when these are considerably altered (see K. Meesters 1940; K. Herter 1953; A. Saxena 1960; and others). They are also capable of "relative" learning; i.e., without retaining the characters of the pattern absolutely, they can learn the relationship between two different patterns. If they are trained, for example, to prefer a pattern of fine black-and-white stripes to one with coarser stripes, and are then offered the fine stripes together with still finer ones, they will choose the latter, even though they are seeing this pattern for the first time. This relative learning, which requires the formation, so to speak, of engrams of a "superior" level, is displayed by many vertebrates when the characteristic differences can be arranged in a monodimensional sequence as finer-coarser, lighter-darker, smaller-larger, less-more, and so on.

Fishes, like humans, are susceptible to simultaneous color contrast. If they are trained for instance to prefer "red" to "blue," and are then offered two grey patterns, one surrounded by a "green" border (the complementary color to red) and the other by a yellow one, they will choose the former (see K. Herter 1953). Fishes, like humans, are also subject to optical illusions; for example, the Zöllner, Ebbinghaus, and Ehrenstein illusions (G. Dücker 1967). Finally, fishes also show signs of emotion, normally by changing color or spreading their fins or gill lids. These signs of emotion also appear in conflict situations, when an experimental neurosis is set up. This sometimes occurs if the fish has successfully learned to discriminate a positive and a negative sample in two pairs of pattern and is then once or twice offered two negative samples only.

If we assume sensory reactions, memorization, the ability to generalize, morphological similarities, and homologies of corresponding structural and functional features in sense organs and brain to be more or less reliable criteria for conclusions by analogy, then we may concede perceptions, mental images, and feelings to be present in all vertebrates, though these phenomena become less complex as we descend the scale from warm-blooded creatures to fishes, which have much simpler brains.

In the classes of invertebrates, the central nervous system and also the sense organs are quite differently constructed, and their behavior too often shows little resemblance to that of the vertebrates. Are we still entitled to rely upon conclusions by analogy which are much vaguer for invertebrates, and assume that they too experience psychic phenomena?

For some groups at least we may answer this question in quite a positive sense.

The brain of cuttlefishes (cephalopods), a class of the molluscs, is relatively and absolutely large and divided into a number of functional regions. Most species also have well-developed eyes capable of discriminating different shapes, and a fairly good chemical and tactile sense. An *Octopus* can be trained to distinguish different geometrical patterns and it is capable, too, of a limited degree of generalization, for once it has been trained with a sample pair, it can recognize the same patterns if presented on a reduced scale or, within limits, on a larger one (J. R. Parriss and J. Z. Young 1962). An octopus can even successively learn three sample pairs, distinguishing a small from a large square, a vertical from a horizontal rectangle, a white circle from a black one, and can then master the three tasks simultaneously (J. Z. Young 1961). So these larger cephalopods can achieve as much as small fishes like the Cyprinodontidae.

A bee's brain is quite differently constructed from that of a cuttlefish; but though it is no bigger than a pinhead, it contains over 850,000 neurons (W. Witthöft 1967). And bees can be trained to discriminate colors, black-and-white patterns, or different scents (see, for example, K. von Frisch 1919, 1934). In the short time (two to three weeks) during which they leave the hive, bees as well as bumblebees can learn to master up to three or four discrimination tasks simultaneously (M. Schulze-Schencking 1970). They can also generalize to some degree (M. Hörmann 1937) and are susceptible to simultaneous color contrast (A. Kühn 1921). Their positive reaction to various kinds of sugar and to flower scents can probably be taken as evidence of positive feeling tones. It seems that generally all animals are aided by positive feeling tones in satisfying their instincts, and are protected from harm and in particular from pain by negative feeling tones.

In the different classes of worms, the structure of the central nervous system and sense organs is simpler. Annelida, like insects, have a supra-esophageal ganglion, corresponding to the insect's "brain," an infra-esophageal ganglion, and a chain of paired abdominal ganglia. Some of the most highly developed members of this group, the marine Polychaeta, have eyepits consisting of numerous visual cells and protective pigment cells, and also chemical and tactile sense organs. They can learn to turn to left or right in a T-shaped maze; the supra-esophageal ganglion, however, is not alone involved in this activity (see J. Arbit 1964; R. B. Clark 1964; and others).

The lowest of the free-living worms, the turbellarians, have still simpler sense organs and nerve systems. But even they have been shown capable of some degree of associative learning in the sense of conditioned reflexes (see R. Thompson and J. McConnell 1955; J. B. Best and I. Rubinstein 1962; C. D. Griffard and J. I. Peirce 1964; A. L. Jacobson 1965). For Coelenterata and in particular Hydrozoa and Anthozoa, final proof of this is still lacking. The same kind of learning in the sense of conditioned reflexes has been claimed for Coelenterata, which have sensory cells and very simple neurons with only a few branchings but no central nervous system—only a loose assemblage of neurons at the base of the tentacle corona (see D. M. Ross 1965; E. A. Robson 1965).

If we take sense organs and neurons as morphological criteria and processes of excitation as physiological criteria, it seems possible to attribute the capacity for perception and recollection to all multicellular organisms except nerveless sponges and Mesozoa, i.e., to all animals from Coelenterata "upwards." Though such conclusions are rather vague in the lower groups, they are supported by the continuity argument put forward in section H II of this chapter: all animals are parts of the continuous and branching stream of life, in which the generations are linked by the uninterrupted chain of reproductive cells. Besides, it is highly unlikely that the psychic phenomena with their specific correspondence to different physiological nervous processes should have suddenly sprung into being, point-like in time and space, at some time in the course of phylogenetic development, which before had run off purely as a sequence of causally determined material events. If we accept, however, an identistic interpretation which has already been outlined and will receive further treatment in the following chapters, the problem as such does not occur, because in that case mind and matter do not exist as separate units, but only psychical or protopsychical causal relationships within an "ultimate something." These relationships either enter a stream of consciousness of a living being and may then be experienced as phenomena, or else they remain unassociated.

On the basis of the continuity argument and an identistic conception, it is also possible at least in principle to ascribe very primitive conscious phenomena to *unicellular organisms*. Their responses to light, touch, heat, chemical, or electrical stimuli, which are usually termed "sensory reactions," closely resemble those made by the smallest worms which possess a nervous system, for instance, microscopic rhabdocoele Turbellaria and Gastrotricha. However, there is no evidence that unicellular animals possess a memory. At most they show alterations of behavior in the sense of "habituation," which can be interpreted in different ways (see C. J. Warden, T. N. Jenkins and L. H. Warner 1940; H. Machemer 1966). So we can at most assume that in unicellular organisms isolated sensations or more probably similar protopsychical phenomena occur.

As *plants* also represent a branch of organisms deriving from unicellular organisms, and as they too form part of the continuous stream of life (see Ch. 3 B), it seems possible to ascribe to them also some such primitive and unassociated protophenomena. We do in fact find "sensory reactions" to various stimuli among plants, like phototropism, chemotropism, thigmotropism, geotropism, photonastia, and seismonastia. And seismonastic movement of the leaves of *Mimosa pudica* are accompanied by processes of excitation which correspond fairly closely to the type and conduction of the action potentials in lower animals. Many great naturalists and philosophers such as G. T. Fechner, F. Paulsen, B. Erdmann, W. Wundt, Th. Ziehen, E. Haeckel and R. H. Francé have already stressed that we cannot in principle rule out the possibility of very primitive conscious phenomena in plants. Of course all conclusions by analogy are very uncertain here. The most one could assume, in my own view, would be unconnected "protophenomena." It is unlikely that any appreciable differentiation of qualities and modalities of sensations has been developed, because natural selection cannot have caused such integrations of protopsychological elements. One can only expect differentiation of this kind if it is advantageous for excitations released by different types of stimuli to become differently utilized within a central nervous system.

As already mentioned, prestages of life also exist, represented by viruses (Ch. 5 E). If we accept an identistic conception, there is no difficulty in principle in ascribing protopsychical properties to these prebiological forms too, which in some stages exist only as macromolecules. And then we may also ascribe a protopsychic nature to all molecules, atoms, and all "matter" (see B. Rensch 1961). This identistic conception is based on the conviction that causal and logical laws govern and have always governed the entire universe, and that these laws, or rather the relationships of the "ultimate something" governed by these laws, have become integrated in more and more multifarious separate laws, in the sense of an epigenesis (Ch. 6 K), as living organisms have evolved and grown in complexity. In the course of phylogenesis, the living organisms have continuously been obliged to adapt their nervous reactions to these lawful relationships of their environment, and the accompanying psychic phenomena have been adapted in a parallel manner. The increasing differentiation of the nervous system caused an increasing integration of protophenomena to phenomena, to sensations and mental images, which became more and more adapted to the surrounding world. Finally, his conscious phenomena have even enabled *Homo sapiens* to arrive at an understanding of his own existence, his evolution, and the laws that govern the universe.

IV. Psychontogenesis. The whole of hereditary information is transmitted from one generation to the next by chemically definable molecules—the nucleic acids and especially DNA (Ch. 4 E). The human hereditary characters transmitted in this manner also include psychical ones. Research into twins, and analysis of nearly related persons, have proved that the inclination to certain mental diseases, individual tempo, the capacity for abstraction, special talents, etc., are inherited (see, for instance, O. von Verschuer 1954; K. Gottschaldt 1964; C. Stern 1960). The structures and functions of the fertilized ovum or early embryonic stages do not show any indications of psychical phenomena. It is only at the latest embryonic stages, when the nerve cells have developed and the brain has been formed, that reactions have been observed which may possibly be accompanied by weak, uncoordinated psychic phenomena (see, for example, L. Carmichel 1951). Immediately after birth multifarious sense reactions run off and it is very probable that sensations are really "experienced," although they are also uncoordinated at first.

Where then does this "psychical" element come from, which seems to be so fundamentally different from the "material" processes? According to certain dualistic conceptions, a "soul" unites with the embryo, and after death it parts from the body again and is immortal. But where in that case was it before it entered the embryo? And what about the "souls" of animals? These questions were seldom formulated in the past, and even then were not often realistically treated, because they touch upon fundamentals of Christian doctrine. Leibniz alone dealt consistently with them in his *Considérations sur les principes de vie et sur les natures plastiques* (1705, vol. ii; see 1906b). Noting that the characters of plants and animals are already present in the seed or sperm, he wrote: "It is reasonable to conclude that what does not begin to live, does not cease to live, and that death and generation are simply two stages of the same creature which is now multiplied and now reduced." He was obviously convinced, then, that the souls of animals live on after death.

Analyses of physiological processes in the brain have increasingly confirmed the view that *all* psychical processes correspond to nervous processes (Ch. 4 C and 6 C). Hence, the excitations of the neurons would take place in the brain in precisely the same way, whether or not there were any corresponding psychical phenomena. So conscious phenomena could therefore be called "epiphenomena." The psychologist Th. Ribot (1839-1916) very aptly compared the phenomena to the rays from a bedside lamp, which light up the dial of a clock without in any way affecting how it goes. As the effect of natural selection has always been to promote and maintain only those new properties which are advantageous, epiphenomena appearing at any phylogenetic level or stage of embryonic growth would have been "superfluous" features, and as such would have had no survival value. If we try to avoid this difficulty by arguing that psychic phenomena are advantageous, because they affect and direct physiological processes in the brain," we find that this contradicts the law of the conservation of energy. It would mean that an "acausal" psychical element would produce an additional "effect" in the course of the continuous gapless causal processes.

With an *identistic* interpretation, however, none of these difficulties arises. According to such a conception, the germ cells and early embryonic stages already have a protopsychical nature, and higher stages of integration of the protopsychical properties, i.e., sensations and ideas, begin to appear as soon as a central nervous system has developed, making it possible for the phenomena to be "experienced" in a stream of consciousness. This development corresponds to the phylogenesis of phenomena.

If one nevertheless admits the view that early embryonic development is purely a succession of "material" processes, and that an entirely novel "psychical" element becomes associated with it at some later stage, one could find support by saying that "discontinuity" of this kind can be observed in sleep, or in an unconscious or drugged condition.

We must, however, take the following facts into consideration. If the brain lacks oxygen, the conscious processes become less intense, and this may lead to a semiconscious condition. This indicates that the processes in the neurons of the brain must attain a certain degree of intensity for the stream of consciousness to be maintained. As we have seen also, inhibition of afferent impulses by the formatio reticularis diminishes the intensity, leading to drowsiness and finally to sleep. Many impulses which are unimportant for the behavior of animals and man are normally cut off and do not enter the stream of consciousness, as for example most of those initiated in the peripheral areas of the retina. The same is true of many other excitations which are not the object of attention (Ch. 6 F IV). Apparently this system by which impulses are switched off developed first in higher animals with numerous sensory cells, to protect the brain from a superfluity of excitations and to allow it to concentrate on reactions to vital stimuli.

But even after the majority of sensory excitations have been switched off in this way, and the stream of consciousness has been interrupted or halted, it is still theoretically possible for uncoordinated sensations and ideas to occur. In a half-drowsy condition we still experience brief, faint trains of imagery and a few feeble sensations, which seem to float up into a weak stream of consciousness and then sink away again. Other phenomena may possibly exist, but we cannot experience them because they do not enter the center of consciousness.

I. Matter

I. The Problem. The general view is that all life processes are bound up with "matter." As already mentioned, a particular structure of certain chemical compounds and systems, and their replication (reproduction), is a characteristic of life, but the same compounds in isolation do not show any vital functions. The components of the psychophysical substratum, the sensory and cerebral cells, likewise consist entirely of such compounds, which during metabolism are converted and partly excreted, and are then replaced by inanimate material deriving from food. When certain vital processes cease to take place and the individual dies, rapid decomposition ensues. We have further noted that all systems which can be considered as living organisms probably evolved step by step from inanimate matter. So we must inquire what the essence of this "matter" is like, this matter which can combine to form a living organism and can revert to an inanimate state, which can build up nerve cells whose functions correspond to conscious processes and can be decomposed to components which seem to lack any "psychical" character.

If we want to arrive at a reliable idea of what matter is, we start with our psychic phenomena as the only absolutely certain factors, and then inquire how far matter has been analyzed by physics, especially microphysics, and what are its characteristic properties. We shall confine ourselves, however, to the most important facts relevant to the problems of biophilosophy. We shall also examine to what extent hitherto unexplained life processes may be interpreted through microphysical findings, what place the emergence of living organisms occupies in the evolution of matter, and how present-day data fit into a biophilosophical, and especially an identistic and panpsychistic picture of the universe.

II. Epistemological Analysis of "Matter". We have already noted that the only absolutely indubitable reality is that of our phenomena (Ch. 6 D II). In regard to matter, we can state with certainty that we experience relevant perceptions, ideas, and trains of thought. A rose consists for us of visual perceptions of shape and color, of tactile perceptions of its hard stem and soft petals, of a pleasant scent, and also of a sound if it falls to the ground. We have also corresponding visual, olfactory, haptic, and auditory images and trains of thought, which are largely connected with the concept "rose," and the actual or imagined sound of that word. This is the manner in which we experience all "things," a term by which we denote objects distinguished or in some way distinguishable from their environment. We humans perceive and therefore imagine most "things" visually or haptically. But there are also some "things" which have no constant shape, like running water, or no fixed outlines, like mist.

We almost always connect the idea of some "object" with our perceptions; that is to say, multitudinous experiences have led us to assume the existence of "things" which initiate our perceptions. With infants, however, the case is probably different. They only gradually form a concept of the "self," and learn to mark it off from both the world around them and their own body. Later, they assume the existence of "things" as the causes of phenomena relating to the outside world. As already noted, a concept of the self, or at least a prestage of it, is very probably present in higher animals also. To judge by their behavior, they distinguish "things" as independent from their own bodies (Ch. 6 H III). But it is very doubtful if or how far this corresponds to "objectivation" in the human sense. It may be that, like infants, they simply distinguish two groups of phenomena differing in their practical significance for the business of living.

"Objectivation" takes place when certain complexes of perceptions presented in a time sequence either remain the same or alter in a continuous manner, as in the case of an object in motion. We then experience several or many successive phenomena made up of the same or similar components, and we thus receive the impression of *sameness*. On the other hand, if several equal phenomena appear together in a spatial arrangement at the same time (phenomenal simultaneity), we never assume that they all refer to the same "thing" (see Th. Ziehen 1934, § 20).

At first all the properties of perception are considered to appertain to the objectivated "thing." Not only for children, but also for the majority of adults, including most "educated" people, a rose *is* red, its petals *are* soft and scented; in other words, these qualities are all regarded as being qualities of the "material" flower. The same is true of its spatial and temporal properties. This naively realistic identification suffices for ordinary everyday purposes, and it is useful. Therefore natural selection has not operated against it as higher animals and man progressively evolved.

The philosophically reflective man, however, has long ceased to believe that this kind of naive objectivation of phenomena does justice to facts. As far back as Hindu Vedanta philosophy—most clearly expounded in the interpretation by Shankara (7th century B.C.)—it was accepted that phenomena do not reproduce the essence of things; they are only "maya," only appearances. In a fragment from Democritus we find the phrase: "In terms of the *nomos* it is 'color', it is 'sweet', it is 'bitter'; but in reality it is only atoms and emptiness." Plato makes Theaetetos declare: ". . . that black and white or any color you choose is something that has arisen out of the meeting of our eyes with the appropriate motion, and what we say 'is' this or that color, will be neither the eye which encounters the motion, nor the motion which is encountered, but something which has arisen between the two. . . . And so we must think in the same way of the rest—'hard', 'hot', and all of them . . ." St. Augustine, too, plainly understood the nature of the problem (see p. 178).

But it was only when the physiological basis of man's sensory activity came to be examined that any real advance was made. Locke's distinction between primary and secondary qualities was of vital importance in this connection. In *An Essay Concerning Human Understanding* (1690) he classed the spatial properties of perception (size, shape, and number, and also solidity) as primary, because he assumed that they are characters of the things themselves. Colors, sounds, scents, etc., he called secondary qualities, because they come about in our nervous system only after some impulse caused by forces in the outside world has been generated in the sense organs: ". . . the ideas of primary qualities of bodies are resemblances of them and their patterns do really exist in the bodies themselves; but the ideas produced in us by these secondary qualities have no resemblance of them at all" (see 1877 ed., p. 246). Besides, Aristotle had already distinguished primary qualities in a similar way; he stated that only "motion, rest, number, form and size" belong to all perceptions (*De Anima*; see 1847 ed., p. 90).

Not very long after Locke's work, Berkeley's *Principles of Human Knowledge* appeared in 1710. This great philosopher, then only twenty-five, held the view already suggested by Hobbes, that the properties of perception and also their "primary" spatial and temporal qualities tell us nothing about extramental "things", and that only psychic phenomena exist. "Being" is thus identical with "being perceived" (*esse* =

percipi). In his *Kritik der reinen Vernunft* Kant also emphasized that primary qualities are subjective, and that we cannot therefore ascribe spatial and temporal qualities to things. But he did not go so far as Berkeley. He accepted the fact of extramental existence, though the essence of the "thing in itself" cannot be known. Kant considered primary qualities to be in a separate class, insofar as we can have a knowledge of spatial and temporal properties independently of experience, a priori. Unlike color, scent, etc., these primary qualities cannot be dissociated from the phenomena. But this distinction does not always hold good. As Ziehen has pointed out (1934, pp. 140ff.), when imagining a triangle we cannot dissociate from the color of its outline. On the other hand, we can dissociate the spatial properties from a tune. It is true, however, that spatial and temporal properties of sensations are more difficult to dissociate than individual colors. That it is impossible, according to Ziehen, to have a notion of space "before all experience" already follows from the fact that we often make mistakes in transferring haptic spatial sensations into the visual sphere.

Schopenhauer's interpretation of Kant's theory already reveals the shortcomings of a purely subjective view involving a priori knowledge. In *Die Welt als Wille und Vorstellung* (IV, i p. 739) he writes: "Obwohl die Zeit wie der Raum die Erkenntnisform des Subjekts ist, so stellt sie sich gleichwohl, eben wie auch der Raum, als von demselben unabhängig und völlig objektiv vorhanden dar. . . . Es existiert nur in den Köpfen der erkennenden Wesen; aber die Gleichmässigkeit seines Ganges und seine Unabhängigkeit vom Willen gibt ihm die Berechtigung der Objektivität." ("Although time, like space, is the form of knowledge of the subject, yet just like space it presents itself as independent of the subject and completely objective. . . . It exists only in the heads of percipient beings, but the uniformity of its course and its independence of the will give it the authority of objectivity.")

Although the views held by modern philosophers differ considerably, most agree that the properties of phenomena, and in particular their qualities, are primarily subjective. We are sure that the rose itself does not possess color, but light waves reflected from it generate photochemical reactions in the visual cells in our eyes, and specific impulses are passed on through the visual nerves to the brain. The sensation of color is only then coordinated with the physiological activity taking place in the brain, or rather the two processes are identical (Ch. 6 K V). The light waves of various lengths, and the excitation patterns they set up, are only quantitatively different. On the other hand, the processes occurring directly at the level where the psychical phenomena correspond are apparently qualitatively different, because of the varied integration of the underlying energy complexes. And the sensations which correspond to these are also qualitatively distinct; we are aware of a red flower and green leaves. Hence, these qualities of sensation do no more than *indicate* certain chemical and physiological peculiarities belonging to the thing we call a rose.

The position is much the same with the scent of the rose. This is not a property of the blossom itself; it arises when certain molecules (geraniol) leave the petals, are drawn into our nostrils, and release impulses in the olfactory cells, which reach the brain where they are coordinated with a scent quality at the level of the psychophysical substratum. If we accept that the qualities of sensation already correspond to the primary process of excitation in the sense organs, this would provide stronger evidence of "objective" properties in the rose (Ch. 6 C II). Yet wherever the psychophysical substratum may be located, we are still left with the question whether its reaction with the qualitative peculiarities of the excitations are coordinated with the sensation of scent, or whether they are identical with it. We shall be discussing this important problem in Chapter K V of this chapter, which is concerned with the so-called laws of coordination (= parallel laws).

The situation is somewhat different in regard to the *intensity of sensations*. The degrees of intensity show a certain proportionality to the intensity of the stimulus. Weber's law, however, shows that the parallel is not an exact one. We may even notice a slight increase of intensity in the stimulus in low levels of intensity, for instance in the realm of a pale pink color, but fail to do so in the level of high saturation, for instance with a denser shade of red. The degree of intensity of the sensation does, however, allow us to infer more about the nature of matter than the qualities of sensation do.

The primary properties of sensation, the spatial and temporal components, which are invariably present in all sensations, are of greater importance for our knowledge of matter. The *spatial properties* are most noticeable in our visual, haptic, kinesthetic, and most auditory sensations, but are much vaguer in sensations of taste and smell. (These merge with haptic sensations on the tongue, or in the nose as we inhale.) They indicate the spatial relations of qualities concerning the position of separate components in a delineated sensation, and the relations between different sensations with regard to their distance from one another and to their direction.

Spatial properties are unlike qualities in being completely homogeneous. They occur serially, so distance and direction are measurable in units. All spatial positions extend three dimensionally, and they unite to form an idea of uniform space with a variably determinable zero point. The point of reference for position, distance, and direction is normally one's own body.

Already in early childhhod the notions of spatial properties all run together to form a uniform visual, tactile, vestibular, kinesthetic, and auditory space. As all these are mutually consistent, and the spatial properties are homogeneous, the whole soon merges into an idea of uniform "absolute" space. This space can be thought of as limitless by adding mental images of one three-dimensional extension to another and imagining that this process can be endlessly continued. But we cannot in reality imagine anything spatially limitless; for it is only possible to think of a chain of finite components.

It is a matter of the greatest epistemological significance that *absolute space is not primarily given with our phenomena*, but that the idea can only be constructed from spatial properties of sensations belonging to different modalities. However, spatial properties do afford a good deal of evidence from which to deduce the essence of that "ultimate something" behind all phenomena which we call matter. But it is advisable to follow Ziehen in distinguishing "local" properties of phenomena from those "locative" qualities of the "thing in itself" which we may deduce from them. Yet it is probable that the two correspond either largely or completely to one another. Locke's assumption that primary qualities are properties of the things themselves and Kant's description of spatial qualities of phenomena as a priori forms of experience were insofar not unfounded.

The *behavior of higher animals* suggests that with them too spatial qualities belonging to different modalities merge into some kind of uniform idea of space. It would be very interesting if we could determine at what phylogenetic stage such generalization first appears. Up to the present very few experiments have been made in this direction. N. Kohts (1928) placed various wooden objects—a ball, a cone, a pyramid, a prism, etc.—in a sack, and then trained a chimpanzee to grope about for one of them in this sack when shown (optically) an object of that particular shape. The animal also learned to select flat objects with a different number of corners in the same way. M. Tellier (1932) conducted a similar experiment with one of the lower monkeys (a *Macacus sinicus*). This animal was first trained to pick out certain plastic samples by sight and to discard others, and then to grope about in a sack for objects corresponding to those which he had learnt to choose. These results suggest that monkeys have a generalized sense of space (although Tellier's experiments were unsatisfactory because the monkey had tactile experience through learning to choose). We may assume that the same is true of birds. They select a twig by sight, and then land on it without looking to see where to grasp it with their claws. It seems unlikely, however, that lower vertebrates possess such a generalized idea of space. It is possible that a salamander creeping into a hole in a rotten tree stump does not combine visual and tactile spatial sensations, and probably the same is true for invertebrates such as cuttlefishes.

Locality as an immediate "given" property of sensation is no more definable than any other such property. Perception of right and left, above and below, front and back, cannot therefore be further traced back to anything else or "explained." On the other hand, the little-known *physiological* basis of these relationships of sensation (mentioned in C II of this chapter in connection with the local sign theory) is causally determined relationships; and they are capable of explanation. This point is of vital importance for an identistic and panpsychistic interpretation, which we have already referred to as offering a possible explanation for many philosophical problems (Ch. 6 K VII).

All sensations also have *temporal properties*. They are always related to the present. Here too we must inquire whether these are purely subjective properties, or whether they either correspond directly to the extramental world or very closely resemble its temporal qualities which, following Ziehen, we may term "temporitative" properties.

Locke, as already mentioned, attributed temporal properties to the things themselves. To Kant they were subjective, but are grasped a priori. His "thing in itself" is timeless in the same way as Plato's timeless ideas which embody "being." Yet temporal sequence is among the most essential characteristics of all "given" data. Kant asserted that it is impossible to dissociate temporal properties, but when we generalize we are in particular abstracting the temporal properties of our individual sensations: general concepts such as "circle," "red," "dignity," are thought of in a timeless manner.

In certain respects temporal properties correspond to spatial ones. They are homogeneous; they form an unbroken sequence; and they can be imagined as infinitely prolonged, although in a corresponding manner to spatial infinity we are unable really to grasp eternity. But it is an important difference that temporal properties show a mono-dimensional order, that they follow a continuous and coercive course without any change of direction. And the zero point is always a constantly shifting point of time, the present. As all our phenomena are experienced in the immediate present, one might be tempted to think of the temporal quality as a superfluous expression of the nature of the phenomenon itself. But this would not do justice to the facts. Most mental images have a characteristic "intentional" property, a relation to the experienced past; in other words, we recognize something as already familiar to us. And we also have mental images with a tone of futurity, when our ideas are linked with something about to be or expected to occur.

The fact that in our waking state the temporal properties of all perceptions are in accord, and the phenomena succeed one another in an unbroken succession, leads us to assume the existence of absolute time, within which all our phenomena occur. But it has long been known and proved by experiments that our experience of time may vary. When we are excited or the body temperature rises, or when we are concentrating hard and so only take note of the present, time seems to pass more quickly (see, for instance, recent experiments by W. Josenhans 1959, and J. Cohen 1964). This last condition is also a feature of persons who have undergone a leucotomy operation, when some of the nerve paths from the diencephalon to the forebrain have been destroyed. Such patients underestimate "objective," i.e., physically measured, intervals of time (P. Glees 1957, p. 410). Differences of this sort may therefore be governed largely by a decrease in imagery of past and future. Hence there is a certain relativity in the temporal properties of our phenomena, especially in mental imagery, and this means that we cannot completely identify the temporal properties of perceptions with the "temporitative" qualities of the "ultimate something" or the "matter" which we infer from them. We must consider, however, that these phenomenal differences are related to our *assessment* of time experience. The relevant physiological processes in the brain can be partly explained by alterations in their speed and intensity and the interconnection of the phenomena involved. And this may also explain the variations in our assessment of time experience.

During phylogeny, the experience of time has probably developed through successive adaptation to the extramental world. If we attribute sensations to all animals endowed with sense organs and nervous systems, we may suppose them capable of some awareness of the present. As already mentioned however, this will only appear as a special property of the perceptions when successive sensations are associated and memories (i.e., primary mental images) are built up, whose quality of familiarity causes a perception of something past. This may possibly occur in the case of some invertebrates like cuttlefish and higher insects which can be assumed to have at least a temporary connection of consciousness of perceptions and mental images. The lowest living organisms such as sponges and unicellular organisms that might be capable of unassociated sensations would presumably experience these phenomena in a timeless manner. This may also be true of some higher organisms with few senses, such as the ticks, which only react to a butyric acid or heat stimulus emanating from a potential host.

The phylogenetic development of temporal experience will no doubt have taken place somewhat as Condillac's famous statue was successively endowed with sensations (1788, pp. 66ff.). As long as it experiences nothing but the scent of roses, it *is* the scent of roses. Only when this is succeeded by the scent of violets is a temporal quality added: "La notion de la durée est donc toute relative: chacun n'en juge que par la succession de ses idées . . ." (p. 77) ("It follows that the idea of duration is completely relative; we can only judge it by the succession of our ideas . . .") In a corresponding manner the statue experiences spatial qualities only by "la coexistence de ses idées" (p. 195) ("the co-existence of its ideas").

We may assume that future tone in mental images only occurs in some birds and mammals which are capable of acting with "foresight."

When we now try to answer the question how we arrive at any idea of what matter is like, we may summarize the results of our epistemological findings as follows. Experience in early youth leads us to connect perceptions with ideas of objects, and these ideas become more firmly established by our use of words to denote such "things." These associated ideas of objects greatly influence our interpretation of matter; but we must ignore them in discussing the properties of phenomena which can give us information about extramental "matter." Among these properties there are some which quite clearly are not identical with those of matter: qualities of color, smell, hardness, etc. On the other hand, the degree of intensity of our perceptions points to a similar degree of intensity in the "material" effect upon our sense organs, although they do not exactly correspond. The positive or negative feeling tones that accompany perceptions have their origin in ourselves and cannot be a property of matter. But the fact that the spatial and temporal qualities are homogeneous in all perceptions suggests that there are very similar although perhaps not totally identical properties in matter.

We see that in order to arrive at an idea of what matter is, we are proceeding by a *process of reduction*. The "reducts" which are left are regarded by most philosophers as fundamentally different from any "psychical" elements. However, our reduction has not been applied to consciousness itself, but only to certain of its properties (see Ziehen 1934, 1939). So *these "reducts" are in fact protopsychical spatio-temporal complexes*, devoid of qualities. This extramental "something" is called "substance" by many philosophers. But as this term suggests the idea of an "object," it may be preferable to speak instead of an "ultimate something." Its characteristics can only be determined by physical analyses, which we must now briefly consider. We shall see that the present-day physical ideas of matter, too, no longer lead us to the earlier notion of it as "solid things" but only as lawfully determined relationships of energy complexes.

III. Physical Characteristics of Matter. It is universally accepted that the world, and therefore "matter" of some kind, existed before there were any humans, or indeed any living organisms, to be aware of it. If we accept an identistic and panpsychistic interpretation, then this matter is protopsychic in character. The matter of which living organisms are constituted has gradually adapted itself by selection in the course of phylogeny to the peculiarities and laws of the world, in particular with regard to the structural and functional development of the sense organs and the brain. These organs have enabled the most highly developed of these organisms, man, to grasp and understand more fully the logical and causal connections which underlie events in the world. This has become easier in proportion as man has produced what might be called additional "sense organs" which have gained him an insight into dimensions and processes which at first he could not perceive. He has constructed microscopes, electron microscopes, telescopes, various devices for detecting x-rays, ultraviolet and infrared rays, magnetic and electric fields, elementary particles, etc. But all the results gained by these devices cannot reveal the real nature of matter; they can only give certain indications, because we get these data by means of our sense organs and our brain which ponder over the data. However, intensive physical research has already led to certain relatively well-founded views on the nature of "matter."

Physical research is not normally affected by epistemological considerations. It is based on the supposition that it is concerned with something *objective* (though in the sphere of microphysics, the limitation of cognition by our sense organs had to be considered). However, this objective approach, at first quite justifiable, has often led physicists to think of the components of matter as "things," as something "solid." We shall see that this does not appear to do justice to the facts.

As the analysis of matter proceeded, it became increasingly clear that *the qualitative diversity of nature is largely conditioned by quantitative differences.* This means that the qualities are determined in the main by *system properties*, which increase in multiplicity as the material units gain in structural complexity (Ch. 6 K III). It is evident that new relationships come into being, and new properties are produced, whenever two or more different structures combine to form a new and more complex unit. To take an example: Under normal atmospheric conditions sodium is an explosive light metal, and chlorine a greenish gas which is lethal to man. But if they become united as molecule NaCl, a harmless substance (cooking salt) with entirely new properties originates. Tens of thousands of molecules are now known to chemistry as stable units of this kind. But they are all composed of a relatively small number (104) of elements or atoms. The number of *molecules* with extremely different properties but composed of carbon, hydrogen, and oxygen alone is immense, for carbon atoms can combine with one another to an almost unlimited extent.

Until the close of the nineteenth century *atoms*, as their name implies, were taken to be ultimately indivisible units; but we now know them to be structures composed of relatively stable subunits. These *elementary particles* are in turn made up of different numbers of a few, normally three, relatively stable types, combined however in different quantities: protons and neutrons (nucleons) which form the atomic nucleus, and the electrons which surround it. In addition to this, material waves exist, light rays for example, which under certain analytic conditions may be regarded as "corpuscular," composed of photons. Following E. Schrödinger (1954) one can imagine these particles perhaps better as more or less transitory structures of the wave field.

The transformation and splitting of atoms and the analysis of cosmic rays have led in recent times to the discovery of many more elementary particles. Most of these, however, are highly unstable, with a lifetime of no more than a hundred-thousandth to a ten-millionth of a second. More than thirty different elementary particles have now been identified, and if we include all activated states of these particles, the number exceeds a hundred. They can be classified in an ascending sequence of mass as photons, leptons (neutrinos, electrons, μ-mesons), π-, τ- and κ-mesons, and baryons (protons, neutrons, and the superheavy hyperons). At the light end of the scale, one might also include the quanta of the gravitational field, the so-called gravitons, though up to now these are still hypothetical.

Particles all differ in regard to "mass," charge, spin, speed, and magnetic moment, which, however, depends upon the spin. All have corresponding *antiparticles* which are opposite in charge (positrons, for instance, have the same properties as electrons but have a positive charge), are highly unstable, and unite with homologous elementary particles or other components of matter by compensation of their charge. The most stable particles are photons, neutrinos, electrons, and protons.

One very important point must be borne in mind as we examine the nature of matter, i.e., the elementary particles which form its basic components: the purely physically determined characteristics of matter must not be thought of as properties of some kind of minute "things." They are no more than relationships of an unknown "ultimate something"; or expressed still more exactly, *they only constitute a system of relationships*. This is why they can be especially well expressed in mathematical formulas. So we must not assume the existence of a "material substance" which has certain properties. It is only a practical manner of thinking, evolved phylogenetically and suited to the perceptive capacities of higher organisms, to refer properties to some "carrier." Schopenhauer already recognized this, and wrote about matter in his *Welt als Wille und Vorstellung*: "denn diese ist durch und durch nichts als Kausalität. . . . Ihr Sein nämlich ist ihr Wirken: kein anderes Sein derselben ist auch nur zu denken möglich. Nur als wirkend erfüllt sie den Raum, füllt sie die Zeit. . . . ihr ganzes Sein und Wesen besteht also nur in der gesetzmässigen Veränderung, die *ein* Teil derselben im andern hervorbringt, ist folglich gänzlich relativ . . ." (I, i, 4) ("Matter is nothing more than causality. . . . Its true being is its action: nor can we possibly conceive it as having any other meaning. Only as active does it fill space and time. . . . the whole being and essence of matter consists in the lawful change which *one* part of it brings about in another part; the existence of matter is therefore entirely relative . . .")

Thanks to the work of Einstein, physicists have come to realize that "mass," which had been regarded as the "material" basis of matter, is equivalent to energy; that is to say, it can be converted into energy $(E = m \cdot c^2)$. Rest mass thus represents potential energy; it is the "inertia of energy" and can therefore be expressed in units of energy. This means that elementary particles are largely interchangeable, and can be partly converted into radiation, just as radiation can become matter. So elementary particles can be produced from kinetic energy. Energy provides us with what is in fact "the valid proof of the uniformity of all matter" (W. Heisenberg 1959, p. 131).

Another reason for regarding properties of elementary particles only as relations is that photons and neutrinos are without "mass." Photons are the quanta of the electromagnetic field, while a neutrino is "pure spin," "a nothing which rotates."

Finally, *all matter has spatial and temporal properties*. However, as

Leibniz already realized, these can only be grasped as relationships; that is to say, with regard to some other "material" elements. A hydrogen atom, for example, occupies a certain position in relation to other atoms. In the same way a light ray has a certain position in relation to a body or another light ray, and it has also a certain direction. These statements lose nothing by the fact that we cannot precisely determine the position of electrons, for spatial relations to other particles are certainly present.

Heisenberg's "uncertainty principle" only means that we cannot determine both the position and the momentum of a particle simultaneously. The wave-corpuscle-dualism is characterized by a similar difficulty: when we establish the corpuscular properties of light, the wave properties disappear, and vice versa. The dualistic description only means that one has not yet found any adequate "image" for it (see C. F. von Weizsäcker 1949). But waves, like corpuscles, have spatial properties. Presumably we must imagine "mass" in some way as a continuum like a "field" (see the discussion in Ziehen 1939, pp. 250ff.). (My own opinion is that we ought in principle to be able to find an "image" for anything which has spatial properties.) We can further establish that all "matter" has temporal properties; for it is a system of relations which either exists in the present or existed in the past, and movement and spin also imply a temporal process.

We find, then, that matter possesses very few basic properties: energy (rest mass in the form of potential energy), charge, spin, speed, and spatial and temporal properties. This suggests that all these properties could be grasped and expressed in a unitary formula. Einstein did this to a great extent in his brilliant *special and general theory of relativity* (see Einstein 1916 and especially 1920; M. Schlick 1920; G. Eder 1960; G. Süssmann 1965; and others). With reference to a biophilosophical conception it is especially significant that according to these theories spatial and temporal properties depend on the state of movement of the system in question; in other words, absolute spatial and temporal measurements are only possible within a strictly delineated stable system. Spatial properties also are indissolubly linked with energy, because mass (energy) increases with increasing speed. "Wirklich ist nur die Vereinigung, die Einheit von Raum, Zeit and Dingen, jedes für sich ist eine Abstraktion." ("Only the combination, the unity of space, time and things is real; taken singly each is an abstraction.") (M. Schlick 1920, p. 37) According to the theory of relativity, then, if "things" vanished, time and space would do so as well.

A universally accepted and definitive *world formula*, which would explain why different elementary particles have a certain mass and charge, does not yet exist. But Heisenberg's complex formulas combining the relativity and quantum theories can at least be taken as an approach to such a formula. They incorporate the coordinates of space and time (γ_1, γ_2, γ_3, γ_4) in the form of certain matrices, as well as the wave function ψ (for calculating the probability of a certain state) which depends on these four coordinates, the universal constants; l (elementary length = diameter of an electron), c (velocity of light), and h (Planck's constant).

It is of great importance that all "matter" is now regarded as having the same basic uniform structure, that subunits are presumed to be present even in the protons of the atomic nucleus (nuclear quanta, Quark), and that all "forces" are seen as differing expressions of the same. Elementary particles can be the "source" of fields extending into space, and fields can be the source of particles. A static electromagnetic field, for instance, corresponds to virtual photons.

IV. General Picture of "Matter". No doubt the most important outcome of our brief discussion of matter is the realization that *epistemological and purely physical analysis yield corresponding results on certain essential points.* Both fields of research can give us nothing more than certain indications about the nature of matter. The earlier idea that it was a substance composed of solid atoms must be entirely discarded. Of the formerly accepted dualism between "force" and "matter" only "force" remains, in the sense of "fields of energy." The "solid matter," the "mass," is found to be equivalent to "energy." Space and time too are now conceived in quite a different way. Epistemological analysis showed that we cannot simply accept an absolute time and absolute space, for primarily there are only spatial and temporal *properties* of sensations and ideas. These properties are very likely related to an "ultimate something" in a similar fashion (as "locative" and "temporatitive" properties) or possibly in the same fashion. This "ultimate something" (in Ziehen's sense) must not, however, be thought of as a "thing" having a certain solidity. We grasp its properties only as relations. The constantly shifting point of present time already makes us aware of this relationship. Modern physics has reached very similar conclusions. The concepts of energy and momentum, too, are ultimately concepts of relations. And spatial and temporal properties have no existence per se; they are inextricably bound up with energy (mass). The former idea of "solid matter" has been largely dispelled.

But the epistemological and the physical conceptions differ in their interpretation of space and time. Our phenomena tell us that space is Euclidean and not "curved." As we must take it that selection has been operating throughout our phylogeny to adapt our perceptions to the extramental world, we conclude that the locative relationships in that world correspond to the Euclidean system. This train of thought underlies one of Ziehen's main arguments, in which he opposes Einstein's theory of relativity (see 1939, p. 300). In my opinion, however, the phylogenetic adaptation of our perceptions and the functions of our brain must have been limited to those spheres where it had some biological significance; in other words, it must have related to what we can apprehend with the naked eye or with our other sense organs. And in this realm the findings of physics are in line with the idea of Euclidean space. At the same time, it is quite possible that entirely different circumstances may prevail in other spheres only accessible with the aid of astronomical instruments, or even in wider ones still. In other words, the universe as a whole, instead of being Euclidean, may be quasi-spherical in Einstein's sense, and then limitless but not endless. Indeed, fluctuations in the movements of Mercury at the perihelion suggest that the general theory of relativity does give an apposite picture of space in the universe. (It must be noted, however, that R. Dicke's calculations of the degree to which the sun is flattened make the theory somewhat less convincing.) On the other hand it is true that astronomical observations up to the unimaginable distances of 100,000 or possibly even 3.5 milliard light-years are also in accordance with a Euclidean conception of endless space (K. Walter 1964). Hence, the problem of space cannot yet be regarded as completely solved by physics.

Another point to be considered is that when relationships are expressed as formulas, the use of symbols and figures may mask the differences in the properties of the extramental world for which they stand. Ziehen has rightly remarked that in physics also false theories have been expressed in exact mathematical formulas. As temporal and spatial relations are homogeneous, it is possible to use the time coordinate as a fourth dimension. However, we must bear in mind that the epistemological analysis shows that time occupies a special position, because the zero point, the present, is a continuously shifting point, and also because the course is isotropic. The theory of relativity apparently takes this difference into account; time, unlike space, has an isotropic steady course, and therefore corresponds to an inertial system. In consequence of this peculiarity temporal dimensions in systems at high velocity expand, while spatial ones contract.

These physical theories have an important meaning for biophilosophical problems. For example: the time expansion just referred to and implicit in the theory of relativity has led to the much-debated assumption that living organisms in spaceships moving at high speeds would age more slowly. We have to wait to see if this consequence can be proved by experiments, perhaps with short-lived lower animals or plants.

The physical statements on the nature of matter are also of significance for our assessment of biological processes, as single microphysical events may have an effect upon macrophysical physiological processes. It could be calculated that even one light-quantum in a human visual cell can release an impulse, and that 10-14 quanta are enough to produce a sensation (see H. Autrum 1948; W. Trendelenburg et al. 1961). One energy quantum can also produce a gene mutation, and this can lead to far-reaching morphological and physiological alterations (the "Treffertheorie"). That does not, however, imply that *acausal* microphysical events intervene in the life process, as P. Jordan (1945) has claimed. We shall return to this problem when discussing the laws in K III of this chapter.

One might well have hoped that microphysical findings would help to explain the particular position occupied by the psychophysical substratum; but no relevant indications have as yet been found.

Another factor of epistemological importance is that we now think quite differently about the "solidity" of the matter from which the living organism is made up. "Solid substance" has now given place to fields of energy, and these can only be comprehended in terms of relationships. This conception makes it easier to incorporate the findings of physics in a panpsychistic, identistic picture (see also similar suggestions in

A. Eddington 1939 and J. Jeans 1943). By linking energy (mass), space, and time indissolubly together, Einstein's view represents a certain approach to Kant's idealistic theory of cognition. But Kant went too far in supposing that space and time would not exist in the absence of a percipient organism. On the contrary, we can well assume that such spatial and temporal relationships (at least in the sense of locative and temporatitive relations) have been in existence as long as matter has. But epistemology and physics agree that neither time nor space exists without matter.

Finally, the physical analysis of the basic properties of matter has clearly illustrated that almost all qualitative differences in molecules and atoms can be traced back to *quantitative* differences of equal components. The elementary particles alone have certain irreducible basic properties: mass, charge, spin, velocity, and spatial and temporal properties. So all the more complex structures which they make up—atoms, molecules, liquids, rocks, living organisms, and their decomposition products— owe their diversity to *systemic properties*.

The establishment of several irreducible basic properties and universal laws, treated in more detail in the next chapter, is also of great importance for an epistemological conception. If we base our considerations on an identistic, panpsychistic interpretation (suggested by various facts; see Ch. 6 L), then these physical basic properties must also correspond to basic protopsychical characteristics. Living organisms can become conscious of these protopsychical and psychical properties as soon as they have reached the stage of "material" integration represented by a central nervous system. New psychic properties could come about in the same measure as new "material" properties appeared, owing to new system relations in higher levels of integration. In other words, sensory qualities appear, by integration of protopsychical qualities, in a manner we have sketched when discussing sensation (p. 190). Because basic irreducible properties and lawful relations are also present on the "material" side, the much-debated contrast between the multifariousness of phenomenal qualities and the mainly quantitative differences of matter disappears.

According to an identistic conception, the difference between the "psychical" and the "material" can be interpreted in the following way. We experience phenomena as qualitatively diverse and immediately given data. How the phenomena have been developed phylogenetically, we can only assess to some extent by conclusions reached through analogy, when we analyze the structure and behavior of animals. However, the causal components which can be established by reduction from our psychic phenomena, what Ziehen called the "reducts," and which we generally describe as "material" (as "things"), are capable of physical analysis at all stages of integration. This is how we arrive at a knowledge of the basic physical properties and universal laws which we have mentioned.

It is therefore possible to state that in the evolution of matter, which I consider to be also the evolution of consciousness from protopsychical elements to self-consciousness, we can distinguish the following stages: elementary particles of protopsychical nature, i.e., simplest fields of energy (neutrinos, photons, Quark, possibly gravitons) (= 5th state of aggregation) ⟶ plasma consisting of elementary particles occupying the greater part of the universe (= 4th state of aggregation) ⟶ atomic gaseous celestial bodies arising by gravitational forces, especially hydrogen nebula (= 3rd state of aggregation) ⟶ emergence of further species of atoms (according to F. Hoyle with increase in temperature in the sequence: He—C, O, Ne—Ng, Si, S, Ar, Co—Fe, Cr, Nr, Co, Mn, V, etc.), by their mutual reactions, origin of molecules and also liquid and solid bodies (= 2nd and 1st states of aggregation) ⟶ prestages of living organisms capable of replication ⟶ lowest genuine organisms, integration of protopsychical properties to sensations ⟶ higher organisms with differentiation of a psychophysical substratum, sensation, memory, stream of consciousness.

This kind of epigenetic evolution, in which each step forms the basis for the next, may have alternated with phases of involution, for it is quite possible that the universe has never had a "beginning" (see K. Krejci-Graf 1960; J. L. Hughes 1963; and others). In any case, the idea of a primeval explosion, to which many astronomers cling, can be doubted. The date, given as 10 milliard years (assigned to the quasi-stellar radio source 3 C g) can no longer be sufficient, for some stars in the Milky Way are probably twenty milliard years old (K. Walter 1964).

We humans are incapable of really imagining either eternity or infinity; for all we experience is limited in space and time. We can only think of spatial and temporal dimensions as endlessly extended; but we cannot follow this thought through to the "end." Besides, when we speak of the world "beginning," the nonexistence of matter, space, and time is absolutely unimaginable. And the hypothesis that the world came into being by expansion is founded mainly on the shifting of red by the Doppler Effect. It remains to be seen whether some other explanation of the shifting of red (possibly increasing loss of energy) may not prove more likely. Hoyle's "steady state" theory of the universe seems more satisfactory to one not versed in astronomy.

K. The Laws of the World

1. The Nature of the Laws. The early Greek philosophers, observing that events often follow an even course, already assumed that universal laws exist (see H. Diels 1906; W. Kranz 1949). Heraclitus (536-475 B.C.) spoke of a "divine law" ($\theta\epsilon\tilde{\iota}os$ $\nu\acute{o}\mu os$) which directs all events. Democritus (470-380 B.C.) clearly recognized the existence of the causal law. In a fragment (possibly attributable to Leucippus) he states: "Nothing happens without a cause; everything is the result of reason (ξ $\lambda\acute{o}\gamma o\nu$) and necessity."

Only much later did philosophers such as Giordano Bruno, Hobbes, Descartes, Spinoza, Leibniz, and others begin to investigate the problem of natural laws in more detail. Hume realized that we can grasp the principle of causality only through experience, and he also stressed the lawfulness of psychic processes (*A Treatise of Human Nature*, 1739). Kant, on the other hand, regarded causality as a purely human form and prerequisite of thought, a peculiarity of thinking a priori not derived from experience ("reine Erkenntnis a priori"). Causal relations, he stated, are arrived at through concepts which form the objective basis underlying the possibility of experience ("Begriffe, die den objektiven Grund der Möglichkeit der Erfahrung abgeben"). This transcendental deduction, expressed in different versions by Schopenhauer and many later philosophers met with many objections, however. Following Hume, it can be proved that our knowledge of the causal law is derived solely from experience. Kant's purely idealistic view, formulated before any reference to phylogenetic development was possible, could not admit such knowledge. The a priori quality of which he speaks does, however, indicate that he was in some way aware of an "objective," transsubjective existence of laws, as a precondition of any possible experience. In this context it is well worth reading Ziehen's fascinating discussion with Kant in the philosophers' Heaven (1913, pp. 213-225).

As all the laws of nature rest upon an inductive, empirical basis, they should, strictly speaking, be expressed as theories, although theories of the highest probability (Ch. 2 B). We have also pointed out in sections F II and H III of this chapter that all psychic phenomena have evolved corresponding to the phylogenetic emergence of sense organs and brains in a gradual process of adaptation to the extramental world and its laws. As a final stage, the most highly developed of all living organisms, *Homo sapiens*, has adapted his increasingly complex thought to the causal and logical laws of the universe in such a manner that he has learned to act in accordance with them and to comprehend them (in the sense of transgredient cognition). He has also gradually succeeded in reducing the multiplicity of his phenomena and thus the multiplicity of the extramental processes underlying them, to a relatively few basic principles. In this way he has been able to grasp the fact that many particular laws and rules are ultimately the effect of a universal causal principle, and that many logical and mathematical laws are dependent upon a limited number of basic logical principles. However, he cannot "explain" these universal laws; that is, he cannot trace them further back to other principles. He can only state that they constitute transsubjective "given" facts (Ch. 2 B).

Leaving out of account the historical, ethical, religious, or juridistic laws determined by man's social life, the natural laws appear to fall into three classes: causal laws; laws which coordinate the different phenomena with certain physiological ("material") processes (Ziehen's "parallel laws"); and logical laws. We shall have to see if it is really necessary to recognize these laws of coordination.

As we have noted previously, the analysis of our phenomena shows that what exists is not a solid substance with its properties but ultimately only a system of relationships. Within the framework of immanent and transgredient knowledge, we can define laws as relationships of relations remaining constant in the course of time (see Ziehen 1927a), which we suppose to be independent of living organisms which may apprehend them, as they are "objective units of validity" ("objektive Geltungseinheiten" Husserl). In my opinion we must also regard the established world constants (h, e, c), the laws of conservation, and the principles of symmetry as special laws (in the widest sense), because they cannot be traced back to the causal or the logical laws. The laws of probability should rank as part of the logical laws (Ch. 6 K VI).

As all these basic laws govern life processes as well, we shall now discuss them briefly. Biological laws, however, belong mainly to the sphere of the causal laws. We mentioned this in Chapter 2 B and shall be discussing it later at greater length. These laws have a special particularity insofar as they are systemic laws, sometimes of great complexity. Some of them also have to be considered in connection with the problematic coordination or parallel laws. *Homo sapiens*, the most highly developed living being, is usually also credited with a faculty which allegedly contradicts the laws of nature which govern all his other functions, including the physiological processes in his brain: the so-called freedom of the will. As we have seen in section G of this chapter, it is doubtful that such a freedom does in fact exist.

II. Universal Constants and the Principles of Symmetry. At the present time, microphysics is going through a somewhat stormy development, and a number of special interpretations and theories cannot yet be regarded as definite knowledge. So it may be sufficient to refer briefly to some findings which are important for the understanding of the world's diversity and therefore also for problems of biophilosophy. In any case, a more detailed treatment of the microphysical problem would not be appropriate for a biologist or philosopher.

The theory of relativity and the quantum theory have revealed some properties of microphysical processes which had hitherto proved to be irreducible principles independent of the causal law. Like all physical laws, they hold good within a rectilinearly moved, uniform system of relationships.

Two laws in particular which appear to set certain limits to causal processes may be considered as well established.

1. The velocity of light ($c = 2.99793 \cdot 10^{10}$ cm/sec, in vacuum) represents the limit for the motion of energy, and so the maximum velocity of any signal. No effect can be propagated faster than light.

It was deduced from this that when two light rays from a fixed point travel in opposite directions, neither of their "apexes" will be removed from the other at a speed faster than that of light, although one would have expected the total velocity to have been twice that figure (see C. F. von Weizsäcker 1949, p. 244). Einstein tried to resolve this contradiction by a criticism of the concept of simultaneity. But the mathematical correctness of this solution cannot dispel epistemological doubt. Certainly, there is no means of establishing a greater absolute speed than c, because all the information can only be conveyed at a speed whose maximum equals the velocity of light. Yet it seems to me quite possible that the two apexes may increase their distance from one another at a speed greater than c. In my opinion maximal velocity for the movement of energy and increase in the distance between two points are two quite different concepts. The latter is based on a logical conclusion.

2. There can be no process in which the energy involved is less than the quantum h ($h = 6.7 \cdot 10^{-27}$ erg sec). The constant l (= elementary length) is also dependent upon this.

3. Electromagnetic phenomena are based upon a constant elementary electric charge ($e = 4.8 \cdot 10^{-10}$ electrostatic units). The unit of spin, the angular momentum, of the elementary particles is bound up with the second and third of these rules.

4. A particular gravitational constant (G) is the basis of gravitational phenomena. (According to P. A. M. Dirac, it is possible that this constant is steadily decreasing in inverse proportion to the age of the world; see P. Jordan 1937, 1938, 1947.)

The *principles of symmetry*, usually expressed as the PCT-theorem, which have a significance for a universal "world formula," are also independent of the causal law. P-invariance assumes parity between left and right, and implies that a world must be physically possible where right and left are interchangeable. C-invariance denotes that besides all elementary particles the existence of antiparticles with opposite charge must be possible. It follows that a world must be physically possible where all electric charges are reversed, and all particles are replaced by their antiparticles. These antiparticles have already been proved to exist. They only last a fraction of a second, however, because they are destroyed by a particle of opposite charge. T-invariance supposes a corresponding symmetry in the course of time; that is to say, it must be possible for the course to be reversed.

All known elementary processes appear to be strictly subject to these three types of invariance. Lately, however, exceptions have been noted. It is probable that no total symmetry exists, and this fact may be one of the reasons why the universe is liable to permanent alterations.

From an epistemological point of view it only remains to be noted that although reversal in time may be theoretically imaginable, absolute and positive isotropy, a continuous and inevitable progress toward the future, is precisely a very characteristic feature of phenomenal time, and very likely of the extramental world also, insofar as it exists as a comprehensible universe.

Other universal laws which have a causal effect, although they cannot be directly inferred from the causal principle, are expressed by the *principles of conservation*. They are also of vital importance in biological processes. They state that in all processes certain total quantities remain unaltered, that they are independent of time. This implies the conservation of mass or energy, momentum, angular momentum, and movement of the center of gravity.

In regard to biological processes, the most important principle is that of the conservation of energy. Expressed in a more general manner it means that the total energy involved in a closed physical or chemical system, i.e., the sum of all the various types of energy, remains unaltered. A more exact formulation of this principle would be: "At all intermediate stages, the sum of the kinetic energy and the potential energy is constant, and is equal to the potential energy at the beginning." (F. Fraunberger 1960, p. 228) The conservation principle also applies to nuclear reactions and is consistent with the general theory of relativity. Nor does it conflict with G. Lemaître's hypothesis of world expansion (1946), as this supposes all the energy in the universe to have been originally present in highly concentrated form. The principle of conservation of energy does, however, conflict with P. Jordan's hypothesis that expansion began with a pair of neutrons, and that new matter is constantly being produced. According to Jordan, this idea is still consistent with the principle of the conservation of energy, for he inserted the necessary potential gravitational energy with a negative sign. We must, however, bear in mind that these are rather bold mathematical suppositions which will have to be verified. Besides, an objection to any theory of expansion is that the main basis of these hypotheses, the shift toward the red, admits other explanations as well.

III. The Causal Principle. While discussing evolution and in particular psychogenesis, we noted how the living organism inevitably adapted all its reactions, including its cerebral processes, to the conditions of its environment which are determined by causal laws. In the course of phylogenetic development, the higher organism became able also to grasp the principles of identity, difference, and similarity; and its comparative, analytic, and synthetic functions became progressively more integrated. Finally, *Homo sapiens* could also employ his logical powers, which were adapted to the universal logical laws of the extramental world, to arrive by transgredient apprehension at an *understanding* of the causal law. In this way our experience and thought to some extent mirror the "factual" relationships of that extramental world, insofar as we can apprehend them either directly or indirectly.

The assessment of causality is particularly important for biophilosophical considerations, for the most important part of biological research is research into causes, and the presupposition of general causality as a heuristic principle has always proved to be correct. Sometimes doubt has been cast on this latter claim. Attempts have been made to interpret those life processes which could not be analyzed, by reference to vital forces not open to causal research and regarded as final principles. More recently also, acausal factors allegedly detected in the sphere of microphysics, have been assumed to produce some degree of "freedom" in processes of development and in cerebral processes and thought processes. Finally, it was asserted that acausal "psychical" factors influence "material" life processes. So we shall have to discuss whether these ideas are justified in the present state of biological research.

First of all we must clarify the definition of causality; but we shall limit ourselves to essential points, without attempting to discuss the vast literature on the subject at any length.

We have already noted that Kant's theory of the a priori and necessary nature of causal thinking is unacceptable. But he is correct in stating that all changes obey the law of cause and effect, and that the principle of sufficient reason is the basis of possible experience. Yet we must not interpret the principle of sufficient reason, as Leibniz and Kant did, as "self-evident." With Hume and Mill we must assume that knowledge of the causal law is acquired through experience. But when inferring a propter hoc from post hoc, and anticipating a causally determined effect, logical thinking is also involved. On the other hand, we can never derive a causal law from logical laws alone.

Ziehen (1939, p. 304) has given us what is perhaps the most concise modern exposition of the causal principle: "1. Alle physikalischen Geschehnisse sind inhaltlich, örtlich und zeitlich nach Kausalgesetzen eindeutig bestimmt. 2. Inhaltlich können sie letztlich als Bewegungen (im weitesten Sinn) aufgefasst werden." ("1. The content, place and time of all physical events are clearly determined by causal laws. 2. In regard to content, they can ultimately be conceived as motion (in its widest sense)." In this manner the necessity of temporal sequence, the unbroken continuity and invariability of the law, are sufficiently characterized. As all physiological processes are ultimately physical or microphysical in character and therefore causally determined—in opposition to some contrary interpretations noted below—the definition thus applies to all "material" events. It is therefore also possible to speak of a causal structure of all "matter" (matter here in the sense of Ziehen's "reducts").

It is well known, however, that many atomic physicists doubt that the causal law is invariably valid. These reservations began with the quantum theory. Because it is impossible to predict the behavior of individual elementary particles from the distribution of intensity of diffracted rays, the mathematical formulation of the laws of quantum mechanics had to introduce a probability function. Along the lines of quantum mechanics W. Heisenberg deduced the uncertainty principle (his Principle of Indeterminacy), which states that the more accurately we measure the impulse of a particle, the more uncertain does the determination of its position become. This principle of indeterminacy places an absolute limit on physical determination, and thereby on causal determination and the possibility of prediction.

Heisenberg, Born, P. Jordan, and other atomic physicists believe that *microphysical acausality* has to be assumed. But the matter is still open to question. The indeterminacy is not necessarily attributable to the microphysical processes themselves. "Die Beschränktheit unserer Erkenntnismittel zwingt uns das Surrogat der Statistik auf, und man legt dies fälschlich den Redukten zur Last." ("The limits of our means for acquiring knowledge impose statistics upon us as a substitute; and for this we mistakenly blame the reducts.") (Ziehen 1939, p. 305; see also M. Hartmann 1937.) It is quite possible to assume that the microphysical processes, like all processes in the universe, are subject to causal and logical laws alike. In the realm of microphysics we cannot determine the causal relations, but we can grasp the logical relations, that is, the relationships of probability.

Strict and uninterrupted causality invariably holds good in the macrophysical sphere, and this represents an integration of many microphysical components. How then could this universal causality have evolved in the course of development from plasma to solid bodies unless it already applied to microphysical processes? It is only because the causal principle is invariably valid that prediction is possible in the macrophysical sphere. Prediction, however, is only a consequence of our knowledge of the causal law. It must not figure in a definition of it. So if predictability is limited in the microphysical sphere, that does not constitute a proof of acausality. Besides, the assumption of different kinds of causality, a "weaker type" as well as strict determinism (F. S. C. Nothrop, in W. Heisenberg 1959), does not seem to be logically justifiable. But we must consider that *two kinds of "determination"* exist: by the laws of causality and those of probability.

It is worth noting that many physicists and mathematicians by no means abandoned the view that the causal law is universally valid. Neither Planck nor Einstein did so. Planck declared in one of his lectures (1933, p. 247): "In dem Weltbilde der Quantenphysik herrscht der Determinismus evenso streng wie in dem der klassischen Physik, nur sind die benutzten Symbole andere, und es wird mit anderen Rechenvorschriften operiert." ("Determinism governs the world of quantum physics as strictly as that of classical physics; it is only the symbols and rules of calculation that have changed.") Later, he stated in rather more guarded terms: ". . . Dass die Durchführung einer streng kausalen Betrachtungsweise . . . auch vom Standpunkt der modernen Physik aus keineswegs ausgeschlossen ist, wenn sich auch ihre Notwendigkeit weder von vornherein noch hinterher beweisen lässt." (". . . a strictly causal approach . . . cannot be ruled out even from the standpoint of modern physics, though its necessity cannot be proved either before hand or retrospectively.") (p. 253) Einstein, writing to M. Born in 1944 (see 1955), declared: "Du glaubst an den würfelnden Gott und ich an die volle Gesetzlichkeit in einer Welt von etwas objektiv Seiendem." ("You believe in the dice-playing God, and I in perfect legality in a world of something objectively existing.") The physicist W. Weizel (1954) wrote in a similar manner: "Das Zustandekommen statistischer Gesetzmässigkeit, wo auch immer man sie findet, lässt sich mur auf einer an sich gesetzmässigen Grundlage finden. . . . Völlig falsch ist es aber, sogar die positive Behauptung aufzustellen, dass das Kausalitätsprinzip im atomaren Bereisch ungültig ist." ("Wherever we meet with them, statistical laws can only have evolved on a basis of lawful principles. . . . It is quite wrong, however, to assert that the principle of causality is invalid in the atomic sphere.")

Biological findings also have often been used as evidence that the causal principle is not universally valid in nature. Several authors have pointed in particular to the purposive development of individual structures and organs, to the capacity of many species to regenerate lost parts, and to the ingenious and efficient structure and function of living organisms. But we have seen that these processes are in principle explicable along causal lines (Ch. 4 F and 5 D). Mutation, gene combination, and natural selection have operated to produce "meaningful," i.e., advantageous, structures and functions; and their embryonic development was caused by the same evolutionary factors. Only after these ontogenetic processes have taken place, however, can they be considered as being purposive. In precisely the same way former causal sequences in the realms of physics or the evolution of the world can be regarded as final processes, when we look back on the course of development. When a gaseous sphere condenses to form a planet, and the earth becomes a place where life becomes possible, we may call this a purposive process. Retrospectively, every causal process can be called a final one.

Causal laws normally relate to processes, that is, to alterations in *time*. This is true of all kinds of laws such as the laws of falling bodies, refraction in a lense, and selection. It is only in some special cases when the causal relationships are held in equilibrium, i.e., when we have to do with static laws, that temporal properties are absent. This is the case, for instance, when the arms of a balance are still, or a body sinking downward reaches a depth where its specific gravity equals that of the water surrounding it, or the cell fluid in a unicellular organism is isotomic in relation to the surrounding water and no osmosis occurs in either direction. As long as causal relationships depend in this way on "conditions," we cannot speak of cause and effect. M. Verworn (1912) and his successors go too far, however, when they seek to replace causality altogether by *conditionalism*, which applies only in such borderline cases. It still remains true that the causal principle is normally expressed in a temporal context. Besides, in cases where this is not so, where we have only potential energy or "dispositions," the laws acquire a concrete sense only when we think of additional causes releasing a succession of states, in other words, the beginning of a movement (see also Ziehen's treatment of conditionalism, 1939, p. 308).

Since mass (energy), charge, spin, gravitation, temporal and spatial properties are all irreducible ultimate basic properties, the world is heterogeneous in character; and this in turn has led to the universal uninterrupted law of causality being split up into many *special causal laws*. As we have seen, these largely consist of *system laws* which developed epigenetically as the emergent astral bodies went through successive states of aggregation (Ch. 2 B). As long as "matter" consisted of a nebula of ionized gases, or plasma, the only relevant laws were those directly determined by the universal constants and the principles of symmetry (for example, the Maxwell distribution of velocity and the laws of fusion; the laws of aerodynamics, saturation pressure, etc., apply to normal [ideal] gases). Only when a fluid and stable state of aggregation is reached can the laws of falling bodies, communicating tubes, refraction in lenses, etc., manifest themselves. Yet all these laws existed implicitly from the beginning in the causal principle, before any fluid or stable state of aggregation had been attained.

It is worth inquiring whether the particular biological *law of selection* has not a much wider sphere of application than is usually believed. Natural selection is at work in competition between variants, species, and higher categories of plants and animals, and between human individuals, special groups, and peoples. But selection also operates when the gravitational forces of the largest of several masses prevail, or when elementary particles combine with antiparticles, or when atoms and molecules form chemical compounds only with certain others to which they have an affinity, or when in a mixed solution only equal atoms or molecules form a crystal. In the realm of living organisms selection also takes place when in the "stream of order" only what is lacking and appropriate is added—for example, specific amino acids in protein synthesis on the ribosomes—or when the necessary compounds for certain specific structures or organs are chosen in the course of metabolic processes. It seems possible, therefore, to speak of a *cosmic law of selection.*

IV. Biological Laws. Like many physical and chemical laws, biological laws are largely *system laws,* and they are based in part on physical and chemical processes. As living organisms are highly complex systems, these causally determined processes are subject to a great deal of superimposition, interrelationship, and mutual interference. Besides, living organisms are "open systems" of only relative constancy. It follows, then, that in most cases we cannot establish biological laws but only rules.

Biological laws and rules have manifested themselves successively with the evolution of our planet, in the same manner as the laws of evolution in Chapter 5 B. This also holds for the specific rules which are only valid in special phyla, classes, and orders of organisms. They were all potentially existent in the laws of the inanimate universe, and particularly in the causal laws. In view of this epigenetic type of manifestation, it is quite possible that living organisms may develop, or rather must have developed, in the same way on other celestial bodies where the same physical and chemical conditions have been operative in the course of their evolution. But any more considerable differences due to the size of the celestial body, the amount of available oxygen and water, the temperature range, the conditions of light, and the succession of seasons would have led to organisms of quite a different kind (see relevant discussion in B. Rensch 1954 or 1960a, and Ch. 4 E). Although about one hundred million galaxies exist, each with millions or milliards of suns, some with planets and large moons, the distance between us and the nearest astral body which could be inhabited by higher organisms or even beings which resemble humans is unfortunately so great that communication of any kind, perhaps by radio waves, is quite unlikely.

In contrast to almost all physical laws, biological laws and rules can only rarely be expressed as equations or by numerical formulas, because they are too complex and embrace too many exceptions. It can best be done with Mendel's first and second rules, and with certain physiological laws based on relatively simple physical or chemical processes, as for example the laws of osmosis, the mass effect in biochemical processes, the ratio of volume to velocity in the circulation of the blood, or the ratio between reaction speed and temperature. Equations expressing the growth ratios of animals, plants, and their organs and structures are nearly exact approximations (e.g., equations of growth, especially allometric growth). It is to be hoped, however, that the biological cybernetics now developing will give mathematical expression to a number of physiological laws, as for example B. Hassenstein (1964) has done in his study on the dependence of the degree of reaction on stimulus intensities.

As already mentioned, attempts have been made in the sphere of biology to establish *"final principles."* Aristotle had already taught that, in addition to effective causes, each thing embodied a principle directed to the realization of an aim (an *eidos* or "form"). He stated that the principle underlies the individual development of plant and animal species and causes them to act in a purposive manner; as, for example, spiders and ants do. This principle, by virtue of which the transition of possibility to reality takes place and which effects change (kinesis) and directs the development toward an aim or goal, Aristotle named entelechia.

This conception has persisted in various forms through the centuries. Galen (A.D. 129-199) assumed that "spiritus" (pneuma) control the supply of blood, the nourishing of the tissues, the constancy of body temperature, and the activity of the brain (see Th. Ballauf 1954). In the scholastic period, the basic principle of life was generally held to be psychical in character. G. E. Stahl (1694-1734), too, assumed the existence of an "anima" which enlivens the organic matter. Stimulated by observations on the considerable capacity of regeneration of polyps, J. F. Blumenbach (1752-1840) spoke of a "nisus formativus" (see K. E. Rothschuh 1953). The great physiologist Johannes Müller (1801-1852) was convinced of the existence of a "vital force" which cannot be traced back to chemical or physical factors. R. Virchow (1821-1902), the founder of cellular pathology, had a similar conception. Schopenhauer identified the vital and developmental forces and named this force the "will," thereby using the term in a greatly extended sense. E. von Hartmann traced the life principle back to the "unconscious."

H. Driesch (1905, 1909; see 1928) sought to give a new foundation to vitalism. He had discovered the remarkable regulative powers possessed by the germs of sea urchins when dissected or joined together, and he thought that no causal explanation for this or for the ontogenetic differentiation of structures and organs and for animal behavior was possible. He therefore assumed the existence of a specific immaterial and autonomous life principle to which he gave the Aristotelian term entelechy. This principle, he stated, exerts a general formative and regulatory influence upon the life processes and determines the wholeness of an individual by suspending and presuming certain reactions. On the other hand, entelechy does not affect the causal processes. But he did not explain how these "Werdebestimmer" ("determinants") were to exert an influence without having an "effect." All such influence would have run counter to the principle of the conservation of energy. (Driesch may perhaps have envisaged something in the nature of an occult phenomenon, for he is known to have believed in the possibility of such phenomena.)

We can omit a characterization of the various forms of neovitalism. J. Reinke (1905) wrote of "system forces" and "dominants," and R. Woltereck of active autogenous "propulsive" forces ("geschehenmachenden Mächten") said to control the individual's self-determination, "self-intendance" and self-preservation ("Selbstdeterminierung, Selbstintendierung, und Selbsterhaltung"). Another theory which it seems unnecessary to discuss in detail is "holism," propounded by J. St. Haldane (1931) and elaborated by A. Meyer-Abich (1948). According to this theory, nature exists in a state of perpetual creation as a subsisting and unfolding living entity. By a somewhat obscure deductive method of "holistic simplification" it is said to be possible to reveal the peculiarities of simpler chemical and physical levels of integration by eliminating the characteristics of the higher biological levels. Holism was significant only insofar as it stressed the meaning of the wholeness of organistic structures and functions.

The explanations of vitalists, finalists, and holists show that they merely constructed a "principle" for those causal life processes which had not yet been analyzed. But they failed to supply any positive statements: the terms they used, which only circumscribe the lack of adequate knowledge at a particular time, were employed for far-reaching speculation.

All these hypotheses about "forces" could be put forward as long as many of the phenomena of life could not be analyzed. Recent decades, however, have witnessed a causal explanation for many "marvels of life"; this development is outlined in Chapter 4 and 5. Extremely rapid advances have been made in almost every branch of biology, and we may expect these to lead to a causal explanation for other hitherto unsolved problems. At any rate, the idea that biological processes can be causally explained has always proved an effective heuristic principle.

At the same time, however, these vitalist and holist views and the more pertinent organismic theory have the merit of stressing that the structures and functions of the living organism represent a whole, and that it is necessary to study the systemic laws operative in the complex totality of such organisms. The problem of how "Gestalten," organismic structures as a whole have arisen has not yet been adequately solved, because correlations within a living organism are so intricately interwoven. So speculations about unknown "totalizing" or seemingly final factors continue to be made. The concept of Gestalt theory, that the whole is not simply the sum of its parts, has often been cited in this connection. As we have already noted, however, such simple summation has scarcely ever been claimed in this strict sense. It is self-evident that in molecules as in total organisms the integration of components is bound to produce new properties (see also Ziehen 1930).

In more recent discussions of the problem of finality it is fairly generally accepted that the life processes follow a causal course. But it is rightly pointed out that purposive ontogenetic development poses certain problems. Thus Ch. P. Raven (1960) wrote: "Every developmental process must be traced backwards to its causes; at the same time, however, it points forwards to an end. . . . While the single elementary processes are the same as those studied by physics and chemistry, and are connected by causal relationships, moreover they are so to speak embedded in an ordered pattern." The problem, therefore, is how living systems could evolve and survive, and how a "stream of order" could emerge. An attempt has been made to answer these questions in the chapters on the origin of life (5 E), on genetics and individual development (4 E and F), and on the problem of entropy (5 F): living beings could come into being after self-replicating DNA and peptides had been formed under prebiological conditions and a "stream of order" emerged. At the same time the organisms continuously disposed of superfluous entropy by giving off the heat generated by metabolism.

Evolution by mutation, gene combination, and natural selection could always operate on the various species of plants and animals in such a way that the alterations of each stage of the individual cycle remained viable. This is how the "strategy of the genes" (C. H. Waddington 1957) operated. It was possible along purely causal lines. Only when we look at an individual cycle *after* this causally determined process of evolution has run its course, are we in a position to regard individual development as a "final" process. Yet, after the course of events, all causal processes can be retrospectively regarded as "final" ones. This applies to the emergence of a new planet in the same way as to the evolution of a new species or the whole phylogenetical tree of plants and animals.

It appears possible, however, that genuine "final" processes take place in *planned actions* by higher animals and man. The aim seems to determine the whole chain of actions, in the manner of a "presumptive cause" (F. Baltzer 1955). But in reality these processes are causal in character. The perceptions and mental images in a situation of choice lead to chains of association concerned with different possible lines of action, and the brain tries these out experimentally, so to speak, to see if they will lead to the desired goal. Once the most suitable one has been selected, motor impulses then set the appropriate muscles in action. So the final stage is determined by a process of selection, in the sense that the dominant idea of attaining the goal directs the associations. Deflections by chains of associations away from the goal probably induce negative feeling tones, whereas positive feeling tones are present whenever the goal is brought nearer (see Ch. 6 F IV, G, and 7 A). "The harmony between nature and reason does not come about because nature is reasonable, but because reason is natural" (G. Klumbies 1956). Hence, following Spinoza, we can denote the *causae finales* as *figmenta humana*.

According to H. Driesch, the mode of behavior of animals and man is directed by a particular form of entelechy which he termed "psychoid," but in fact it scarcely differs from what is normally called "Psyche." Almost all dualists assume that the "psychical" processes affect the "material" body, in the sense of *psychosomatic influences*.

We have already pointed out, however, that in all probability psycho-physical dualism does not rest upon the existence of two fundamentally different modes, the psychical and the material, but is based on the fact that an "ultimate something" exhibits relations of two kinds: (1) extra-mental ones corresponding to the causal and logical laws of the universe and governing also the processes in the brain as well as inanimate matter; and (2) those which are limited to the organism's stream of consciousness. The problems of such a restricted kind of "dualism" based upon a panpsychistic, identistic conception will be discussed in the following section.

V. Phenomenal Coordinating Laws (Parallel Laws), or Realistic Indentity of Qualities and Extramental Existence? As we mentioned briefly in Chapter 6 H IV, the assumption of psychophysical interaction is hampered by the fact that such interaction is inconsistent with the law of the conservation of energy. Psychophysical parallelism is not open to this objection and yet it is hard to comprehend if one does not assume, with the Occasionalists, an intervention by God in each special case or, with Leibniz, a pre-established harmony after the fashion of synchronized clocks. On the other hand, if we accept an identistic view, which is supported by psychophylogenesis, psychontogenesis, the analysis of the psychophysical substratum, and the analysis of phenomena and matter, such difficulties do not exist. But we still have to consider the connection between phenomenal qualities such as "red," "sweet," "warm," "soft," and their relationships to causal relations.

Ziehen (1913, 1934, 1939), eminent both as a philosopher and as a scientist, has provided us with a clear and accurate treatment of the problem. He started with the phenomena (which he terms "gignomena") as the only absolutely certain factors, and he analyzed these by a process of reduction; he distinguished the causal components from the qualities and feeling tones of sensations and ideas which are not subject to causality, and he traced their local and temporal properties back to the locative and temporitative properties of a neutral "ultimate something." In this way he divided phenomenal properties into "reducts" and parallel components (qualities, etc.); but he emphasized that both are conscious components and that they are experienced only as relationships.

Ziehen regards the relationships of the "ultimate something" on the one hand as subject to the causal laws, and on the other, as dependent upon laws which certain sensations, ideas, or trains of thought invariably coordinate with particular causal processes in the brain. He refers to these latter coordinating laws as *parallel laws*. The physiological processes in the brain run off in a causally determined sequence. The parallel components are not postponed, but are coordinated instantaneously. The existence of an extramental world subject to the causal cannot be doubted. And as the reduction process was applied to certain properties alone, and not to consciousness as a whole, Ziehen's "reducts" are not to be regarded as unconscious matter.

Thus Ziehen held a hylopsychistic, panpsychistic identism. But he assumed a certain contrast between the extramental and the phenomenal world, as his parallel laws are confined to the latter. The extramental world is the realm of "reduct things," which he defines in the following terms (1939, p. 339): "Das Reduktding ist die kausalgesetzliche räumlich-zeitliche *Gestalt* des letzten Etwas mit Bezug auf seine Reduktkomponenten. . . ; es ist also das *Produkt* des Zusammenwirkens der Kausalgesetze." ("The reduct-thing is the spatio-temporal, causally determined *Gestalt* of the ultimate something, with reference to its reduct-components. . . ; it is therefore the *product* resulting from the cooperation of the causal laws.")

As already indicated, I too advocate a panpsychistic, identistic conception. But I believe that its application can be further extended. While it is characterized as "conscious," Ziehen's world of "reducts" still has something "material" as opposed to "psychical" about it, although he emphasizes the necessity of getting away from the idea of an "object" which we normally associate with perceptions and mental images, so that we are hardly ever aware of pure phenomena as such.

We must then base our considerations on phenomena free from all such associations. We must experience phenomena as a child does, which does not yet distinguish between phenomena and objects, or as we experience them in dreams. Following Ziehen, we could say in that case that the dream image of a red light is coordinated with a causal biochemical or biophysical process which we usually suppose to be the revival of an engram. In my view, however, we could equally well say that the mental image "red" *is* the process taking place at the level of the psychophysical substratum. So we do not need the assumption of hypothetical parallel laws.

As already indicated in earlier writings (1961a, 1964, 1969), and in D IV, F II, and IV, and G II of this chapter, the following interpretation seems to be largely free of contradiction. The system of protopsychical relationships which we usually call "matter" and which is manifest at different stages of integration as elementary particles, atoms, molecules, viruses (complexes of proteins and self-replicating polynucleotides), unicellular and multicellular living organisms, had led in the course of evolution to the production of nerve cells. In these cells the protopsychical properties are integrated in such a way that sensations as new systemic properties arose. Once nerve cells had combined to form more complex multicellular nerve centers, a stream of consciousness could come into being and the animal individual could then "experience" sensations as phenomena. The impulses left engrams behind them, and when corresponding patterns were repeated, the engrams were elicited, and memories, i.e., mental images, arose, which could be incorporated in the stream of consciousness also. Even when these mental images are not evoked, we may assume their generalized conscious character to be present, but this cannot be positively stated until the mental images enter a stream of consciousness and become "experienced."

In my view *the qualities of sensations and mental images are not "co-ordinated" with causal relationships but are identical with them.* Phenomenal qualities like "red," "sweet," "warmth," or "pain" *are* at the same time causal reality. So we do not need any "coordinating laws." Indeed, Ziehen himself has stressed that there is no connecting link between causal and parallel components. My conception, put forward as a working hypothesis, can be termed *realistic, panpsychistic identism* built on a positivistic basis.

According to this interpretation, the many differing qualities of our sensations, mental images, and complex thought processes are not "parallel components" in contrast to totally different "causal components" of phenomena, which are largely quantitatively characterized. We have already noted that the irreducible basic relationships which we call energy (mass), momentum, spin, and charge are subject to many laws;

and as they combine, new relationships are built up by system laws, and qualitatively diverse "material things" come into being. "Matter," in my view, is a system of relationships of generalized consciousness. Only after certain stages of nervous integration have been reached can it be experienced as phenomena within a stream of consciousness.

Among the many *psychological laws* some exist which seem to have no physiological basis. These correspond to Ziehen's noncausal, "endo-parallel" laws (1939, p. 109; but not 1913). These laws manifest themselves in such ways as the appearance of complementary colors, the combination of colors giving a three-dimensional color body, as well as in intentional relations in voluntarily directed chains of association and concepts of value. If these noncausal laws represent a special category, then "parallel components" would also have to be recognized as different in character from the "causal components" of phenomena. So far, however, there is no evidence that the facts which are characteristic of psychological laws are not based on causally determined nervous processes. It seems very probable that complementary colors can be traced to opposite reversible processes in the same nervous system. In the case of volitional action, our experience may derive its specific character from the fact that a dominant idea is directing the train of association, our mental imagery is in almost continuous relationship with the idea of our self, and certain sensations of tension often enter into the complex of phenomena (see below). Concepts of value may acquire their phenomenally specific character through positive feeling tones.

W. Wundt (1894) had declared that the law of conservation does not hold good for mental processes as it does for material ones. Ziehen agreed with him in this. He pointed to the largely qualitative character of parallel components as evidence against such a law. In my opinion the qualitative character is only a property of our phenomena which does not allow us immediately to deduce the law of conservation, or the causal law from it. This is not the same as saying that these laws do not govern mental processes. It is worth noting, too, that the mensurable spatial and temporal properties of the so-called "causal components" and the corresponding "parallel components" appeal to be identical. (See also the discussion of qualities in section VII following.)

VI. Logical Laws. Most philosophers regard logical laws as solely laws of human thought. Such a conception, however, puts them in an unlikely exceptional position; whereas all other laws are valid irrespective of space and time, or are like the numerous causal laws in being temporal manifestations of a universal law, logical laws would then have originated at a very late stage of development on our particular planet, when man with his capacity for logical thought had appeared. In spite of this, these laws are said to have universal validity, and Kant called them "principles a priori." This aprioristic character, which Laplace (1814) also emphasized, already suggests a validity independent of man. In all scientific considerations, we must suppose that the basic principle "if $a = b$ and $b = c$, then $a = c$" is valid for three atoms in space which no one had ever apprehended, no less than for the perception of three congruent circles within our field of vision.

E. Husserl (1900) already recognized that logical principles possess a timeless and "objective validity" apprehended independently of our thought by an "immediate grasp of their essence" ("Wesensschau"). But Ziehen (1934, § 22) was the first to draw attention to the fact that in the course of phylogenetic development the process of thinking had obviously adapted itself to the extramental logical laws of the universe.

Logical laws are the most general, "formal" conditions, i.e., conditions devoid of all content, of everything "given," of all phenomena and all the relationships of the extramental world which can be transgrediently deduced from the phenomena. The world is not only causally but also logically *determined*. Logical relationships can be expressed in mathematical terms, and we may regard all arithmetic and algebra as applied logic. This implies that mathematical statements are valid in the extramental world. In the course of his phylogeny, man has done no more than reach the stage of recognizing their existence and finding corresponding formulations for them. This was made easier by the fact that many statements could be checked and developed by means of purely abstract examples and paradigmatic conclusions. The axioms which underlie these, as well as the consciousness of their validity, their "certainty," depend upon their being consistent with the laws of the universe.

Progressive adaptation to the logical laws has already taken place in the course of phylogenetic development of animals, for all hereditary reactions inconsistent with them and therefore disadvantageous were gradually eliminated in the competitive struggle. This was particularly true of reactions dependent upon cerebral activity. A comprehension of likeness and difference, the basis of logical action, was developed at a fairly early stage; learning experiments with vertebrates and higher invertebrates have revealed this ability. These peculiar *functions of differentiation* can only be experienced and cannot be further defined. In an epistemological context we must therefore accept them as "given," as functions adapted to the system of universal laws. With regard to the physiological bases, however, we can try to find causal explanations such as those briefly outlined in F II of this chapter. But if we accept an identistic interpretation, these physiological interpretations would be relevant for epistemological theory as well. Adaptation to the logical laws also took place in a nonhereditary fashion in man, who acts with understanding and foresight. Errors of thought or action based on formal or material miscalculation leading to failure, as well as successful performances which are in accordance with the logical order of the world, are retained in the memory; and these memory traces help to direct future thinking and acting.

The *principles of probability* occupy a special position among the logical laws. They appear to be based on irreducible principles which are illustrated by the fact that if we toss a coin, it will on average fall heads or tails an equal number of times, and that a series of heads five times in succession will happen more often than a series of heads twenty times. And yet each throw is determined by purely causal factors involving force and direction of throw, center of gravity, and gravitational forces. But the above laws of averages and probability of shorter or longer series cannot be derived from the causal law. This influence of the laws of probability is responsible for the special position of "statistical laws, especially in the sphere of microphysics, where only the effect of logical laws and no longer that of causal laws can be observed.

VII. Survey of the Laws and the Problem of Qualities. We have seen that the great diversity of particular physical, chemical, biological, and psychological laws can be traced back to relatively few basic laws. These basic irreducible principles deduced transgrediently from our phenomena have given rise to the special laws in an epigenetic manner, as the appropriate stage was reached by the system of relationships which we call "matter." We may enumerate at least the following basic laws.

1. The uninterrupted, effective *causal law* determines every process in the world, in conjunction with the law of probability, and also microphysical processes which we can only formulate as statistical laws. The causal law is also the basis of all biological laws, which are mostly system laws and are often so interwoven that they can only be expressed as "rules." There is no necessity to postulate any acausal or vitalistic laws. Nor are there any "final" laws as such, though causally determined processes of development may be regarded as "final." Borderline cases exist, where causal relations are balanced by conflicting influences, and the time factor which is normally characteristic of cause and effect is lacking.

If we accept a panpsychistic, identistic conception and equate the psychophysical processes with the phenomena, there is no need for any particular coordinating or *parallel laws*, which would coordinate the phenomena with specific physiological processes in the brain and possibly in the sense organs as well. But if we equate physiological and psychical processes, then psychological laws also and Ziehen's "endoparallel laws" are all incorporated in the other laws, and in particular in the causal law.

2. The *universal logical laws* are also valid in the sphere of extramental matter. In the course of phylogenetic development, the brain functions of higher animals and man, and therefore also human thinking, have been adapted to these laws. The laws of probability, like the already mentioned laws of averages, have to be included in the logical laws, as they concern logical possibilities. The so-called statistical laws are both causally and logically determined.

3-6. The *universal constants* which cannot be traced back to the causal law can also be formulated as special laws: the velocity of light, the quantum of energy, the elementary charge, and the gravitational constant.

7-9. The *principles of conservation*, too, are not determined by the causal law. These are the laws of the conservation of energy, momentum, and angular momentum.

10-12. The established *principles of symmetry* can also rank as universal laws. The PCT-theorem determines the possibility of physical processes in which right and left are reversed, the formation of antiparticles, and the doubtful possibility of a reversal of time.

The combined effect of all these laws on the stages of manifestation of the relation complexes which we call "matter" produces specific physical, chemical, biological (and psychological) system laws. Their impact depends upon the state of aggregation and the stage of development reached by the astral bodies and living organisms.

The fact that physical analysis reveals ultimate irreducible properties of matter, energy (mass), charge, and spin, and that this matter is governed by twelve different basic laws, explains how a vast diversity of physical and chemical phenomena could arise at the present late stage of manifestation which the earth has reached. New complex systems were continually creating new interconnecting relationships, i.e., new properties. So we can now appreciate the limits of the monistic principle, discussed in Chapter 2 D: *the "material" world cannot be characterized in purely quantitative terms.* On the contrary, an ultimate irreducible diversity of lawful relationships forms the basis for the extraordinary diversity of physical, chemical, biological, geological, meteorological, and cosmic phenomena.

We had stated that a good deal of evidence exists suggesting that this "material" world is protopsychological in character (see pp. 190, 221). This brought us to a panpsychistic interpretation, and thereby somewhat nearer to an explanation of another problem: the development of the qualities of our sensations and mental images. *It is probable that the multifarious qualities which determine the content of all our thought, of sensations as well as of auditory word images and visual images, have become differentiated by system laws corresponding to the relevant stage of complexity of protopsychical matter* (Ch. 6 D III, IV, and F I). (Such a process of progressive differentiation would also be compatible with a theory of psychophysical parallelism.) As we will see in section L I, it is possible to make the contrast conceivable which seems to exist between the so-called "psychic" sphere, characterized by irreducible qualities, and the "material" sphere, characterized by mainly quantitative differences. The different possibility of analyzing the development of both spheres is the decisive factor here.

The realization that everything which occurs in the world (including man's processes of will; Ch. 6 F IV, G) is entirely determined by laws makes it possible to envisage a peculiar conception already fore-shadowed when discussing the laws of evolution (Ch. 5 D IV). *It is possible to think of the whole course of the world as a stable spatio-temporal configuration along which we ourselves trace our coercive course in the fourth dimension.*

Attempts have been made to cover all the world's lawful relationships by consistent mathematical formulas, or even by a single all-embracing formula. In principle this would seem to be possible. But Einstein's brilliant formulas and Heisenberg's exciting search for a "world formula" have shown how extremely difficult it is to solve the problem. Even if one succeeded in finding a definitive formula of this kind, however, the endless variety of simultaneous physical, chemical, and biological processes on our own planet and in a possibly infinite universe would preclude the possibility of doing anything more precise than sketching a few main lines of development. We shall never attain to what Laplace (1814) attributed to such a "universal intelligence":

"Une intelligence qui pour un instant donné connaîtrait toutes les forces dont la nature est animée et la situation respective des êtres qui la composent, si d'ailleurs elle était assez vaste pour soumettre ces donées à l'analyze, embrasserait dans la même formule les mouvements des plus grand corps de l'univers et ceux du plus léger atome; rien ne serait incertain pour elle, et l'avenir comme le passé serait présent à ses yeux."

("If for a given moment an intellect comprehended all the forces that animate nature, as well as the respective situations of the bodies of which it is made up; and if such an intellect were vast enough to be able to analyze these findings, it would then include within a single formula the motions of the greatest bodies in the universe and also those of the slightest atom. Nothing would be uncertain to an intellect of this order; both past and future would be present in its eyes.")

L. The Panpsychistic, Identistic, and Polynomistic Picture

I. Brief Summary. In order to arrive at a "true" interpretation adequate to both phenomenal reality and extramental existence, I based all my considerations on the *positivistic principle**. The starting point of my philosophical discussions was, therefore, the indisputable reality of *all* phenomena and their properties. I have examined these in the light of man's innate powers of comparison, analysis, and synthesis, and have drawn conclusions from the facts thus gained, with the help of the laws of logical thought which inevitably developed as man adapted himself to the logical laws of the universe. In this process, I exceeded the bounds of immediate experience only in the sense of "transgredient" processes of thought. I avoided all such transcendent assumptions and postulates as form the basis of many philosophical systems, appearing as "intuitively grasped" or "self-evident" or "revealed" truths. At the same time, however, I strove to make use of any findings of science, (most of which are gained without reference to epistemology), which seemed relevant to philosophical considerations and could be accepted as reliable and often experimentally verifiable facts.

We have seen that both epistemological findings and well-established scientific facts gained by very different methods can be combined in a *panpsychistic, identistic, and polynomistic view*. This kind of interpretation has the advantage of being relatively simple and largely free of contradiction, and of offering a uniform solution to many problems of philosophy and the sciences.

Of course we must always bear in mind that these findings and conclusions correspond to the actual state of our knowledge and our human cognitive faculties. Any attempt to frame an adequate universal view is rendered difficult because from childhood on we have all been too prone to consider what our senses perceive as a direct image of the extramental world; whereas they can at most provide us with hints about the nature of extramental existence, hidden from us as behind the veil of "maya." On the other hand, we must not forget that the rays of our human intelligence are able to penetrate that veil and allow us at least a shadowy glimpse of ultimate truth.

Our biophilosophical picture, the result of our investigations, can be briefly summarized as follows. Epistemological and psychogenetic statements lead to a panpsychistic, identistic conception according to which

---

* The term "positivistic" is used here in the sense discussed in Ch. 6 B. Auguste Comte, who coined it, took it in a much narrower sense, using as primarily "given" scientific and sociological facts alone. So did many later writers.

there is *no contrast between mind and matter.* We must recognize that all "matter" is protopsychical in character. When atoms and molecules came into being, the protophenomena of the elementary particles acquired new relationships and in this way new properties caused by system laws emerged. Sensations and recollections, i.e., primary mental images, only became possible at the complex structural level at which nerve cells developed. And self-consciousness and a concept of the self could only result if these sensations merged into a stream of consciousness within a larger central nervous system. Finally, human self-consciousness representing the highest phylogenetic stage yet reached enabled logical thought processes to arrive at an understanding of the universe.

Hence, the difference between extramental existence and self-consciousness is only a question of the degree of complexity and the relevant stage of integration reached by "matter" and its properties. So there is no contrast in principle between the qualitative stages of matter, which are largely determined by quantitative differences, and the phenomena, which are qualitatively distinct. In each case qualitative properties evolve as a result of new relationships set up within the relevant configuration by system laws. Each, too, is based upon the same system of relationships which we call matter as long as it does not appear within a stream of consciousness as phenomena. However, *we can establish phenomena as qualitatively highly differentiated only at a final stage of phylogenetic development, as phenomena of Homo sapiens.* Only to a very limited extent can we appreciate the various phylogenetic stages of phenomenal integration which have led from the inanimate to the animate and to higher animals. *In the sphere of what is extramental and inanimate, however, we are in a position to examine all the different stages of integration.*

What we call matter, then, is fundamentally a system of relationships governed by a number of basic laws. *It is precisely this polynomistic basis of all being that explains the diversity of the properties possessed by both matter and phenomena, in other words, both extramental relationships and those which enter into a stream of consciousness.*

It follows that there is no difference in principle between phenomenal reality and being, but only between two systems of relationships: those which enter a stream of consciousness and those which remain extra-

mental. The latter relationships are called "matter," the inframental ones "phenomena." This interpretation is *realistic* in that the physiological processes in the psychophysical substratum are identified with the phenomena. Berkeley's oft-quoted "esse = percipi" is true in the sense of "percipi = pars entis" or "percipi = esse, partim." *The cognitive capacity of man and higher animals is purely the result of the co-activity of universal laws at a certain stage of integration reached by the system of relationships representing being, the stage of complexity attained by a highly developed brain.* The universal laws, insolubly bound up with space and time, can possibly be regarded as infinite and eternal. They are "given" facts which cannot be traced back to something else, and they cannot be further "explained."

II. Summary of Some Essentials on Which Panpsychistic Identism Is Based. Philosophical systems are so extraordinarily diverse that it has been necessary to discuss both epistemological and scientific findings in some detail, because only by so doing could I show why I reject certain philosophical conceptions and exclude them from the present consideration. As the factual material is so ample, the arguments for the biophilosophical conceptions I have developed have been spread over several chapters. It therefore seems useful to summarize a few main points here. Without going into the history of the problem in any detail, I shall also refer to some philosophers whose views are similar to my own at least in certain aspects.

My panpsychistic and identistic conception is based upon the following facts:

1. The only reality of which we can be absolutely certain relates to experienced phenomena, which include sensations, mental images, feelings, and volitional processes, and thought processes as a whole.

2. Man does not consist of two separate components—matter and mind, or body and soul, but represents an indivisible psychophysical unity. Many philosophers and psychologists have rightly maintained this to be so, though they have often differed in some of the reasons they put forward. Apart from classical philosophers this conception is held by Maupertuis, Diderot, E. von Hartmann, Fechner, Wundt, Riehl, Höffding, Spencer, Bain, Spaulding, Mach, Schuppe, Erdmann, Ziehen, G. E. Müller, and others (see also R. Eisler 1910).

3. When we trace the phylogenetic development of conscious phenomena backward from man to the lowest animals, we can base our considerations only on conclusions of analogy. In the case of higher animals at least, these are thoroughly reliable. But at no lower phylogenetic level is there any indication that the psychical element might suddenly have emerged as something fundamentally new, during the process of gradual and progressive alteration of the lines of descent. The existence of intermediate forms between the animate and the inanimate (like viruses), and the probability that life has gradually developed out of lifeless matter, suggest that molecules, atoms, and elementary particles are protopsychological in character.

4. The course of individual development, psychontogenesis, points to the same conclusions. The claim that at some stage in the development of the human embryo a "psychical" element is added to it is an arbitrary assumption (partly postulated on the basis of religious convictions). As mental characters are also inherited, but the material basis of this process is to be found in the DNA molecules, a protopsychical character can obviously be ascribed to these molecules, their atoms and elementary particles. During the process of increasing biochemical complexity and structural integration, as the fertilized ovum develops into the embryo and the brain is formed, the protopsychical characters will also lead to greater integration: to sensation and then also to mental images.

5. If we assume that the "psychical" element, whatever its origin, is added to the "physical" matter at some stage in the development of the individual or a line of descent, and that it later induces physical "effects," such psychosomatic influence would be incompatible with the law of the conservation of energy. The same objection can be raised with regard to all theories of mutual psychophysical interaction and corresponding explanations of volitional behavior.

6. Apparently, endlessly vast spaces in the universe are entirely devoid of life. The pointlike appearance of a previously nonexistent "psychical" element in the brain of living organisms is a great deal more hypothetical than the assumption that integration took and takes place among already existent protopsychical properties of the systems of relationships which we call matter.

7. Even from the point of view of physics, matter can no longer be

regarded as something "solid," but as complexes of energy. (It seems to be particularly difficult for many scientists to imagine that felt "solidity" is not a quality of matter itself, but arises in our sense-organs and brain.)

8. The psychophysical substratum, the nerve substance, consists of molecules and atoms which derive from food and which are exchanged in the course of metabolism; protein molecules are partly exchanged in neurons in the course of some weeks. The conscious processes correspond to biochemical and biophysical processes. It is probable, then, that the molecules, atoms, and elementary particles involved are protopsychical in character.

9. If ectodermal tissue of the human embryo which would normally have developed into skin—the soles of the feet for instance—is transplanted to the region above the anterior chorda-mesoderm, this latter causes it to develop into brain tissue, "so that we stand and walk on portions of the body which, if they had developed at another part, we could have used for thinking" (H. Spemann 1943, p. 167).

10. It is impossible to point to any fact which would prove that matter is *not* protopsychical in character.

Since all considerations show that my interpretation combines a panpsychistic and an identistic conception, it may be well to emphasize in this context that I do not hold the view that matter has "material" and also "psychical" properties, as the Greek hylozoists already assumed. The complexes of relations which we call "matter" *are* of psychical nature; they are *protophenomena*.

Many philosophers have held hylopsychistic, panpsychistic, and identistic views. But in each case these correspond only in part to the interpretation outlined in the present work. Spinoza regarded the psychical and the material as "attributes" of the same basic "substance." In Proposition VI of Chapter II of his *Ethices* (1677, see 1914) he wrote: ". . . quod quidquid ab infinito intellectu percipi potest tanquam substantiae essentiam constituens, id omne ad unicum tantum substantium pertinat, et consequenter quod substantia cogitans et substantia extensa una eademque est substantia, quam jam sub hoc jam sub illo attributo comprehenditur. Sic etiam modus extensionis et idea illius modi una eademque est res, duobus modis expressa." ("Whatever can be perceived by infinite intellect as constituting the essence of substance, invariably appertains to one substance alone; and consequently thinking substance and extended substance are one and the same thing, which is now comprehended through this and now through that attribute. Thus also the mode of extension and the idea of that mode are one and the same thing, but expressed in two manners." (Translation by T. S. Gregory, 1955) It is worth noting, however, that Spinoza's brilliant system is purely rationalistic and is not grounded on experience in its wider sense.

The idealistic theories of Berkeley and Kant have also contributed to my panpsychistic view to some extent. These, however, are not sufficiently applicable to a period of the earth's history before there were any humans or any living organisms on it. And in some other respects also, especially with regard to cognition, they are at variance with present scientific findings. But in one of his early works, *Träume eines Geistersehers*, Kant wrote the following remarkable passages:

"Weil . . . meine Seele . . . von jedem Element der Materie nicht unterschieden wäre, und die Verstandeskraft eine innere Eigenschaft ist, welche ich in diesen Elementen doch nicht wahrnehmen könnte, wenn selbige in ihnen allen angetroffen würde, so könnte kein tauglicher Grund angeführt werden, weswegen nicht meine Seele eine von den Substanzen sei, welche die Materie ausmachen, und warum nicht ihre besondere Erscheinungen lediglich von dem Orte herrühren sollten, den sie in einer künstlichen Maschine, wie der tierische Körper ist, einnimmt, wo die Nervenvereinigung der inneren Fähigkeit des Denkens und der Willkür zustatten kommt. Alsdann aber würde man kein eigentümliches Merkmal der Seele mehr mit Sicherheit erkennen, welches sie von dem rohen Grundstoffe der körperlichen Naturen unterschiede, und Leibnizen's scherzhafter Einfall, nach welchem wir vielleicht im Kaffee Atomen verschlucken, woraus Menschenseelen werden sollen, wäre nicht mehr ein Gedanke zum Lachen."

("Because . . . my soul does not differ from any element of matter . . . the power of reasoning is an internal property which I could not perceive anyhow, although it might be found in all these elements. From these considerations no valid reason can be brought forward, why my soul should not be one of the substances of which matter consists, nor why its peculiar manifestations should not originate in the place which it occupies in such an ingenious machine as the animal body, where the combination of nerves favours the inner faculty of thinking and will-power. In that case, however, there would remain no peculiar characteristic of the soul by which it could be surely recognised and distinguished from crude elementary matter, and the jocose suggestion of Leibniz would not be laughable any more, that in our coffee we swallow, perhaps, atoms which are to become human souls.") (Translation by E. F. Goerwitz 1900)

Among more recent philosophers who hold a panpsychistic identistic view or at least approximate to it, E. Mach (1905, p. 6) may be cited: "Die Gesamtheit des für *alle* im Raume unmittelbar Vorhandenen mag als das *Physische,* dagegen das nur *einem* unmittelbar Gegebene, allen anderen aber nur durch Analogie Erschliessbare vorläufig als das *Psychische* bezeichnet werden. Die Gesamtheit des nur einem unmittelbar Gegebenen wollen wir auch dessen (engeres) *Ich* nennen." ("The totality of what is directly present for *all* men may be termed the *'physical'*; on the other hand, what is immediately present as a 'given' factor to *one* individual alone and discernible to all others only by analogy, may provisionally be called the 'psychical.' We shall call the totality of these immediately 'given' facts the individual's (narrower) self.")

Those thinkers who come closest in thought to my own view are Ziehen and M. Schlick, who both uphold a panpsychist identism. As I have noted, however, Ziehen differentiates between his causally determined components, the "reducts," and parallel components; this does not correspond to my more realistic identification of physiological processes in the nervous substratum with the phenomena. M. Schlick (1925, p. 267) came nearer to my opinion: "Die Welt ist ein buntes Gefüge zusammenhängender Qualitäten; ein Teil von ihnen ist meinem (oder irgendeinem anderen) Bewusstsein gegeben, und diese nenne ich subjektiv oder psychisch; ein anderer Teil ist keinem Bewusstsein unmittelbar gegeben und diesen bezeichne ich als objektiv oder extramental— der Begriff des Physischen kommt hierbei überhaupt zunächst nicht vor." ("The world is a richly varied configuration of interdependent qualities; some of these are given factors in my (or another's) consciousness, and I call these subjective or psychic; others are not directly given to any consciousness and these I term objective or extramental—the concept of the physical does not arise in this connection.") But Schlick differs from my more realistic conception when he goes on to state, somewhat later (p. 293): "Das Bewusstsein kann nicht das Ansich der Gehirnteilchen sein, denn wo das erstere fehlt (im Tod und Schlaf) bleiben die letzeren vorhanden." ("Consciousness cannot be the essence of the brain particles for they are present even when consciousness is absent, as in death or sleep.") Against this I would say that "brain particles," like all matter, can be regarded only as systems of relationships, and the question is only whether these are merged in a stream of consciousness.

III. The Significance of Polynomism. When discussing monistic tendencies (Ch. 2 D) we noted that prescientific interpretations of being as well as many philosophical theories reveal an effort to trace the entire multifarious human experience back to some uniform fundamental principle. Or else they assume a uniform basis for either matter or mind or possibly for both, which are normally considered as being completely different.

We have seen, however, that this reduction to a uniform basis is applicable only in the realm of scientific research, and even there it has its limits. On the other hand, the diversity of phenomena seems to be irreducible. "Ein formelhafter metaphysischer Monismus gibt nicht Rechenschaft mit seinem Satze, dass alles Sein in Wahrheit *eines* ist; es bedarf notwendig irgendeines pluralistischen Prinzips zur Ergänzung." ("Metaphysical monism, with its thesis that all existence is fundamentally *one*, cannot be considered as adequate; it needs amplification by some pluralistic principle") (M. Schlick 1925, p. 305)

Physical analyses of the nature of matter have proved that a number of irreducible basic principles exist: energy, charge, spin, velocity, spatial and temporal properties. Matter is ultimately a system of relationships subject to various laws: the causal law, the universal constants which may be expressed as laws, the principles of symmetry, the laws of conservation, and the logical laws. This *basic irreducible multiplicity* of the "material" world postulated by science is the reason why such an extraordinary wealth of different "objects" and processes could arise by progressive stages of integration of their components and the new relationships set up by system laws.

A panpsychistic, identistic interpretation offers at the same time an explanation for the diversity of phenomena. *As the system of relation-*

ships which we call "matter" is regarded as being psychical in charac-
ter, and the irreducible physical properties are therefore also protophe-
nomenal, it follows that the different "material" stages of integration and
the relevant processes of the psychophysical substratum, which are de-
termined by the laws of the universe, must inevitably lead to very diverse
phenomena. The difficulty of comprehending this lies in the already mentioned fact that all stages of integration reached by matter from the elementary particles, atoms, and molecules to the structures of the living
organisms can be investigated; whereas the phenomena can only be apprehended at the final phylogenetic stage of *Homo sapiens.* We can
assume by analogy that mental processes took place in phylogenetical
animal prestages also. Especially in regard to the important initial integration processes, however, when protophenomena were combined to
form sensations and mental images, the conclusions are very vague (Ch.
6 H). Yet it is a necessary and legitimate task to trace back man's highly
complex phenomena on the basis of experiment and observation of animals, to successively simpler phenomena experienced by animals at
lower phylogenetic levels. We must even try to go back as far as Protozoa and prestages of life. With higher animals such consideration yields
fairly certain results. But higher animals have gradually developed from
lower forms; we have to bear in mind that from the beginning life is a
continuous process, although divided into innumerable lines of descent.
And it is also important that life probably emerged successively from
lifeless matter. Psychogenetic considerations of this kind are also significant in that their results supplement the general view of life and fit
without contradiction into a biophilosophical conception obtained by
totally different methods.

IV. Advantages of the Biophilosophical Picture Now Developed. According to the panpsychistic, identistic interpretation outlined here, the system of relationships which we call "matter" is psychical in character and
represents either protophenomena or phenomena, according to the stage
of integration reached. Instead of the contrast between mind and matter
assumed by dualists, I make only a distinction between extramental
protophenomena and phenomena which are associated within a stream
of consciousness and are experienced by an individual. This implies that
thought, as we saw, is not different in kind from being; *thought is a part*

of being. In my opinion this interpretation provides a relatively simple solution to the most widely discussed problem of epistemology, a solution derived directly from analysis of the phenomena and from scientific facts. It also simplifies a philosophical conception.

My view is also *realistic* in that it identifies the physiological processes in the psychophysical substratum with the phenomena. This realistic version of panpsychistic identism is very largely consonant with the realism of many scientists. Though it does not ascribe the qualtities of phenomena to the "objects" perceived, it considers them to be identical with the relevant physiological nervous processes.

This kind of view has no need of a "carrier" of psychical processes or any "substance," and *it replaces Kant's "thing in itself" by a system of ultimately psychical relationships which, if present in a central nervous system, can enter into a stream of consciousness*. This seems to me to establish *a firm link between the humanities and the sciences*. Besides, such a view may also help to reconcile materialistic and idealistic interpretations. If "matter" is psychical in character, then *this form of identism corresponds in many respects to materialism*.

Another advantage of equating the phenomena with corresponding psychological processes is that *all scientific analyses and explanations of the processes of the psychophysical substratum are also epistemologically significant*. It becomes conceivable that the DNA molecules transmit the hereditary factors of mental abilities and weaknesses, and that sensations, mental images, and more complex phenomena develop successively before and after birth. We can also understand how living organisms could have gradually emerged from lifeless matter, and how further complication, histological differentiation, and progressive development in animals led to increasingly higher levels of phenomenal integration. Finally, this view explains why many "sensory" excitations do not become conscious (experienced) and why processes which have been conscious may cease to be so, as actions become automatic. We can moreover understand how trains of excitations running off outside the stream of consciousness, after passing into other trains of excitations which are coordinated with consciousness, produce "intuitive" ideas. It remains to be seen if my conception proves of heuristic value in solving further problems such as the localization and revival of engrams.

Various philosophical systems have already been described as realism or neorealism. Some have points of contact with my interpretation; but most of them assign quite a different meaning to the word "realism." In contrast to purely idealistic or spiritualistic views, they normally assert that some "real" extramental element exists besides the phenomena, something which cannot become the content of consciousness itself and can only be conceived as extramental reality (see O. Külpe 1912-1922, etc.). There is no need to discuss all the relevant literature on the various types of realism. In most cases these views already differ from my conception, because the authors could not yet be aware of modern biological and physiological findings or they have not concerned themselves with the biological knowledge of their time.

Chapter Seven

Ethical Consequences
of the Conception
Now Developed

We have seen how essential it is, when dealing with problems of natural philosophy, to exclude all transcendent ideas, all "self-evident" truths or religious convictions. My conception, based on epistemology and science, led to the assumption that all being and all events are governed by the ceaseless interplay of relationships determined by universal laws. *Each human individual, therefore, can be regarded as a partial complex of this universal system of relationships, a partial complex which is inescapably integrated into the grandiose order of the universe.* From a purely scientific point of view this interpretation may be a satisfying one.

In practical life, however, some difficulties present themselves. As all the processes in the brain are governed by laws, we have seen it is most unlikely that the will is free. This conception leads to important consequences regarding the conduct of life, education, the administration of justice, and ethical standards.

Furthermore, panpsychistic identism cannot hold that the soul can be separated from the body. But this is at variance with many religious convictions. It leads to the question whether and how far religious ideas of any kind are compatible with our biophilosophical views, or how near the two attitudes can approach to one another.

As the philosophical conception of the large majority of people is determined more by practical consequences than by epistemological or scientific principles, I shall now consider certain ethical and religious problems insofar as they have a direct relationship to the views I have expounded. As already mentioned in my introduction, my treatment should not be regarded as being more than an outline. These are problems of far-reaching significance, and some aspects of them go beyond the scope of the present work. I shall therefore cite only a few particularly relevant views selected from the extensive literature on the subject.

A. Problems Concerning the Lack of Free Will

In Chapter 6 G we noted that volitional thought is determined by universal and individual hereditary factors—particular gifts and failings, strength or lack of will power, constitution, tempo, and age—and by nonhereditary factors including experiences of every kind, particularly education, feelings, affects, illness, and fatigue, as well as actual perceptions at the moment in question. Although it cannot be stated with certainty, because of the complexity of the associative processes involved, it appears extremely unlikely that the will is free. Volitional behavior, in particular in animals and young children, is often quite clearly determinate. Moreover, we have seen that such freedom of brain processes could only have appeared in the world after man or at least higher animals had originated, and would then continually infringe the laws of the universe. Yet to deny the freedom of the will seems to jeopardize a number of valuable ethical concepts such as responsibility, duty, conscience, guilt, and indeed the ideal of freedom. This is why many philosophers, whose views are otherwise deterministic, have accepted at least a limited degree of free will—or, expressed more appositely, they have tried to "rescue" it in some way or other (see also B. Rensch 1963).

If we deny that the will is free, does this necessarily lead to sterile fatalism, to a purely passive attitude toward life? Are our ethical standards then indefensible? Have we no right to blame and punish criminals? In my view such fears are groundless. All that is needed is a change in our understanding of the determinate nature of volitional processes, and a new attitude toward wrongdoers. No radical change is required in our conduct of life, education, administration of justice, though at certain points some modification is overdue.

As to *fatalism*, there is a great deal to prevent us from sinking into apathy: the continuous activity of our brains; the urge toward self-preservation, which we share with all animals; the instincts of food getting and reproduction; and the striving for precedence which also has a hereditary instinctive basis. Other powerful impulses as well, attended by positive feeling tones, operate against apathy and inactivity. We seek to live our own lives, to have our share in life's riches, and to be in harmony with our surroundings; we take pleasure in helping others, and we set ourselves an aim in life. Yet many civilized people believe that some destiny directs all that happens. Even the gods of ancient Greece were subject to moira. The Arabs hold that kismet ultimately determines one's fate. Christians speak of God's inscrutable decrees. But none of these peoples have abandoned themselves to a passive life. They have attained to great historical achievement, and have produced abiding cultural values.

No one concerned with philosophical problems need be restrained from maximum achievement by the thought that he is only a particular and complex part of a universal system of relationships subject to equally universal laws, and that therefore his own thought processes are not free. Nor is it normally possible to hinder him in this way; for all the hereditary and nonhereditary factors we have enumerated direct his associations and determine his thought. He can only be hindered if, when he realizes that the world is determinate, he draws false conclusions and adopts erroneous fatalistic ideas. Although these false conclusions, too, would be determinate, this rarely happens unless his physiological activity is already diminished.

Such considerations already indicate the decisive significance of education of all kinds—at home, school and work—and of daily experiences, for man's thought and actions. The moral and ethical standards instilled in youth or later life, the individual's values and ideals, all become determinants of his thinking. In the "play of motives" these established complexes of ideas, which apparently correspond to cytologically consolidated engrams, generally prove stronger than conflicting ideas in helping one to come to a decision.

The *concept of freedom* is of particular importance in this connection. Are we to refrain from aspiring to freedom of body or mind because in the last instance freedom does not exist and everything that happens is determined? This would be a false conclusion, for the concept of freedom is itself an important determinant in our thought. It prevents us from taking a passive attitude toward events or breaking off the "play of motives" because the universal laws are what ultimately decide every issue. It helps us to go on until we reach the best possible solution, in thought and action, for our physical and mental well-being. We are spurred on by the "feeling of freedom" which is a special kind of complex phenomenon (see pp. 205, 232). In this way the legitimate drive toward freedom, the drive to be free of all avoidable determining factors limiting our thought and action, is in itself determinate in character.

All these considerations deeply affect problems connected with the *administration of justice*. If our will is not free, the concept of "guilt" would seem to be meaningless; but without the reproach of guilt no wrongdoer can be punished. Yet human society has to be protected from wrongdoers; so punishment must not be motivated by a sense of retribution. A man may be more or less driven to criminality, partly because of his hereditary disposition, some mental limitation, lack of will power, or compelling sexual drive. P. A. Jacobs, M. Brunton and M. M. Melville (1965), and also M. Casey (1966), who based their work on special State Hospitals, found that sex-chromosome abnormalities were thirty times more frequent in men detained in State Hospitals for those requiring special security than among the rest of the population. Types with an additional Y-chromosome were particularly prone to crimes of violence. Nonhereditary factors such as lack of education, bad company, and alcoholism are also significant determinants. As early as 1905, observations of this kind led F. von Liszt to urge that the wrongdoer should be regarded as a patient, and that no reproach of guilt should be attached to him and no ethical standards applied. There should be "punishment" but only in the sense of taking measures to protect others from possible harm, to act as a deterrent and lead to amendment and reintegration into society.

This attitude has made some progress among many civilized peoples. But the terms "guilt" and "punishment" have been retained, especially among juristically minded laymen who expect legalistic reactions from the criminal and still regard punishment as retribution (see P. Bockelmann 1961, p. 37). There is no doubt that the concept of guilt is an important determinant in human behavior. But the judge must realize that his sole purpose in attacking guilt is to impress certain concepts upon the wrongdoer, and others as well, which will affect future thinking. He should always bear in mind how it has come about that the criminal is in the dock while he himself is on the bench. "Alle Schuld ist Schicksal und alles Verdienst nicht minder" ("All guilt is fate, and so is all merit") (H. Groos 1939, p. 46). Justice, then, becomes solely the "Inbegriff aller durch organisatorischen Zwang gesicherten Normen" ("the essence of all the norms ensured by organized compulsion . . ."). (B. Rehfeldt 1962, p. 41)

There are still many countries of which it cannot be said that justice is based upon recognition of the absence or restriction of free will. In some, however, it has been taken into account that crimes are largely determined by inherited or noninherited causes, and increasing efforts are being made to reform the wrongdoers and to reintegrate them into society. The Communist countries in general, and the USSR and China in particular, have apparently gone farthest in this respect, and they seem to have had a fair measure of success. Convicts in these countries are "re-educated" in social thinking (see, for example, B. L. Bodard 1959). These methods are often discredited in the West where they are labeled "brain-washing." From the physiological point of view, however, all efforts to impress ethical and religious ideas on children by fixing the appropriate engrams, even in early youth, might also be called brain-washing.

Besides, there are practical difficulties in the way of modern justice. In most cases it is impossible to assess a hereditary taint with sufficient accuracy. And it is particularly hard to reintegrate into society persons with certain hereditary predispositions, such as a predisposition to sexual crimes. In these cases the only method is to persuade them to consent to castration, by which the stimulating hormones are eliminated. Moreover, it would be futile to explain to the criminal that his will is not free (see B. Rehfeldt 1962). It is also extremely difficult and requires great psychological skill to eradicate firmly established ideas at variance with the law. And minor criminals are not detained for long enough to allow such treatment. Discussion of the administration of justice is being actively carried on at the present time; to assess the wide differences of opinion it is useful to compare the writings of E. Mezger (1926), R. Lange (1955), W. Welzel (1960), and A. Kaufmann (1962) who stress man's "freedom" and "dignity," with the philosophically more neutral comments of H. Leferenz (1948, 1959), P. Bockelmann (1961), and B. Rehfeldt (1962), and those of F. von Liszt (1905), J. Lange (1929), F. Bauer (1957), and M. Danner (1969), which advocate wide reforms. The work of F. Bauer in particular is of great interest on this subject.

B. The Inquiry into the Purport of Existence

The invariably experienced sequence of cause and effect, and the realization that a consistent causal principle governs the universe, have made it normal for man to plan his actions; that is to say, he runs over the chains of probable events in his mind and then selects a kind of action appropriate to his aim. Man's daily life, as well as all his future, is directed by a succession of aims of this kind; and corporations and institutions of all types—political, economic, technical, scientific or religious —often extend their plans far beyond the span of a single generation. So it is natural for the question of a final aim, the purport of existence in regard to both individuals and mankind as a whole, to arise.

The question could only emerge after man had at least recognized the causal course of everyday events, and had begun to reflect on his own existence. The Australopithecines and men of the *Homo erectus* type can certainly have been no more concerned with it than the animal pre-stages of the hominids had been. It would not be right to say that the preservation of their species constituted the "purport" of existence for them; for they did not live *in order to* reproduce their kind. They existed *because* they did so in sufficient numbers to survive in the struggle for life. It was only after the man had begun to think about death, and to form an idea approximating that of the "soul," that the problem of the purport of existence dawned upon him. This may possibly have taken place at the Neanderthal stage. No doubt it long remained a minor consideration for man, for there are many primitive people today who hardly reflect upon it.

We do not know what direction the future evolution of *Homo sapiens* will take. It is not to be expected that his brain will improve structurally, since natural selection has ceased to favor further progress of this kind. It seems more likely that there may be a decline, brought about by continuous undirected mutations in the absence of natural selection toward mental advance. And even if mankind takes eugenic measures to maintain progressive development, the significance of such action could only be temporary; it could not constitute an ultimate aim.

Any question of the purport of human existence is obviously bound up with the future of our planet. Very little definite information about this was available until the last few years. It is only recently, since astronomical processes have been investigated with the help of computers and a more intensive study of certain clusters of stars has been carried out, that astronomers have been able to demonstrate the probable evolution of the astral bodies. Judging from the Hertzsprung-Russell diagram, the sun has reached a relatively stable stage, in which nuclear fusion in its interior is transforming hydrogen into helium. When the hydrogen is exhausted, however, the outer layer will expand to such an extent as to destroy all life upon earth. It is estimated that this may take place after about eight milliard years or at least within a calculable period (see R. Kippenhahn and A. Weigert 1967; H. Haffner 1967; and others).

If living organisms have developed or may develop elsewhere than on our planet, which is possible, as our Milky Way system contains about ten milliard suns, and approximately a hundred million additional galaxies exist, then their destiny will resemble our own. Organisms can only come into being if the star in question reaches a firm state of aggregation, if certain biochemical conditions are fulfilled, and if the surface remains within a definite range of temperatures. A longer or shorter period of evolution will then ensue, after which the planet will very probably be destroyed again.

Science, then, can offer no proof of an ultimate aim of existence for the organisms which emerge and vanish again as life rolls on in a continuous stream, nor a purport of existence for the highest species, *Homo sapiens*. We humans too are no more than temporary, finite, highly complex systems of the protophenomenal "matter" of which the world is composed, a "matter" representing a system of certain relationships subject to universal laws.

Leaving aside any religious ideas, we may say that neither the existence of the individual nor that of humanity has any "aim." *Sub specie aeternitatis* it has no more "purport" than the extramental world or the totality of existence has. *The "purport of existence" can therefore be looked for in finite aims alone.* And man has always created aims of this kind for his personal life and that of social communities. There is always some satisfaction in pursuing and attaining certain nearer or more distant aims.

This kind of limited "purport of existence" has been interpreted in many different ways. As man is a social animal, and his well-being is bound up with the organization and achievements of society, his aims have often been formulated as *ethical ideals and standards*. The aim here has usually been "happiness," a maximum of satisfaction and pleasure for each individual, as expressed by many other philosophers since Epicurus, including Francis Bacon, Hobbes, Descartes, Leibniz, Mill, Spencer, Fechner, and many more. But pleasure and satisfaction have been differently interpreted. Aristotle found supreme happiness in the cognition of eternal truths, the Theoria. Hobbes wrote: "The greatest good, or what is called happiness or an ultimate goal, is not to be found in this life. . . . The greatest good is unimpeded progress from one goal to further ones." (1658; see 1949 Ch. ii, 15) Descartes, Spinoza, Leibniz, Fichte, Herder, Comte, E. von Hartmann, Wundt, and later philosophers up to Bertrand Russell saw it in a striving after wisdom, man's "ennoblement," the creation of new values, in every endeavor toward cultural progress. However these aims of existence have been defined, the empirical bases remained man's innate urge toward positive feeling tones, toward pleasure, and the indispensable restrictions imposed by life in human society. But they never amounted to "ultimate aims."

Kant alone opposed this kind of view, with his demand for a formal ethic based not empirically but on an absolutely existent moral law grounded in "transcendental freedom." He held that the consciousness of duty ought to be the mainspring of our action. His "categorical imperative" provided each individual with an aim, an ultimate goal.

No matter what view guides our way of life, appreciation of ultimate truths, creation of new values, cultural progress, empirically based ethical ideals and norms, or some highly debatable absolute moral law, all these views are bound to lend existence some kind of limited "purport." This preserves at least those who are not philosophically minded from falling into the pessimism that results from too much reflection over the absence of a genuine "ultimate purport" in the life of the individual or the development of the world.

However, most people do not concern themselves with those problems at all, because their lives are directed by religious convictions which provide an aim in the life of the soul after death. But religious interpretations themselves differ greatly. As the next chapter will illustrate, some of them even approximate the biophilosophical conception which is developed in the present work.

Chapter Eight

Relationships to Problems of Religious Philosophy

It may be emphasized again that in this chapter I have limited myself to a bare outline. But it seems essential at least to touch upon the main relevant problems, in view of the fact that the theologians of today are seriously engaged in coming to terms with the findings of modern science.

A. The Roots of Religions

Homo neanderthalensis of the Mousterian period already placed gifts in the graves of his dead. So we may assume that some kind of life after death was envisaged, even at the cultural level of more than 150,000 years ago. It may be that primitive religious notions were connected with the practice. In the later paleolithic period, some 15,000 years ago, when *Homo sapiens* created the cave dwellings of southern France and northern Spain, it is likely that ideas of magic were already current, especially those connected with the hunt.

In line with growing tradition, animistic conceptions developed which we still find among underdeveloped peoples. Inanimate natural objects such as the sun, moon, stars, or planets came to be regarded as alive, but the idea of a soul need not necessarily have been attached to them. Although the dead were honored, and a primitive form of ancestor worship grew up, whatever element was considered to live on—what, for instance, the Jagga negroes describe as the "shadow"—did not always amount to a "soul" now freed from the body (N. Soederblom 1916, 1964).

The concept of the human soul—and by analogy something similar, the "souls" of animals—may have developed at the beginning of the advanced cultures of some six to eight thousand years ago. It then became possible not only to think of the soul as living on after death but to imagine similar independent spirits and demons, and later other gods as well, gods of lightning and thunder, rain and wind, illness, death, and also other unintelligible processes in nature. Animism led in this way to polytheism.

The Ionic natural philosophers also thought of nature as largely animated and "plena deorum." Thales of Miletus (624-545 B.C.), Anaximander (610-547), and Anaximenes (588-524) had already reflected on growth and decay in nature and sought to establish certain basic principles in line with monistic ideas. This led to the assumption of a primal matter ($\alpha\rho\chi\acute{\eta}$) of which all other matter was said to be composed. In this way the spheres of philosophy and religion had begun to draw apart.

A similar development had apparently taken place half a millenium earlier in India. The concept arose there of the impersonal brahman, the all-one, out of which everything takes form, and into which it returns again. The people, however, also personified this brahman, and it became worshipped as the god Brahma. In India, therefore, religious and philosophical ideas have remained in closer contact. The early Chinese philosophers of the sixth to third centuries B.C. such as Tsang Hsi, Lao Tse, Lieh Tse, Chuang Tse, and others arrived at a similar concept of an ultimate unity, the "tao," the source of everything (see H. Hackmann 1927).

But a growing understanding of the causal and logical relations of all events did not necessarily lead to a transmutation of religious ideas into philosophical ones. Religions still exist at the present in all parts of the world, and they have a far greater influence on most men than philosophical ideas have. So religion must have other roots besides ignorance about the processes of nature; these roots must be sought in man's general mental character and requirements. I shall not attempt to deal with the literature on this topic, but simply mention a few of what I consider to be the most important factors involved here.

The idea that the "soul" parts from the body at the moment of death was no doubt influential at an early period; for it seemed to explain the difference between a dead body and a living one, a difference which would not yet be understood in biological terms for some thousands of years. Fear of what unknown future the soul might face presumably also played a considerable part. At the same time, these ideas were combined with the desire for a life after death, and this has probably been the most powerful factor in the development of many religions. Added to this, there was often the longing for compensatory justice, for it was reassuring to know that some kind of "other world" existed, where good deeds would be rewarded and evil ones had to be expiated. Many Egyptian frescoes and reliefs of the Old Empire already portray this idea in an attractive manner; they show the heart, in the underworld after death, being weighed in a balance against a feather as the symbol of justice. This idea of compensatory justice was very probably one reason why the religious ideas of Ancient Egypt remained relatively stable for three millenia. Corresponding convictions have helped the spread of Christianity, Islam, and Buddhism among those who were less fortunate in life than the relatively few powerful or rich men.

Besides these largely rational factors, emotional ones also played a part in the origin of religions. Men in need or distress, in misfortune or sickness, or threatened by death, craved for the consolation and protection of some overruling power. They discovered this in the gods to whom they could pray. These gods, like the saints and bodhisattvas, also represented suprahuman ideal figures, and it seemed worth aspiring to approach or emulate them.

There were also purely psychical factors of importance such as the "religious shudder" which may be linked to a sense of the "sublime." E. Burke (1757) had aptly remarked that the feeling of the sublime arises when man is confronted with something immense and stupendous, such as a natural feature, a high mountain or a vast building, or some abstract power, or something limitless. All this fills him with a vivid consciousness of his own puny size and weakness. That is why most temples, pagodas, and churches are uncommonly large buildings. The enormous size of many statues depicting gods, too, has an imposing effect. The gigantic bronze statues of Buddha at Kamakura, and in the Todaiji temple at Nara in Japan, are 11.4 and 15 meters high; the reclining Buddha in the Wat Po in Bangkok is 49 meters in length. A "religious shudder" can also be intensified by darkness. The interiors of many temples and churches are dark or dimly lit; Burke noted that "almost all heathen temples were dark" (1757, Part II, Sec. iii). Other unaccustomed sensory impressions also play a part: a particular tone in the priest's voice, solemn music, echoing chants, the scent of incense or joss sticks. All this creates a mood which transports those who are oppressed, troubled, or fearful into an unfamiliar and unreal sphere. They feel a "pious shudder," but they also find reassurance and escape from their prosaic rational existence.

Different factors, then, have contributed to the development of religions during the past millenia. Many philosophers and researchers into religion have recognized this, though most have stressed only one or another component. The chief factors have apparently been a lack of understanding of much that occurs in nature, a feeling of impotence in the face of these natural events, a dread of largely incalculable forces (see Hume, Holbach, Ebbinghaus, etc.), the mystery of death, the desire for an afterlife, and the hope of compensatory justice after death. They also include belief in a universal "moral order" (see Kant, Fichte), the longing for safety in the care of some higher and beneficent divine being that can accomplish "miracles" (see Schleiermacher's "feeling of dependence" and E. von Hartmann's "need of redemption"), and finally a sense of sublimity induced by these ideas and the "pious shudder" which accompanies them. As all these factors have affected man—and indeed they still do—it is understandable how the development of all early cultures was largely influenced by religion. Most of the imposing buildings which these cultures have left behind them are therefore temples and shrines, like the temples of Ancient Egypt; the ziggurats of Babylon; the temples of the Greeks, Romans, Hindus, and Jainas; the Shintoist shrines; the Buddhist temples and pagodas; the temples and temple pyramids of the Mayas, Aztecs and Incas; the mosques of Islam; and Christian cathedrals and churches.

B. Divergent Development of Religion and Philosophy

Belief and knowledge, religion and philosophy, are radically different in regard to their intellectual basis; yet both spring from man's endeavor to comprehend and influence the course of events. The factors we have outlined in considering the origin of religions soon brought about a cleavage between the two spheres of thought. It may seem at first as if the discipline known as religious philosophy might constitute a bond between them. But this is not so, chiefly because the term is used for two different branches of knowledge. What I might call "pure religious philosophy" aims at grasping the genuine philosophic content of the different religions, without reference to the truth of any religious premises, "eternal truths," or similar maxims. But the term "religious philosophy" is often used to cover all kinds of investigations already based upon firm religious convictions. Religious thoughts and their development are only examined and interpreted by philosophical methods, or are linked with philosophical ideas. Such studies may of course be instructive and valuable, but they are not strictly speaking philosophical, as they represent a mixture of knowledge and belief. Belief, however, means accepting processes to be true which cannot be confirmed by experience and indeed are often contradicted by it.

Religion has usually drawn apart from pure science by personifying the Divine and establishing a body of dogma. Surely, dogmas form an essential part of every enduring religion; for the believer must know what he is to believe and how he is to conduct himself. He must also be convinced that these truths are sure and abiding ones. But these conservative elements of religion may lead to difficulties if they clash with new views that accompany every cultural advance. So dogma has to be supplemented and interpreted in new ways in the course of time. Christianity has done this to a great extent, while Buddhism has remained relatively stable.

All religions, including those with a more philosophical basis, show a tendency to personify the power which they accept as the principle governing the order and course of the world. But the inclination to personify has manifested itself outside the context of religion as well. Many features of language reveal this: we speak of "mother earth," the "mother church," the "mother liquor," and so on. The German miner calls his crutch-handled stick "Krückmann" and his food-wallet "Henkelmann." The Englishman calls his ship a "man of war" and the Malay his toe "anak kaki" (child of the foot). It was therefore natural for primitive peoples to personify the forces of nature and to regard them as spirits, demons, or gods. In the most advanced religions, the conception of a personified god no doubt developed as an ideal figure capable of creating and directing the regulated course of nature, the motions of the stars, the alternation of day and night, the passing seasons and the stages in man's life as well as the wonderfully adapted structure of men, animals, and plants, and the incomprehensible workings of fate. A figure of this sort could only be imagined as an idealized human: only a personal god might be propitiated by sacrifices, and he was the only one to whom one could pray as to a suprahuman father. This was the only kind of being whom one could love.

Hinduism, too, teaching that the ultimate aim of man's existence is to be merged in the All-One, also represented the brahman in human form as Brahma, Shiva, Vishnu, Durga, and other gods. Many Buddhists pray to Buddha as if he dwelt as a human in Nirvana, though in Nirvana, all forms, sensations, and ideas are said to have ceased to exist (see Buddha's words, and H. Oldenburg 1914 and other relevant literature).

The faithful cherish a belief that the gods can also break the laws of causality and logic. In most religions gods can perform miracles and help man in situations which are hopeless in regard to their causal conditions. But this has led religions into conflict with everyday experience, and with the findings of science. We therefore find most religious ideas based on feelings rather than on rational thought. We have seen, however, that feeling tones are only a property of sensations and ideas. They can influence and direct a train of thought, but they cannot lead to knowledge of the extramental world. Valid knowledge can only come through reason.

The history of cultures shows that religious and scientific convictions can persist side by side, despite the inconsistency of their different experience. As most people have no particular philosophic bent, and this state of affairs will probably continue for a long time to come, they will generally find more satisfaction in one of the religions than in rational experience and scientific knowledge. The reassurance and strength for the task of life which prayer brings to many is indeed a positive factor. The difficulties arise, however, when efforts are made to unite dogmatic ideas with certain rational findings, and conflict between the two ensues. The danger which then presents itself is that purely theoretical discussions encroach on practical life, and scientific truth is denied and freedom of thought stifled. As the history of cultures shows, this has too often been the case, especially with Europeans, who are particularly intolerant. We shall refer briefly to these matters in our treatment of individual religions.

C. Panpsychistic Identism in Its Relation to Religions

The writings of many philosophers offer a combination of knowledge and belief. Besides giving well-documented facts, they imply the existence of a personal god as creator of the world, a divine "will," or an afterlife of the "soul," as valid. But such assumptions, which are incapable of proof, are incompatible with the view I have developed in the present work, based strictly on given phenomenal facts and their logical treatment. Yet this does not mean that philosophers tinged in this way with religious views may not possess other aspects which are not dependent upon religious convictions and are quite consonant with my conception. Fundamentally, however, a deep division separates them from any consistent "pure" philosophy exempt from prejudice.

This is particularly true of the *double truth theory*, first put forward in the 14th century by the great Franciscan philosopher William of Occam, since revived from time to time, and occasionally held in our own day. G. Hennemann (1967), for example, has written (p. 77):

"Dass die von der Naturwissenschaft in Raum, Zeit und Materie objektivierte Welt die einzige Wirklichkeit sei, ist ein (heute auch von der Naturwissenschaft selbst erkanntes) Vorurteil, dass uns den Zugang zumal zur religiösen Wirklichkeit verstellt. So ist *Himmelfahrt*, um deren Wirklichkeitskern es uns jetzt geht, keine Ortsveränderung im Naturgeschehen und in diesem Sinne real, sondern sie bedeutet ihrem Wirklichkeitsgehalt nach eine Wesensänderung in der Seele, von der man auch entsprechende Wirkungen ablesen können muss. . . . Das aber kann nicht begriffen, sondern nur geglaubt werden."

("That the only reality is the world objectified by science in space, time and matter, is a prejudice (admitted even by science today) which blocks our access to religious reality. The *Ascension*, whose reality is the point at issue, is not factually 'real' in the sense of representing a change of place in the natural world; its reality consists in a fundamental alteration within the soul, and it must be seen to have corresponding effects. . . . This cannot be comprehended; it can only be believed.")

In the epistemological sphere, the religious ideas which emphasize the *destiny* determining all existence have more in common with the conception I have advanced. The Greek belief in moira as the supreme and absolutely determinant power, or the islamic idea that all existence is irrevocably predetermined (kismet), or the Christian doctrine of predestination, do not contradict the conclusion that the whole course of the universe and of each individual human life is determinate and that the will is not free (see Ch. 6 F IV and G). And the monistic version of the Indian *doctrine of brahman, the All-One*, into which the atman (the individual being) re-enters after death, corresponds at least to a certain degree with my conception of being as a uniform and lawful system of relationships which is psychical in character. After all, every *panentheistic theory*, insofar as it regards god as *impersonally* present in all things and all events, can be largely combined with nomistic and panpsychistic views. In regard to Christian doctrines we must also bear in mind that Western philosophers were *obliged* for many centuries to retain a conception of god, even when they did not assume the existence of a personal god, as Christian belief claims (see below).

The practical *ethical demands* made by most religions are very largely the same in every case, and they hardly differ from those which could be based on a panpsychistic, identistic conception or any other philosophical system exempt from religious ideas. But the arguments given for these ethical rules and precepts are of course totally different. Man is a *zóon politikón*. Like all social creatures, he can only exist if the relations between individuals are regulated by certain rules. These rules are partly determined by everyday life, and they have developed and become established by a kind of natural selection from among possible types of conduct. As chaos would ensue if every member of a community were at liberty to lie, slander, steal, rob, injure, or kill his fellows, customs have grown up to prevent this as far as possible. Offenders are punished, even if only by being exposed to disapproval for some slight misdemeanor, or by being hindered from rising in order of precedence. Larger communities have had to embody such customs in laws.

Ethical maxims, however, also derive meaning from the fact that concepts such as responsibility, propriety, and justice act as standards of conduct. If someone acts counter to these concepts based on "rational" ideas, then the members of the social group blame or condemn him. On the other hand, acting in accordance with ethical demands may increase one's self-reliance and make life easier by giving it a firm direction.

Compassion, too, has played an important part in the development of ethical standards. It is based on man's sensitivity and his capacity to put himself imaginatively in another's place, and so to feel and suffer with him. Such empathy leads to a display of charity which gratifies the man who acts in this way and helps him to live a reasonably peaceful existence. But more philosophically minded men who are convinced that every event is determined by chains of eternal laws will have a more profound understanding of their fellows' conduct and this will influence their own ethical behavior.

All these factors, which we have briefly touched upon here, have led to the formation of a system of *natural ethics*, whose practical demands are roughly the same as those which are based on divine commandments, the idea of transmigration of souls, or other religious ideas, or on the assumption of an absolute moral law existing a priori. They have relatively little to do with the question of the purport of existence.

Christianity. A great deal has been written in recent centuries on the relation of Christianity to scientific and philosophical ideas. To discuss this literature in any detail would go beyond our present considerations. It may be enough simply to indicate how far this religion shows convergences and divergences with respect to a panpsychistic and identistic conception. The initial difficulty is that it is almost impossible to state the nature of Christian belief in unequivocal terms. There are different churches—Roman Catholic, Greek Orthodox, Lutheran, Reformed, and Anglican—and countless sects, each with its own widely differing tenets. Besides, almost all religious assumptions and expositions of Bible texts have undergone considerable modification in the past few centuries, as the history of dogma shows (see A. von Harnack 1931; M. Werner 1959; and others). Moreover, very far-reaching changes of interpretation are taking place at the present time, and extremely varied views are being put forward (see, for example, the Jesuit K. Brockmöller's book).

The differences even concern the idea of god. Man has usually regarded god as a personal deity, but he has not infrequently thought of god as impersonal being. As these radically different lines of thought have a particular bearing on my biophilosophical conception, I shall consider them briefly.

I shall begin with more orthodox ideas. If one compiles the attributes and abilities of god as illustrated in the New Testament, one finds the following descriptions. God is "everlasting," "eternal," "invisible," "wise," "just," "good," "patient," "forebearing," "faithful," "gracious," and, on occasion, also "angry"; he "sees," "hears," "wills," "speaks," "commands," "promises," "comforts," "judges," "pardons," "is not mocked," etc. Except for the first three, these are all human attributes; that is to say, attributes applicable only to living organisms with nerve cells, brains, and sense organs, and not for instance to a protopsychical

All-One. Many theologians and philosophers in the past have been aware of this; and in the Middle Ages these anthropomorphic expressions were already sometimes regarded as merely symbols to indicate an inscrutable being (see below). "Gott schuf den Menschen nach seinem Bilde, das heisst vermutlich, der Mensch schuf Gott nach dam seinigen." ("God made man in His own image, presumably means that man made God after his own.") (G. Ch. Lichtenberg)

Most Christians, however, regard God entirely in human terms, as a man and like a father. The Virgin and the saints are also generally thought of as human beings. Pictorial representations and the doctrine of the bodily Assumption have, of course, contributed to this way of thinking. As already mentioned, personification corresponds to a human tendency. Besides, one can only revere and love a person and not an abstraction; and genuine prayer is impossible except to a personal being. One of the most valuable things about great religions, however, is the opportunity they afford of drawing strength and consolation from prayer. But the course of modern philosophy has shown that few of those who really reflect on these problems are convinced that such personification is convincingly imaginable.

In addition to these more orthodox ideas, however, there are other basic Christian convictions which are incompatible with my own biophilosophical conception. One of the most significant in this connection is the idea that body and soul are separate entities, and that after death the soul quits the body and can live on "in heaven." As already noted, epistemological analyses of the phenomena and scientific statements on psychophysical processes, psychophylogenesis, and psychontogenesis have shown this kind of dualism to be most unlikely (see Chap. 6 C and H). Besides, if this "other world" has no spatial properties, it cannot be reconciled with the idea that one will see the souls of relatives and friends again there, for in that case one would have to experience a spatial impression of their features. The concept of the soul has also involved making a distinction between "mind" and "soul." Higher animals are said to have some kind of "mind," but only humans are considered to have "souls." From the psychological point of view, however, the two concepts are inseparable. We have seen how all our phenomena are built up from sensations and mental images, and they are embedded in a constantly changing stream of consciousness, in which it is quite impossible to discriminate between components of "mind" and of "soul." In scientific publications it is therefore usual to make no such distinction.

The Christian conception of the creation, too, has proved inconsistent with the idea that man has gradually evolved from animal ancestors. This is why the theory of evolution has been so strongly attacked. It was only when it became clear that an overwhelming body of evidence pointed to man's animal origin that progressive theologians began to accept the theory and to cite evidence from the Bible and other works of Christian literature to the effect that there is no cogent reason against it (see, for instance, H. Volk 1955). But even today it is only the evolution of man's body which is officially acknowledged. The evolution of the soul, psychophylogenesis, still presents great difficulties. It may perhaps be possible for Christian doctrine to incorporate this, if the beginnings of human symbolic language are interpreted as taking place "at God's behest."

The attitude of modern Roman Catholic theologians toward evolution appears to be very fully covered in a noteworthy book by W. Bröker (1967), who is well versed in both biology and theology and who inquires into the purport of evolution. He points out that since it obeys certain laws, evolution of the universe as well as of life can be conceived as "formal purport" ("Gestaltsinn"). Evolution can thus be regarded as "a natural revelation of God."

"Das Bild des Schöpfergottes wird dadurch, dass die Welt, sein Werk, als eine evolutive erkannt wird, grösser und inhaltlich reicher." (p. 108) "Unter diesem Sinnaspekt der Schöpfung in Evolution als eine Kundgabe des noch gewaltiger und noch genialer zu denkenden Schöpfergottes erscheint Evolution als geradezu eines göttlichen Schöpfers würdig, da sie in der Erkenntnis zur grösseren Ehre Gottes beiträgt." (p. 103) [Wenn dann Christus] "das Ziel der Schöpfung ist, dann unterstreicht die für menschliches Empfinden unendlich lange Zeit der Vorbereitung seines Kommens durch Evolution seine Bedeutung. Eine solch gewaltige Anstrengung des Kosmos ist der Würde Christi angemessen." (p. 110)

("The image of God the Creator is greater and richer if we conceive the world, His work, as evolutive in character. . . . If we regard creation as in a process of evolution in this way, as a manifestation of a Creator-God of even greater genius and power, then evolution appears worthy of a divine creator. Recognition of it then redounds to His greater glory."

[If Christ is] "the goal of creation, then His coming gains in signifi-cance because the time of preparation for this event, through the medium of evolution, seems to us humans so infinitely long. This stupendous effort on the part of the whole universe is appropriate to Christ's dignity.")

Bröker, however, is content to call his work merely a "contribution to the discussion" and a "prologue to a conclusive synthesis."

In my view the theory of evolution and Christian doctrine are most likely to draw closer together in regarding man as the latest product in the history of creation, as God's image, representing a final unsurpassable stage. It is true that in his somatic structure *Homo sapiens* does in fact represent a final stage, reached after 70,000 to 100,000 years; and unless selection in regard to build and brain structure takes place, he is not likely to alter much in the foreseeable future. It is always possible, how-ever, that he may learn to direct his own phylogenetic development to some extent. And with regard to the scope of his knowledge, which ex-tends from atomic forces to the origin of stars, from the evolution of living organisms to the course of civilization, present-day may may ap-pear almost "god-like" in comparison to any other living beings (see B. Rensch 1965a, Ch. 5).

Christian doctrine is difficult to reconcile with any cosmology based on the established laws of the universe, because the implications must appear unsatisfactory to the faithful. They would mean either that a personal God has created the world and fashioned its laws, and this world then runs on without his aid; or else that God is continually obliged to intervene in its development and in all microphysical, physical, chemical, biological, geological, and cosmic events, but always in accordance with causal and logical laws. On the other hand, he might sometimes disrupt this determinate activity and produce "miracles."

One particular problem arises whenever faith and scientific knowledge are brought up against one another over any issue; reason, which Chris-tian teaching tells us is God's gift to man, leads the researcher to scien-tific and philosophical results which contradict religious convictions. Surely this action on the part of God would be inconceivable.

The practical effect of orthodox Christian views has often been to lead to events which offer a sad contrast to the ethical standards of most men. Bertrand Russell (1957) is probably right in calling Christianity the most intolerant of all religions. We have only to recall the many wars against "pagans" and the destruction of their cultures such as those of the Mayas and Incas, the persecution of all who dared to have scruples over doctrinal niceties, the Inquisition with its barbarous tortures and burnings, or the spiritual agony of those threatened with hell fire. Intellectual progress has often been obstructed, and the list of thinkers whom the Christian church has persecuted is a long one, beginning in the ninth century with Johannes Scotus Erigena and continuing with Albertus Magnus, Roger Bacon, Giordano Bruno, Galileo, Campanella, Fichte, La Mettrie, Holbach, D. Fr. Strauss, and others. Even Kant's theistic work *Die Religion in den Grenzen der blossen Vernunft* (*Religion within the Bounds of Mere Reason*) (1794) came under censure from Frederick William II. His Order in Council denounced it as a misuse of philosophy and a degradation of the fundamental doctrines of Holy Writ. The professors of philosophy and theology at the university of Königsberg were all forbidden to lecture on the subject. But even today a certain intolerance which should be incompatible with Christianity often mars both family and professional life (see also B. Russell 1957, and G. Szczesny 1958).

Other religions, of course, have also been guilty of religious persecution. "Every religion has its popes and cardinals, its idolatry and witchhunts. The cards and the game are the same; it is only the names that are different." Our comment on these words of S. Radhakrishnan is bound, if we are honest, to be an admission that in Europe the game is played with unusual vigor and ruthlessness.

Since the beginning of Christianity, however, there have been other interpretations besides the orthodox one, and some of them have more in common with our biophilosophical conception. Often on a basis of simple piety but mainly influenced by philosophical views, God has been conceived in a largely *impersonal* manner as immanent in all created beings, or even in the whole universe, and identified in some way with the universe or to a certain degree with the laws that govern the world. Some brief remarks may be sufficient to characterize these *"panentheistic"* ideas. In the ninth century Johannes Scotus Erigena, influenced by Neoplatonic doctrines, declared that all creatures emerge from God and re-enter His being; our life is God's life in us ("Nam et creatura in Deo est subsistens, et Deus in creatura mirabili et ineffabili modo creatur, seipsum manifestans . . .") (B. Geyer 1928) David of Dinant, writing in the twelfth century, dared to equate God with matter. In the fifteenth century, Nicolaus Cusanus taught that God is the substance of all things and the threefold root of all being ("causa efficiens, formalis, finalis"). Giordano Bruno, burned as a heretic in Rome in 1600, held a similar view. He thought of God as the immanent cause of the universe, the universal reason, and acting nature ("natura naturans").

In Sir Thomas More's *Utopia* (1516) we read that men revere ". . . a certain godly power, unknown, everlasting, incomprehensible, inexplicable, far above the capacity and reach of man's comprehension, dispersed throughout all the world, not in bodily size but in action and power" (see 1947, p. 143). Angelus Silesius (1624-1667) gave the same ideas poetic expression in his *Cherubinischer Wandersmann* (see 1905): "Ich weiss, dass ohne mich Gott nicht ein Nun kann leben; werd' ich zunicht, er muss vor Not den Geist aufgeben." (I, 8) "Eh' ich noch etwas ward, da war ich Gottes Leben . . ." (I, 73) "Ich bin nicht ausser Gott, und Gott nicht ausser mir . . ." (I, 106) ("I know that without me God cannot live one instant; for if I perish, he too must give up the ghost. . . . Before ever I *was*, I was God's life. . . . Without God, I am not, nor does God exist without me . . .") Jakob Boehme (1575-1624) declared: "Denn Gott ist selber das Wesen aller Wesen und wir sind Götter in ihm, durch welche er sich offenbart." ("For God is the essence of all beings; and we, through whom he manifests himself, are gods in him.") (C. Richter 1943)

This pantheistic interpretation reaches its greatest heights in Spinoza's identification "deus sive mundus." In a letter to H. Oldenburg (1675; see *Works* 1871, vol. 2, 21st letter), he explains: "that God is what one might call the immanent cause of all things; he is not a force working from without." Spinoza expressly abstained from personifying god in any way. The following passage is taken from a letter to W. van Blyenbergh (1665, 36th letter): ". . . in philosophy, however, where we clearly understand that we cannot ascribe or impute to God those attributes that make men perfect, any more than we could ascribe to man what makes an elephant or a donkey perfect." Another letter (the 60th) contains this statement: ". . . for I believe that a triangle, if it could speak, would maintain that God is eminently triangular . . . " In the letter to Queen Sophie Charlotte of Prussia (1702; see 1906), Leibniz

offers the following definition: "The ultimate basis of all things, common
to everything and embracing all the well-knit elements of nature, is what
we call *God*. God must therefore be an absolute and absolutely perfect
substance." In his *Théodicée* (1710, see 1879) he also calls God the
cause of all being; at the same time, however, god is individualized as
the central monad and lawgiver, who has created the pre-established
harmony of bodies and souls. Oken (1809, p. 15) wrote: "Alle Dinge
sind nichts als Vorstellungen, Gedanken, Ideen Gottes" ("All things
are nothing else than ideas and thoughts of God").

Hegel (1832) taught that the object of all religion was the absolute,
which can only be imagined but cannot be apprehended. The theologian
Schleiermacher (1884) thought of religion as "das unmittelbare Bewusstsein von dem allgemeinen Sein . . . alles Zeitlichen im Ewigen . . ."
("the immediate consciousness of universal being . . . all that is temporal during eternity . . ."); in other words, the feeling of union with
eternity and of dependence upon it. D. Fr. Strauss also defined religion
as a feeling of dependence upon the universe. G. Th. Fechner (1879), a
typical panentheist, identifies God with world consciousness and the
universal law of the world. Though their views were formulated as
generalized statements of religious philosophy, these last four philosophers clearly had Christianity principally in mind. Similar views held
by P. Natorp (1908) and G. Störring (1920) will be referred to briefly
in section D.

These philosophers and theologians have all been cited only as
examples of how Christian doctrine can also be interpreted in a panentheistic sense, which means that God is completely impersonal, and the
worship of God is equated with a rational and emotional grasp of man's
place in relation to the universe and its laws. This view is quite compatible with my biophilosophical conception (except for "worship").
*When we are referring to the universal being and the universal laws
which direct everything including human thought and action, it is after
all a mere question of terminology whether we speak of "the power that
governs us" or of "God."*

What makes it difficult to be precise about the original elements of
Christianity is the fact that we have no historical records of it from
the first few centuries of our era. The Dead Sea Scrolls (Qumram
texts) make no mention of Christ although their date is somewhere

between 200 B.C. and A.D. 70. (The C_{14} test gave an average of A.D. 33; see J. M. Allegro 1959, and others.) Some Protestant theologians have even stated as their considered opinion that Christ never lived at all. H. Raschke's linguistic studies (1954, 1966) led him to the opinion that "Jesus" (in Hebrew, jeschuoth = the savior; in Greek Christos = the helper) did not originally refer to a person, and that the miracles were later associated with Jesus as a result of a poetic play on words then common in connection with Hebrew and Greek place names. The Revelation of St. John, the oldest of the New Testament books, makes an equally impersonal statement (xix, 13) ". . . his name is called the Logos of God." This suggests to Raschke that the earliest form of Christianity was a cult of the Essenes and in fact a "neoplatonic philosophy after the manner of Philo and the Apologetics . . ." It is noteworthy that Tertullian already complained of the "adulteration" involved in rendering "logos" as "word." To the Stoics, however, whose panentheist theories were prevalent about the time of "Christ's birth," "logos" meant the divine law of the world. If we were now to return to early Christianity as Raschke interprets it—which would mean a second reformation!—it would have much in common with the opinion of many philosophers and also with my own biophilosophical conception. Unfortunately, theologians find it extremely difficult to examine this interesting question historically in an impartial spirit, for their views are understandably colored by their beliefs, and they must consider the possibility of official protests from their own churches.

Islam. This religion, founded in 622 by Mohammed, is monotheistic in a much stricter sense than Christianity with its doctrine of the Holy Trinity. To the Mohammedan there is only one God, Allah. Jesus, like Abraham, Moses, and Mohammed, is honored only as a prophet. Allah, however, is completely personified, and endowed with human attributes in the same way as the God of orthodox Christianity. Looking through the Koran we find Allah also described as "all-knowing," "all-hearing," "all-seeing," "all-powerful," "all-bountiful," "strict," and so on. He "speaks," "ordains," "is angry," "punishes," "rewards." None of this shows any points of contact with my biophilosophical conception.

Hinduistic Conceptions. The Indian religions have undergone various changes and Hinduism in particular is split up into many sects. Here as elsewhere the conception of the gods is an anthropomorphic one. In this case, however, philosophical theories developed from these religious ideas at quite an early stage. These theories have some basic ideas in common with my panpsychistic and identistic conception. This is especially true of the Upanishads, which were added as an explanatory supplement to the Veda or holy writings, and in particular their panentheistic interpretation by Shankara (in the seventh century) and the school of Advaita which still flourishes in our day. In the present context it is perhaps sufficient to refer to some of the more significant points of contact with natural philosophy (see Shankara 1957; S. Nikhilananda 1951; T. M. P. Mahadevan 1953; M. M. Chatterji 1947; P. Deussen 1922; S. Radhakrishnan 1929, 1931; H. von Glasenapp 1948; B. Rensch 1966; and others).

According to philosophical Hindu belief, man's nature like that of every animal and plant is the atman which has its origin in the brahman, the All-One, into which it will enter again after death. So everyone must constantly bear in mind that each living organism is part of the same eternal and divine brahman (*tat tvam asi*), to which no violence must be done (*ahimsa*). The Katha Upanishad contains the following passage (II, ii, 9): "As the same non-dual fire, after it has entered the world, becomes different according to whatever it burns, so also the same non-dual atman, dwelling in all beings, becomes different according to whatever It enters. And It exists also without." (Translation by Swami Nikhilananda 1951) These ideas correspond very well with my conception of the unity of all being, which is likewise assumed to be protopsychical in character. Besides, they are in line with the increasingly probable assumption that life emerged phylogenetically from lifeless matter.

According to Shankara, honoring the divine in personified form as Brahma, Shiva, Vishnu, Kali, etc., represents a lower stage of understanding. In his *Viveka-Chudamani* (Crest Jewel of Discrimination) he states: "I, the atman, am Brahma. I am Vishnu, I am Shiva, I am this universe." But the brahman is impersonal, it is pure consciousness, without form or qualities, invisible, soundless, tasteless, odorless, unchanging and eternal. This again corresponds very largely to a possible conception of "matter" as a system of relationships and protopsychical in character. The real nature of the brahman (which corresponds in a certain degree to the term "being") is veiled by maya, the transformation by our sense organs. And this precisely corresponds to epistemological analysis. But Shankara appears also to be aware of the significance of phenomena as the most reliable basis, for he states that we can doubt much, but not the fact of our own existence. These Hinduistic doctrines conflict much less than do Christian ideas with such findings of science as the theory of evolution, the theory of relativity, and modern cosmology.

The old belief in reincarnation after death, in metempsychosis, which is connected with the atman-brahman relationships, is of course inconsistent with my biophilosophical conception. But the idea that all one's good and evil deeds are rewarded or punished in the next incarnation, placing one higher or lower in the scale of living organims, has an important ethical significance.

Buddhism. Almost the same may be said of Buddhism, which split off from early Hinduism (see H. Oldenburg 1914; P. Deussen 1922; etc.). During the Conference of Religion in the Age of Science held at Star Island (U.S.A.) a leading Burmese Buddhist, Th. M. Th. S. U Chan Htoon (1958), said in his address: "There is no principle of science, from biological evidence to the general theory of relativity, that runs counter to any teaching of Gotama Buddha." In the course of a discussion between us at the University of Münster, the quotations from Buddha which he gave in answer to my relevant questions made it clear that Buddhism sanctions every kind of scientific knowledge, and that it accepts the causal principle as a universal factor. Buddhism does not aim at proclaiming a "divine revelation." The Buddhist idea of successive world epochs, too, is in line with what seems to me the most probable interpretation of the universe, namely that celestial bodies originate and vanish in eternal alternation.

Buddhism, like Hinduism, presupposes the existence of an absolute moral law which corresponds more or less with Kant's categorical imperative. We have seen, however, that there are reasons for doubting this assumption (Ch. 7 B). Nevertheless, it has considerable ethical value in that Buddhists believe in metempsychosis and therefore in a kind of "moral causality." The ultimate goal, Nirvana, resembles the brahman of Hinduism, and it thus has something in common with my own view of a protopsychical "All-One." In general, however, we European scientists do not share the Buddhist view that final truths are to be attained by profound meditation. We are convinced that they are to be discovered by scientific research and logical deduction.

There are some Buddhist customs, however, which any people would do well to adopt. At a certain age many of our young men are given to excesses of various kinds: hot music, hard drinking, antisocial behavior; moreover, they are conscripted for military service where they learn to handle weapons designed to kill. It would be a good thing if instead they had to spend at least a few months as monks, as they do in Thailand, living a frugal life, practicing self-discipline, and learning to meditate intensively on the vital problems of existence.

There is one particular virtue which we in the West conspicuously lack and which the peoples of southern and eastern Asia could teach us: religious and intellectual tolerance. In Japan, for example, one sees a Shinto shrine on the fringe of almost every larger Buddhist temple area, and a small Buddhist temple beside almost every larger Shinto shrine. There are also some temples, such as the one in Nikko, where these two very different religions are even merged to some extent. Over the whole of India one may still come upon pillars and rocks on which the edicts of the great ruler Priyadarsin (Asoka) were carved in the third century A.D. He introduced Buddhism into India, but at the same time ordered complete religious tolerance (see E. Hardy 1902; W. Schumacher 1948; and others). In his 7th Rock edict we find the passage: "It is the will of the pious king Devanampriya Priyadarsin that all religious communities shall be able to settle in any place whatever; for they all aspire to self-discipline and purity of heart."

There are also certain affinities between my biophilosophical view and *panentheistic theosophy*, which tries to combine basic ideas from every major religion. It is significant that all attempts of this kind originated in India, and that most of its exponents are still to be found there. The great emperor Akbar (1542-1605) already invited Christian missionaries, Brahmans, and his own Islamic Ulemas to his palace at Fatehpur Sikri, where they had to expound the advantages of their religions to him, and to engage in discussion between themselves. Akbar himself was not an adherent of any one of these religions. Combining what he found acceptable from them all, he created a new religion based on sun worship, but he did not impose this upon his people. On the contrary, in 1593 he issued an edict of complete tolerance, announcing also full parity between all the peoples and races within his great empire (see F. A. von Noer's biography of this Great Mogul, 1880, 1885, etc.).

It is Ramakrishna (1836-1886), however, who is regarded as the real founder of modern theosophy; through his intensive study of various religious doctrines he came to identify himself in visions with Kali, Brahma, Allah, Christ, and Buddha (see R. Rolland 1931; Swami Gambhirananda 1958, and others). He arrived at a theosophical interpretation which was widely disseminated by his adherents in India, America, and Europe. Like many of the panentheistic doctrines of Christianity and Hinduism, his teachings approach my nomistic ideas to a certain degree. A sympathetic feature of theosophy is that it induces Buddhist monks, Hindu believers, Christian priests, and indeed all interested in religion to meet with one another, for example at Ramakrishna House in Calcutta. Buddha, Christ, and Ramakrishna are represented side by side in a chapel there, each with a cushion before him on which the faithful can kneel in contemplation.

D. Scientific Essentials for Any Religion of the Future

Spinoza had already aspired to reconcile religious and philosophical ideas. And since his day, many other philosophers and some liberal theologians have made the same attempt. Among more recent philosophers, P. Natorp (1908) limited religiousness to a "sense of eternity"; G. Störring (1920) identified God with the nonspatial fundament of the world ("unräumlicher Weltgrund"); and M. Scheler (1921) described religion as an immediate experience of the absolute. Ziehen (1928) elaborated a special theory, "nomotheism," based purely on epistemology and the findings of science.

Scientists too have attempted to bring about the union of science with religion. One need only recall E. Haeckel's panpsychistical writings on "monist religion" (1893, cf. 1922), "cell-souls" (1909), and "God-Nature" (Theophysis) (1914), in which he interprets god in a panentheistic sense, but denies the possibility of miracles and the existence of individual souls surviving after death. Haeckel's views have been strongly attacked, particularly on ideological grounds. Some of the hypotheses with which he supplemented his facts also seemed to be insufficiently supported. But it is now possible to state that most of these hypotheses have since been confirmed. These include his rejection of any "vital" force or purposive tendency in phylogenetic development, and his ideas on psychogenesis and the dependence of self-consciousness on the presence of a complex central nervous system. The Union of Monists, founded in 1906 on his initiative, still propagates his views in *Die freigeistige Aktion* (Hannover). Contributions to this journal, especially those of G. von Frankenberg, show a spirit of mutual understanding by their readiness to recognize all religious aspirations which do not conflict with scientific facts. The International Humanistic and Ethical Union operates on a similar but wider scale.

Views such as these bring man, with his responsibilties vis à vis nature and society, back into the center of quasi-religious endeavors; this had already been claimed by Comte. J. Huxley in particular has stressed this point: "Man's most sacred duty . . . is to promote the maximum fulfilment of the evolutionary process in his earth" (1954, p. 16). And Th. Dobzhansky (1956, 1962, 1966) holds a similar view, though he expresses it with some restraint. Modern man is becoming more and more able to influence his own phylogenetic future by eugenic measures and by adapting the increase of the population to the reduced natural selection brought into being by the advances in medical science.

E. Concluding Remarks

My panpsychistic, identistic conception is not consistent with any belief in individual "souls" and their survival. Death is followed by decomposition, and this means disintegration of innumerable relationships and a return to the protopsychical "All-One," the brahman. This is in agreement not only with Indian thinkers but with many European philosophers and scientists as well. Only those on whom contrary ideas have been imprinted from childhood will find these views unsatisfactory. However, as the young Haeckel wrote to his fiancée (1859-1860): "Die Früchte vom Baum der Erkenntnis sind es immer wert, darüber das Paradies zu verlieren." ("It is always worth losing Paradise to gain the fruit of the tree of knowledge.")

Nor is there any need to fear that such a biophilosophical conception may endanger our moral conduct. There are many "unbelievers" in every advanced country today, in Europe and America no less than in India, Japan, and the Communist lands. These irreligious people too direct their lives according to ideals which are acceptable and advantageous for human society. Ethical standards need not be rooted in religion. They also spring in every human society from a primal pleasure in doing good, in compassion and sympathy with one's fellows, an apparently inherited instinct toward recognition in the social group, a satisfaction in great undertakings possibly exceeding the lifetime of the individual, and, among judicious people, contentment at having followed a harmonious and straightforward course of life.

Finally, any ethical attitude must be based on a clear philosophical conception which, with the advance of human knowledge, corresponds more and more to the reality of being. It must surely be an elevating and reassuring idea to know oneself at one with the universe, that wonderful system governed by eternal laws, and at the same time to realize that one represents the most highly developed stage of integration, able to conceive the interrelationships of "matter" and the course of the world, except for its irreducible and hence irrational basis, the existence of phenomena, and laws governing the ultimate something. But in my opinion the present stage of human knowledge obliges us to confess that this philosophical attitude is no longer equivalent to a religion.

Bibliography

Abercrombie, M., and J. Brachet (eds.) 1961-1967. *Advances in Morphogenesis*. 6 vol. New York, London (Academic Press).

Ach, N. 1935. *Analyse des Willens*. Berlin, Wien (Urban u. Schwarzenberg).

Adrian, E. D. 1947. *The Physical Background of Perception*. Oxford (Clarendon Press).

Allegro, J. M. 1959. *The People of the Dead Sea Scrolls in Text and Pictures*. London (Routledge and Kegan Paul).

Alverdes, F. 1933. Die Ganzheit des Organismus. *Zool. Anz.*, **102**, 1-15.

———. 1935. *Die Totalität des Lebendigen*. Leipzig (Barth).

Angelus Silesius 1905: *Cherubinischer Wandersmann* (after ed. of 1675) ed. by W. Bölsche, Jena u. Leipzig (Diederichs).

Arbit, J. 1964. Learning in annelids and attempts at the chemical modification of this behaviour. *Animal Behaviour*, Supp. 1, 83-88.

Aristoteles 1847. *Über Sinn und Sinnliches*. Trans. by F. A. Kreuz. In: *Werke*, vol. 2, Stuttgart (Metzler).

———. 1847. *Drei Bucher von der Seele*. Trans. by F. A. Kreuz, Stuttgart (Metzler).

———. 1856. *Tiergeschichte*. Trans. by Ph. H. Külb. In: *Werke* III, IV, Stuttgart (Metzler).

———. 1860. *Metaphysik*. Trans. by J. Rieckher. In: *Werke* V, Stuttgart (Metzler).

Aristotle, *Psychology*. Trans. by W. A. Hammond. London.

Aster, E. von. 1935. *Geschichte der neueren Erkenntnistheorie*. Berlin, Leipzig (de Gruyter).

Auerbach, S. 1906. Beitrag zur Lokalisation des musikalischen Talentes im Gehirn und im Schädel. *Arch. f. Anat. u. Physiol., Anat. Abt.*, 197-230.

Augustinus. 1930. *Bekenntnisse und Gottesstaat*, ed. by J. Bernhart. Stuttgart (Kröner).

Autrum, H. 1948. Über Energie- und Zeitgrenzen der Sinnesempfindungen. *Naturwiss.*, **35**, 361-369.

Avery, O. T., C. M. Mac Leod and M. Mac Carty. 1944. Studies on the chemical nature of the substance inducing transformations of pneumococcal types. *J. Exp. Med.*, **79**, 137-158.

Bacci, G. 1965. *Sex Determination*. Oxford, London (Pergamon Press).

Bain, A. 1864. *The Senses and the Intellect.* 2d. ed. London.

Ballauf, Th. 1954. *Die Wissenschaft vom Leben. Eine Geschichte der Biologie.* vol. I. Freiburg, München (K. Alber).

Baltzer, F. 1955. Votum in "Finalisme et physicisme." *Actes Soc. Helvét. Sci. Nat.* Porrentruy, 94-98.

Bauer, F. 1957. *Das Verbrechen und die Gesellschaft.* München, Basel (Reinhardt).

Beach, F. A. 1962. *Hormones and Behavior.* New York (Cooper Square Publ.).

Beermann, W. 1962. Genaktivität und Genaktivierung in Riesenchromosomen. Verh. Dtsch. Zool. Ges. Saarbrücken (1961). *Zool. Anz.,* Supp. 25, 44-75.

———. 1965. Operative Gliederung der Chromomeren. (103. Vers. Ges. Dtsch. Naturf. u. Ärzte) *Naturwiss. Rundschau,* **18,** 161.

Bell, D. A. 1962. *Intelligent Machines.* New York (Blaisdell).

Bell, E. (ed.) 1965. *Molecular and Cellular Aspects of Development.* New York, Evanston, London (Harper and Row).

Benary, W. 1922. Studien zur Untersuchung der Intelligenz bei einem Fall von Seelenblindheit. *Psychol. Forschung,* **2,** 209-297.

Bernal, J. D. 1949. *The Physical Basis of Life.* London (Routledge and Kegan Paul).

———. 1965. Molecular Matrices for Living Systems. In: S. W. Fox, *The Origin of Prebiological Systems.* 65-88. New York, London (Academic Press).

Berkeley, G. 1776. *Principles of Human Knowledge* (1710) new ed. London (Dodsley).

Bertalanffy, L. von. 1932, 1942. *Theoretische Biologie.* 2 vol. Berlin (Bornträger).

———. 1937. *Das Gefüge des Lebens.* Berlin, Leipzig (Teubner).

———. 1945. Zu einer allgemeinen Systemlehre. *Blätt. f. Dtsche Philos.,* **18,** Nr. 3-4.

———. 1949. *Das biologische Weltbild.* vol. 1. Bern (Francke).

———. 1950. An outline of general system theory. *Brit. J. Phil. Sci.,* **1,** 134-165.

———. 1932, 1951. *Theoretische Biologie,* 2 vols. Berlin (Bornträger), 2d. ed. Bern (Francke).

Best, J. B. and I. Rubinstein. 1962. Maze learning and associated behaviour in planaria. *J. Comp. Physiol. Psychol.,* **55,** 560-566.

Bier, A. 1944. *Die Seele.* 10th ed. München, Berlin (Lehmann).

Bingham, H. C. 1929. Chimpanzee translocation by means of boxes. *Comp. Psychol. Monogr.* 5, No. 3.

Blois, M. S. 1965. Random polymers as a matrix for chemical evolution. In: S. W. Fox, *The Origin of Prebiological Systems.* 19-38. New York, London (Academic Press).

Bockelmann, P. 1961. *Vom Sinn der Strafe.* Heidelberger Jahrb. V, 25-39.

Bodard, L. 1959. *Chinas lächelndes Gesicht. Erfahrungen und Erlebnisse.* Hamburg (Ch. Wegner).

Böker, H. 1935, 1937. *Einführung in die vergleichende biologische Anatomie der Wirbeltiere,* 2 vols. Jena (G. Fischer).

————. 1936. Was ist Ganzheitsdenken in der Morphologie? *Z. ges. Naturwiss.,* 253-276.

Bonnet, Ch. 1760. *Essai analytique sur les facultés de l'âme.* Copenhagen (Philibert).

Born, M. 1955. Albert Einstein und das Lichtquantum. *Naturwiss.,* **42**, 425-431.

Brachet, J. 1960. *The Biochemistry of Development.* New York, Los Angeles, London, Paris (Pergamon Press).

Brachet, J., and A. E. Mirsky (ed.). 1959-1964. *The Cell. Biochemistry, Physiology, Morphology.* Vols. 1-6. New York, London (Academic Press).

Braun, H. 1952. Über das Vermögen von Papageien, unbenannte Anzahlen zu unterscheiden. *Z. f. Tierpsychol.,* **9**, 40-91.

Bresch, C. 1965. *Klassische und molekulare Genetik.* 2d. ed. Berlin, Heidelberg, New York (Springer).

Broca, P. 1878-1879. *Anatomie comparée du cerveau.* Rev. d'Anthrop.

Brockmöller, K. 1967. *Christentum am Morgen des Atomzeitalters.* Frankfurt a.M. (Knecht).

Brodmann, K. 1909. *Vergleichende Lokalisationslehre der Großhirnrinde, in ihren Prinzipien dargestellt auf Grund des Zellenbaues.* Leipzig (Barth).

————. 1910. Feinere Anatomie des Großhirns. In: Lewandowsky, *Handb. d. Neurol.,* vol. 1, Berlin (Springer).

————. 1925. *Vergleichende Lokalisationslehre der Großhirnrinde.* Leipzig (Barth).

Bröker, W. 1967. *Der Sinn von Evolution. Ein naturwissenschaftlich-theologischer Diskussionsbeitrag.* Düsseldorf (Patmos).

Brücke, E. Th. von. 1934. Gehirnphysiologie, allgemeine und vergleichende. *Handwörterb. d. Naturwiss.,* 2d. ed. vol. 4, 810-850. Jena (G. Fischer).

Bruno, G. 1902. *Von der Ursache, dem Prinzip und dem Einen.* Philos. Bibl., vol. 21. 3d. ed. Leipzig (Dürr).

Budde, F. 1933. Die Lokalisation der äußeren Sinnesempfindungen in den peripheren Organen . . . *Phil. Jahrb.,* **46**, 319-330, 441-449.

Bumke, O. 1948. *Gedanken über die Seele.* Berlin, Heidelberg (Springer).

Bünning, E. 1948. *Entwicklungs- und Bewegungsphysiologie der Pflanze.* Berlin, Göttingen, Heidelberg (Springer).

Bures, J. 1965. Zur Neurophysiologie des Gedächtnisses. In: W. Rüdiger, *Probleme der Physiologie des Gehirns,* 131-151, Berlin (Volk u. Gesundheit).

Burke, E. 1807. *Inquiry into the Origin of our Ideas of the Sublime and Beautiful.* (1757) London (Rivington).

Campbell, A. W. 1905. *Histological Studies on the Localization of the Cerebral Function.* Cambridge (Univ. Press).

Carmichel, L. 1951. Ontogenetic development. In: S. S. Stevens, *Handbook of Experimental Psychology.* New York, London (Wiley).

Carus, K. G. 1846. *Psyche. Zur Entwicklungsgeschichte der Seele.* (Pforzheim), Leipzig (Kröner).

Casey, M. D. 1966. Sex chromosome abnormalities in two State Hospitals for patients requiring special security. *Nature*, **209**, 641-642.

Cassirer, E. 1922, 1957. *Das Erkenntnisproblem in der Philosophie und Wissenschaft der neueren Zeit.* I-III. 2d. ed., IV.

Changeaux, J.-P. 1965. The control of biochemical reactions. *Sci. Amer.*, **212**, 36-45.

Chan Htoon, Th. M. Th. S. U. 1958. Address to the 16th Congress of the International Association for religious freedom, Chicago U.S.A. *The Religion in the Age of Science.* Rangoon (Democracy Publ. Co.).

Chatterji, M. M. 1957. *Viveka-Cudamani or Crest-Jewel of Wisdom of S' ri S'Amkaracarya.* Madras (Theosoph. Publ. House).

Chen, P. S. 1959. Über den Nukleinsäure- und Proteinstoffwechsel der Frühentwicklung bei Seeigeln. *Viertelj. schr. Naturforsch. Ges. Zurich*, **104**, 284-293.

Child, C. M. 1914. Susceptibility gradients in animals. *Science*, **39**.

———. 1928. The physiological gradients. *Protoplasma*, **5**, 447-476.

———. 1941. *Patterns and Problems of Development.* Chicago (Univ. Press).

Chow, K., J. S. Blum and R. A. Blum. 1950. Cell ratios in the thalamocortical visual system of *Macaca mulatta. J. Comp. Neurol.*, **92**, 227-239.

Clark, R. B. 1965. The learning abilities of nereid polychaetes and the role of the supra-oesophageal ganglion. *Animal Behaviour*, Supp. 1, 89-100.

Clifton, C. 1957. *Introduction to Bacterial Physiology.* New York (McGraw-Hill).

Cohen, J. 1964. Psychological time. *Sci. Amer.*, **211**, 116-124.

Cold Spring Harbor Symposia of Quantitative Biology. vol. 24. *Genetics and Twentieth Century Darwinism.* 1959.

Cold Spring Harbor Symposia of Quantitative Biology. vol. 33. *Replication of DNA in Microorganisms.* 1968.

Condillac, M. de. 1766. *Traité des animaux.* Amsterdam (Jombert).

———. 1788. *Traité des sensations.* Nouv. ed. Londres (1st ed. 1754).

Costa de Beauregard, O. 1963. Le dilemme objectivité-subjectivité de la mécanique statistique et l'équivalence cybernétique entre information et entropie. *La Nuova Critica* (Roma) **13-14**, 89-115.

Cowles, J. T. 1937. Foodtokens as incentives for learning by chimpanzees. *Comp. Psychol. Monogr.*, **14**, No. 5.

Crawford, M. P. 1937. The cooperative solving of problems by young chimpanzees. *Comp. Psychol. Monogr.* 14, No. 2.

Czihak, G. 1962. Entwicklungsphysiologie der Echinodermen (Sammelref.). *Fortschr. d. Zool.*, 14, 238-267.

Dacqué, E. 1935. *Organische Morphologie und Paläontologie.* Berlin (Bornträger).

Danner, M. 1969. *Gibt es einen freien Willen?* 2d ed. Hamburg (Kriminalistik-Verl.).

Dantschakoff, V. 1941. *Der Aufbau des Geschlechts beim höheren Wirbeltier.* Jena (G. Fischer).

Davson, H. 1964. *A Textbook of General Physiology.* London (Churchill).

Descartes, R. 1870. *Philosophische Werke*, ed. by J. H. v. Kirchmann. Berlin (Heimann).

Deussen, P. 1922. *Die nachvedische Philosophie der Inder.* 4th ed. Leipzig.

Diels, H. 1906. *Fragmente der Vorsokratiker.* 2d ed. Berlin (Weidmann).

Ditchburn, R. W. and D. H. Fender. 1955. The stabilized retinal image. *Opt. Acta*, 2, 128-133.

Dobzhansky, Th. 1951. *Genetics and the Origin of Species.* New York (Columbia Univ. Press) (1st ed. 1937), 3d ed.

———. 1956. *The Biological Basis of Human Freedom.* New York (Columbia Univ. Press).

———. 1960. Evolution and environment. In: S. Tax, *Evolution after Darwin*, vol. 1, 403-428, Chicago (Chicago Univ. Press).

———. 1962. *Mankind Evolving.* New Haven and London (Yale Univ. Press).

———. 1967. *The Biology of Ultimate Concern.* New York (New Amer. Library).

Driesch, H. 1901. *Die organischen Regulationen.* Leipzig.

———. 1928. *Philosophie des Organischen.* Leipzig (Quelle u. Meyer).

Dubitscher, F. 1937. Der Schwachsinn. In: A. Gütt, *Handb. d. Erbkrankheiten*, vol. 1, Leipzig (Thieme).

Dücker, G. 1967. Untersuchungen über geometrisch-optische Täuschungen bei Wirbeltieren. *Z. f. Tierpsychol.*, 23, 452-496.

Dusser de Barenne, J. G. 1937. Physiologie der Großhirnrinde. In: Bumke u. Förster, *Handb. d. Neurol.*, vol. 2, 268-319, Berlin (Springer).

Eccles, J. C. 1957. *The Physiology of Nerve Cells.* Baltimore (J. Hopkins Press).

Economo, C. von u. G. N. Koskinas. 1925. *Die Cytoarchitektonik der Hirnrinde des erwachsenen Menschen.* Wien, Berlin.

Eddington, A. 1939. *The Philosophy of Physical Science.* Cambridge (Univ. Press).

Eder, G. 1960. Relativitätstheorie. In: W. Gerlach, *Physik.* Fischer-Lexikon, 316-330, Frankfurt a.M. (S. Fischer).

Edgar, R. S. and R. H. Epstein. 1965. The genetics of a bacterial virus. *Sci. Amer.* **212**, 71-78.

Ehrenfels, Ch. von. 1890. Über Gestaltsqualitäten. *Vierteljahresschr. wiss. Philos.*, 14.

Ehrenstein, W. 1947. *Probleme der ganzheitspsychologischen Wahrnehmungslehre.* Leipzig (Barth) (1942) 2d ed.

Einstein, A. 1920. *Über die spezielle und die allgemeine Relativitätstheorie.* 6th ed. Braunschweig (Vieweg) (1st ed. 1916).

Eisler, R. 1910. *Geschichte des Monismus.* Leipzig (Kröner).

Eisler, R. u. R. Müller-Freienfels. 1922. *Handwörterbuch der Philosophie.* 2d ed. Berlin (Mittler).

Ellson, D. G. 1941. Hallucinations produced by sensory conditioning. *J. exper. Psychol.*, **28**, 1-20.

Fechner, G. Th. 1919. *Die Tagesansicht gegenüber der Nachtansicht.* Leipzig (Breitkopf u. Härtel) (1879) 3d ed.

Ferster, Ch. B. 1964. Arithmetic behavior in chimpanzees. *Sci. Amer.*, **210**, 98-106.

Feuerbach, L. 1913. *Das Wesen der Religion* (1845), ed. by H. Floerke. Berlin (Deutsche Bibl.).

Fischer, J. C. 1871. *Die Freiheit des menschlichen Wollens oder die Einheit der Naturgesetze.* 2d ed. Leipzig.

Fischer, R. A. 1950. *Statistical Methods for Research Workers.* Edinburgh (Oliver and Boyd) (1st ed. 1925) 11th ed.

Florkin, M. et E. Schoffeniels. 1966. *Biochimie et biologie moléculaire.* (Desoer).

Flourens, M. J. P. 1842. *Recherches expérimentales sur les propriétés et les fonctions du système nerveux.* Paris (1824). 2d ed.

Fox, S. W. 1965. *The Origins of Prebiological Systems and Their Molecular Matrices.* New York, London (Academic Press).

Frank, H. 1964. Was ist Kybernetik? In: H. Frank, *Kybernetik, Brücke zwischen den Wissenschaften.* Frankfurt a.M. (Umschau Verl.), 9-20.

Fraunberger, F. 1960. Mechanik. In: W. Gerlach, *Physik,* 224-232. Fischer-Lexikon Frankfurt (S. Fischer).

Frey, E. 1951. *Der frühkriminelle Rückfallverbrecher, Basel* (Verl. Recht u. Gesellsch.).

Frick, H. u. D. Starck. 1963. Vom Reptil- zum Säugerschädel. *Z. f. Säugetierkde,* **28**, 321-341.

Frisch, K. von. 1919. Über den Geruchssinn der Bienen und seine blüten biologische Bedeutung. *Zool. Jahrb., Physiol.* **37**, 1-238.

——. 1965. *Tanzsprache und Orientierung der Bienen.* Berlin, Heidelberg, New York (Springer).

Fritsch, G. u. E. Hitzig. 1879. Über die elektrische Erregbarkeit des Großhirns. *Arch. f. Anat., Physiol. u. wiss. Medicin.*

Gall, F. J. 1809. *Anatomie et physiologie du système nerveux en général, et du cerveau en particulier.* 4 vols. Paris.

Galton, F. 1892. *Hereditary Genius* (1869). London (Macmillan).

Gambhirananda, S. 1958. *A Short Life of Sri Ramakrishna.* Calcutta (Advaita Ashrama).

Gaze, R. M. 1958. The representation of the retina on the optic lobe of the frog. *Quart. J. Exp. Physiol.,* 43, 209-214.

Gaze, R. M. and M. Jacobson. 1963. Types of single-unit visual responses from different depths in optic tectum of goldfish. *J. Physiol.* (London), 169, 92-93.

Gelb, A. u. K. Goldstein. 1920. Zur Psychologie des optischen Wahrnehmungs- und Erkennungsvorganges. In: *Psycholog. Analysen hirnpathol. Fälle.* Leipzig.

George, F. H. 1961. *The Brain as a Computer.* Oxford, London, New York, Paris (Pergamon Press).

Gerisch, G. 1965. Spezifische Zellkontakte als Mechanismen der tierischen Entwicklung. *Umschau,* 65, 392-395.

Gerlach, W. 1960. Elementarteilchen. In: W. Gerlach (ed.), *Physik,* 115-123. Fischer-Lexikon. Frankfurt a.M. (S. Fischer).

Geyer, B. 1928. Die patristische und scholastische Philosophie. In: F. Ueberwegs *Grundriß d. Gesch. d. Philos.* Berlin (Mittler).

Gieseler, W. 1957. Die Fossilgeschichte des Menschen. In: G. Heberer, *Evolution d. Organismen,* 2d ed. 951-1109, Stuttgart (G. Fischer).

Glasenapp, H. von. 1948. *Der Stufenweg zum Göttlichen. Shankaras Philosophie der All-Einheit.* Baden-Baden (Bühler.)

Glees, P. 1957. *Morphologie und Physiologie des Nervensystems.* Stuttgart (Thieme).

Glees, P. and H. B. Griffith. 1952. Bilateral destruction of the hippocampus in a case of dementia. *Mon. schr. Psychiatr. Neurol.* 123, 193-204.

Glueck, E. and Sh. Glueck. 1951. *Unraveling Juvenile Delinquency.* Cambridge.

Goldscheider, A. 1906. Über die materiellen Veränderungen bei der Assoziationsbildung. *Neurol. Zentralbl.* 25.

Goldschmidt, R. 1935. Geographische Variation und Artbildung. *Naturwiss.* 23, 169-176.

———. 1952. Evolution as viewed by one genetist. *Amer. Scientist,* 40, 84-135.

Goldstein, K. 1927. Die Lokalisation in der Großhirnrinde. In: Bethe et al., *Handb. norm. u. pathol. Physiol.,* vol. 10, Berlin (Springer).

Goodall, J. 1963. Feeding behavior of wild chimpanzees. *Sympos. Zool. Soc. London,* No. 10, 39-49.

Gottschaldt, K. 1954. Zur Theorie der Persönlichkeit und ihrer Entwicklung. *Z. f. Psychol,* 157, 2-22.

Griffard, C. D. and J. T. Peirce. 1964. Conditioned discrimination in the Planarian. *Science*, **144,** 1472-1473.

Groos, H. 1939. *Willensfreiheit oder Schicksal?* München (Reinhardt).

Grosser, G. S. and J. M. Harrison. 1960. Behavioral interaction between stimulated cortical points. *J. Comp. Physiol. Psychol.*, **53,** 229-233.

Grundmann, E. 1964. *Allgemeine Cytologie.* Stuttgart (Thieme).

Hackmann, H. 1927. Chinesische Philosophie. In: *Geschichte d. Philos. in Einzeldarst. Abt.* I, vol. 5, München (Reinhardt).

Haeckel, E. 1922. *Der Monismus als Band zwischen Religion u. Wissenschaft.* (1893) 17th ed. Leipzig (Kröner).

———. 1914. *Gott-Natur (Theophysis). Studien über monistiche Religion.* Leipzig (Kröner).

———. 1923. *Zellseelen und Seelenzellen.* (1914) Leipzig (Kröner). 2d ed.

———. 1921. *Italienfahrt. Briefe an die Braut.* Leipzig (K. F. Koehler).

Haecker, V. 1925. *Pluripotenzerscheinungen. Synthetische Beiträge zur Vererbungs- und Abstammungslehre.* Jena (G. Fischer.)

Haffner, H. 1967. Sternhaufen und Sternentwicklung. *Arbeitsgem. f. Forsch. Land. Nordrhein-Westf., Nat. wiss., H.* 173, Köln, Oplanden.

Haldane, J. St. 1931. *The Philosophical Basis of Biology.* London.

Hämmerling, J. 1951. *Fortpflanzung im Tier- und Pflanzenreich.* (Samml. Göschen). Berlin (de Gruyter).

Harada, K. and S. W. Fox. 1965. Thermal polycondensation of free amino acids with polyphosphoric acid. In: S. W. Fox, *The Origin of Prebiological Systems,* 289-298. New York, London (Academic Press).

Hardy, E. 1902. König Asoka. In: Kampers et al. *Weltgeschichte in Charakterbildern.* 1. Abt. Mainz (Kirchheim).

Harlow, H. F. 1943. Solution by rhesus monkeys on a problem involving the Weigl principle using the matching from sample method. *J. Comp. Psychol.*, **36,** 217-227.

Harnack, A. von. 1931-1932. *Lehrbuch der Dogmengeschichte.* 5th ed. Tübingen (Mohr).

Hartmann, E. von. 1924. *Phänomenologie des sitlichen Bewußtseins.* Berlin (1879). (Wegweiser Verl.) 1924.

Hartmann, M. 1937. *Philosophie der Naturwissenschaften. Berlin* (Springer).

———. 1943. *Die Sexualität.* Jena (G. Fischer).

———. 1948. *Die philosophischen Grundlagen der Naturwissenschaften.* Jena (G. Fischer).

———. 1951. *Geschlecht und Geschlechtsbestimmung im Tier und Pflanzenreich.* (Samml. Göschen), 2d ed., Berlin (de Gruyter).

———. 1956. Neue Befunde über Befruchtungsstoffe (Gamone) und ihre theoretische Bedeutung. *Naturwiss.* 43, 313-317.

Hartmann, N. 1941. *Grundzüge einer Metaphysik der Erkenntnis.* 3d ed. Berlin (de Gruyter).

———. 1949. *Ethik.* 3d ed. Berlin (de Gruyter). (1st ed. 1925).

———. 1950. *Philosophie der Natur*. Berlin (de Gruyter).

Hassenstein, B. 1964. Forschungsbeispiele aus der biologischen Kybernetik. In: H. Frank, *Kybernetik*, 23-29, Frankfurt a.M. (Umschau-Verlag).

———. 1966. Kybernetik und biologische Forschung. In: L. von Bertalanffy u. F. Gessner, *Handb. d. Biol.*, vol. 1, 629-719, Frankfurt (Athenaion).

Hayes, C. 1952. *The Ape in Our House*. New York (Harper).

Heath, R. G. 1964. Pleasure response of human subjects to direct stimulation of the brain: physiologic and psychodynamic considerations. In: Heath (ed.), *The Role of Pleasure in Behavior*. 219-243, New York, Evanston, London (Hoeber).

———. 1964. *The Role of Pleasure in Behavior*. New York, Evanston, London (Harper and Row).

Heberer, G. 1952. Die Fortschritte in der Erforschung der Phylogenie der Hominoidea. *Ergebn. Anat. u. Entwickl. gesch.* 34, 499-637.

———. 1956. Die Fossilgeschichte der Hominoidea. In: Hofer, Schultz, Starck, *Primatologia*, vol. 1, 379-650, Basel, New York.

———. 1959a. (ed.). *Die Evolution der Organismen*. 2d ed., 2 vols., Stuttgart (G. Fischer) (1. ed. Jena 1943).

———. 1959b. Die subhumane Abstammungsgeschichte des Menschen. In: Heberer, *Evolution d. Organismen*. 2d. ed. 1110-1142, Stuttgart (G. Fischer).

———. (ed.) 1965. Menschliche Abstammungslehre. Stuttgart (G. Fischer).

Hegel, G. W. F. 1832. *Vorlesungen über die Philosophie der Religion*, ed. by Ph. Marheinke. In: *Werke*, vol XI and XII, Berlin.

Heinze, H. 1942. *Psychopathische Persönlichkeiten. Erbbiologischer Teil.* In: A. Gütt, *Handb. d. Erbkrankh.* vol. 4, Leipzig (Thieme).

Heisenberg. W. 1959. *Physik und Philosophie*. W.-Berlin (Ullstein).

Henle, J. 1838. Über das Gedächtnis in den Sinnen. *Wochenschr. f. d. ges. Heilkunde*. 281-289, 297-306.

Hennemann, G. 1967. *Probleme der physikalischen und religiösen Wirklichkeit*. (Erfahrung u. Denken Bd. 18). Berlin (Duncker u. Humblot).

Hering, E. 1905. *Grundzüge der Lehre vom Lichtsinn*. In: Graefe, Saemisch, *Handb. d. Augenheilkde.*, 1. Teil, Leipzig (Engelmann).

Hershey, A. D. and M. Chase. 1951. Genetic recombination and heterozygosis in bacteriophage. *Cold Spring Harbor Sympos. Quant. Biol.*, 16, 471-480 (*J. gen. Physiol.* 36, 39, 1952).

Herter, K. 1929. Dressurversuche an Fischen. *Z. vergl. Physiol.* 10, 688-711.

———. 1953. *Die Fischdressuren und ihre sinnesphysiologischen Grundlagen*. Berlin (Akad. Verl.).

Hess, W. R. 1954. *Das Zwischenhirn*. Basel (Schwabe).

Hesse, R. 1935. *Tierbau und Tierleben* (2d. ed.). vol. 1, Jena (G. Fischer).

Hirsch, G. Ch. *Die Zellorganellen*. In: L. von Bertalanffy u. F. Gessner, *Handb. d. Biol.*, vol. I, H. 19-21, Konstanz (Athenaion).

Hitzig, E. 1874. *Untersuchungen über das Gehirn*. Berlin (Hirschwald).

Hobbes, Th. 1655. *Elementorum Philosophiae Sectio prima de Corpore.* London (Crook). Opera philosophica, Aalen (Scientia 1961).

———. 1949. *De Homine* (1659). Lehre vom Menschen. Trans. by M. Frischeisen-Köhler. Leipzig (Meiner).

Höffding, H. 1911. *Der menschliche Gedanke.* Leipzig.

Hofmann, F. B. 1925. *Physiologische Optik* (Raumsinn). In: Graefe, Saemisch, Hess. *Handb. d. ges. Augenheilkunde.* 2d ed., 3 vols. Berlin.

d'Holbach, P. Th. 1770. *Système de la nature.*

Holst, E. von. 1957. Die Auslösung von Stimmungen bei Wirbeltieren durch "punktförmige" elektrische Reizungen des Stammhirns. *Naturwiss.*, 44, 549-551.

Holtfreter, J. 1934. Formative Reize in der Embryonalentwicklung der Amphibien, dargestellt an Explantationsversuchen. *Arch. exp. Zellforsch.*, 15, 281-301.

———. 1939. Studien zur Ermittlung der Gestaltungsfaktoren in der Organentwicklung der Amphibien. I. *Roux' Arch. Entw.*, 139, 110-190.

Hopf, A. 1964. Localization in the cerebral cortex from the anatomical point of view. In: G. Schaltenbrand and C. N. Woolsey, *Cerebral Localization and Organization.* 5-16, Madison, Milwaukee (Univ. Wisconsin Press).

Hörmann, M. 1931. Über den Helligkeitssinn der Bienen. *Z. vergl. Physiol.*, 21, 188-219.

Hörstadius, S. 1939. The mechanics of sea urchin development, studied by operative methods. *Biol. Rev. Cambridge Phil. Soc.*, 14, 132-179.

———. 1949. Experimental researches on the developmental physiology of the sea urchin. *Publ. Staz. Zool. Napoli*, 21, Supp., 131-172.

Hosemann, H. 1949. *Die Grundlagen der statistichen Methoden für Mediziner und Biologen.* Stuttgart (Thieme).

Hubel, D. H. 1963. The visual cortex of the brain. *Sci. Amer.*, 209, No. 5, 54-62.

Hubel, D. H. u. T. N. Wiesel. 1965. Receptive fields and functional architecture in two non-striate visual areas (18 and 19) of the cat. *J. Neurophysiol.*, 28, 229-289.

Huber, F. 1965. Aktuelle Problems in der Physiologie des Nervensystems der Insekten. *Naturwiss. Rundschau*, 16, 143-156.

Hughes, J. L. 1963. Possible application of the mass energy cycle on a micro- and macroscopic scale in the universe. *Nature*, 197, 441-442; 198, 772-773.

Hume, D. 1913, 1921. *A Treatise of Human Nature* (1739). London (Noon).

Husserl, E. 1913, 1921. *Logische Untersuchungen.* 2 vols. 2d ed. Halle (Niemeyer).

Huxley, J. 1932. *Problems of Relative Growth.* London (Methuen).

———. 1941. *Religion without Revelation.* The Thinker's Library No. 83, London (Watts).

———. 1942. *Evolution. The Modern Synthesis.* London (Allen and Unwin) (2d ed. 1963).

———. 1954. *Evolutionary Humanism.* (Austral. Inst. of Internat. Affairs). Melbourne.

———. 1960. The evolutionary vision: the Convocation Address. In S. Tax, *Evolution after Darwin,* vol. III, 249-261, Chicago (Chicago Univ. Press).

Huxley, J., A. C. Hardy, E. B. Ford. 1954. *Evolution as a Process.* London (Allen and Unwin).

Hydén, H. u. E. Eghazi. 1962. Nuclear RNA-changes in nerve cells during a learning experiment in rats. *Proc. Nat. Ac. Sci. (Wash.)* **48,** 1366-1373.

Hydén, H. and P. W. Lange. 1966. A genetic stimulation with production of adenic-uracil rich RNA in neurons and glia in learning. *Naturwiss.,* **53,** 64-70.

Jacobs, P. A., M. Brunton and M. M. Melville. 1965. Aggressive behaviour, mental subnormality and the XYY male. *Nature,* **208,** 1351-1352.

Jacobson, A. L. 1965. Learning in planarians: current status. *Animal Behaviour,* Supp. 1, 76-82.

Jacobson, M. and R. M. Gaze. 1964. Types of visual response from single units in the optic tectum and optic nerve of the goldfish. *Quart. J. Exp. Physiol.,* **49,** 199-209.

Jasper, H. H. 1954. Functional properties of the thalamic reticular system. In: J. F. Delafresnaye, *Brain Mechanics and Consciousness.* Oxford (Blackwell).

Jeans, J. 1913. *Physics and Philosophy.* Cambridge (Cambridge Univ. Press).

John, E. R. 1967. *Mechanisms of Memory.* New York, London (Academic Press).

Jordan, H. J. 1929. *Allgemeine vergleichende Physiologie der Tiere.* Berlin, Leipzig (de Gruyter).

Jordan, P. 1937. Die physikalischen Weltkonstanten. *Naturwiss.* **25,** 513-517.

———. 1938. Zur empirischen Kosmologie. *Naturwiss.,* **26,** 417-421.

———. 1945. Die Physik und das Geheimnis des organischen Lebens. *Die Wissenschaft,* vol. 95, 4th ed. Braunschweig (Vieweg).

———. 1947. *Die Herkunft der Sterne.* Stuttgart (Wiss. Verlagsges.).

Josenhans, W. 1959. Versuch einer Objektivierung von Änderungen im Zeiterleben und Zeitempfinden. *Naturwiss.,* **46,** 538-539.

Juda, A. 1934. Über Anzahl und psychische Beschaffenheit der Nachkommen von schwachsinnigen und normalen Schülern. *Z. Neurol.,* **151.**

———. 1935. Über Fruchtbarkeit und Belastung bei den Seitenverwandten von schwachsinnigen und normalen Schülern und deren Nachkommen. *Z. Neurol.,* **154.**

Just, G. 1928. Methoden der Vererbungslehre. In: *Methodik der wiss. Biol.,* vol. II, 502-607, Berlin (Springer).

Kant, I. 1922a. *Träume eines Geistersehers erläutert durch Träume der Metaphysik* (Königsberg 1766). In: *Werke*, ed. by E. Cassirer, vol. II, 329-390, Berlin (Cassirer).

———. 1922b. *Idee zu einer allgemeinen Geschichte in weltbürgerlicher Absicht.* (Berlinische Monatsschr. 1784, 386-410), in: *Werke*, ed. by E. Cassirer, Bd. IV, Berlin (Cassirer).

———. 1923a. *Kritik der reinen Vernunft* (2d ed. Riga 1787). In: *Werke*, ed. by E. Cassirer, Bd. III, Berlin (Cassirer).

———. 1922c. *Kritik der praktischen Vernunft* (Riga 1788). In: *Werke*, ed. by E. Cassirer, vol. IV, Berlin (Cassirer).

———. 1923b. *Die Religion innerhalb der Grenzen der bloßen Vernunft* (Königsberg 1794). In: *Werke*, ed. by E. Cassirer, vol. VI, Berlin (Cassirer).

———. 1923c. *Anthropologie in pragmatischer Hinsicht abgefaßt.* (2d ed. Königsberg 1800). In: *Werke*, vol. VIII, ed. by O. Schöndörffer, Berlin (Cassirer).

Kapune, T. 1966. Untersuchungen zur Bildung eines "Wertbegriffs" bei niederen Primaten. *Z. f. Tierpsychol.*, **23**, 324-363.

Karlson, P. 1966. *Kurzes Lehrbuch der Biochemie.* 5th ed. Stuttgart (Thieme).

Karrer, P. 1963. *Lehrbuch der organischen Chemie.* 14th ed. Stuttgart (Thieme).

Kaudewitz, F. 1964. Beiträge der Bakteriengenetik zum Verständnis der molekularen Grundlagen der Vererbung. *Vehr. Dtsch. Zool. Ges. München* (1963). *Zool. Anz.*, Supp. 27, 43-60.

Kaufmann, A. 1961. *Das Schuldprinzip.* Heidelberg (Winter).

Keosian, J. 1964. *The Origin of Life.* New York (Reinhold Publ. Corp.), London (Chapman and Hall).

Kippenhahn, R. u. A. Weigert. 1967. *Das Leben der Sterne. Bild d. Wiss.*, **2**, 99-107.

Klages, F. 1961. *Naturstoffe.* In: H. Kelker, F. Klages, R. Schwarz, *Chemie.* Fischer-Lexikon, Frankfurt a.M., Hamburg (S. Fischer).

Kleist, K. 1934. *Gehirnpathologie. Vornehmlich auf Grund der Kriegserfahrungen.* In: von Schjerning, *Handb. d. ärztl. Erfahrungen im Weltkriege*, vol. 4, Leipzig (Barth).

Klingenberg, M. 1963. *Die funktionelle Biochemie der Mitochondrien.* In P. Karlson, *Funktionelle u. morphologische Organisation der Zelle*, 69-85, Berlin, Göttingen, Heidelberg (Springer).

Klumbies, G. 1956. Kausalität oder Finalität? *Ärztl. Wochenschr.*, **11**, 765-769.

Klüver, H. 1933. *Behavior Mechanisms in Monkeys.* Chicago (Univ. Chicago Press).

Knussmann, R. 1961. Zur Methode der objektiven Körperbautypognose. *Z. menschl. Vererb. u. Konstitutl.*, **36**.

Kochs, W. 1890. Kann die Kontinuität der Lebensvorgänge zeitweilig völlig unterbrochen werden? *Biol. Zentralbl.*, **10**, 673-686.

———. 1892. Über die Vorgänge beim Einfrieren und Austrocknen von Tieren und Pflanzensamen. *Biol. Zentralbl.*, **12**, 330-339.

Koenigswald, G. R. H. von. 1953. Die Phylogenie des Menschen. *Naturwiss.*, **40**, 128-137.

Koffka, K. 1912. *Zur Analyse der Vorstellungen und ihrer Gesetze. Eine experimentelle Untersuchung.* Leipzig (Quelle u. Meyer).

———. 1935. *Principles of Gestalt Psychology.* New York (Harcourt, Brace, World).

Koehler, O. 1933. Die Ganzheitsbetrachtung in der Biologie. *Schr. Königsberger Gelehrten Ges., Nat. Kl.*, **9**, 139-204.

———. 1941. Vom Erlernen unbenannter Anzahlen bei Vögeln. *Naturwiss.*, **29**, 201, 218.

———. 1943. "Zähl"-Versuche an einem Kolkraben und Vergleichsversuche an Menschen. *Z. f. Tierpsychol.*, **5**, 575-712.

———. 1952. Vom unbenannten Denken. *Verh. Dtsch. Zool. Ges.*, **46**, 202-211.

Köhler, W. 1920. *Die physischen Gestalten in Ruhe und im stationären Zustand.* Braunschweig (Vieweg).

———. 1921. *Intelligenzprüfungen an Menschenaffen.* Berlin (Springer).

———. 1921. *The Mentality of Apes.* New York (Harcourt Brace).

———. 1928. *Gestalt Psychology.* New York (Liveright).

———. 1940. *Dynamics in Psychology.* New York (Liveright Publ. Corp.).

Kohts, N. 1928. Recherches sur l'intelligence de chimpanzé par la méthode de "choix d' après modèle," *J. de Psychol.*, **25**, 255-275, 1928.

Koller, S. 1953. *Graphische Tafeln zur Beurteilung statistischer Zahlen.* Darmstadt (Steinkopff) (1st ed. 1939), 3d ed.

El Koran, 1916. Trans. by Goldschmidt. Berlin (Brandus).

Kranz, W. 1949. *Vorsokratische Denker. Auswahl aus dem Überlieferten.* Berlin, Frankfurt (Weidmann).

Krejci-Graf, K. 1960. Das Alter der Erde und der Welt. *Naturwiss. Rundschau*, **13**, 424-427.

Kretschmer, E. 1961. *Körperbau und Charakter.* Berlin (Springer) (1921) 24th ed.

Krüger, F. 1937. *Das Wesen der Gefühle.* 5. Aufl. Leipzig (Akad. Verl. Ges).

Krüger, F. 1967. Zur mathematischen Wiedergabe des Rattenwachstums. *Verh. Dtsch. Zool. Ges. Göttingen* (1966), *Zool. Anz.*, Supp. 30, 587-599.

Kühn, A. 1921. Nachweis des simultanen Farbenkontrasts bei Insekten. *Naturwiss.*, **9**, 575-576.

———. 1965. *Vorlesungen über Entwicklungsphysiologie.* 2d ed. Berlin, Heidelberg, New York (Springer).

Kuhn-Schnyder, E. 1963. Wege der Reptiliensystematik. *Paläont. Z.*, **37**, 61-87.

Külpe, O. 1902. Über die Objektivierung und Subjektivierung von Sinneseindrücken. *Philos. Studien*, **19**, I. Teil, 508-556.

――――. 1912, 1920, 1922. *Die Realisierung.* 3 vols., Leipzig (Hirzel).

――――. 1923. *Einleitung in die Philosophie.* 11th ed. Leipzig (Hirzel). (1st ed. 1895).

La Mettrie, J. O. de. 1909. *Der Mensch eine Maschine.* Trans. after original edition of "L'homme machine" 1748, by M. Brahn (Philosoph. Bibl. Bd. 68), Leipzig (Duerr).

Lange, J. 1929. *Verbrechen als Schicksal.* Leipzig.

Lange, R. 1955. Grundfragen der deutschen Strafrechtsreform. *Schweiz, Ztschr. f. Strafrecht*, **70**, 373-397.

Laplace, P. S. 1814. *Essai philosophique sur les probabilités.* Paris.

Lashley, K. S. 1929. *Brain Mechanisms and Intelligence.* Chicago (Chicago Univ. Press).

Leferenz, H. 1948. Die rechtsphilosophischen Grundlagen des § 51 StGB. *Der Nervenarzt*, **19**, 364-372.

――――. 1959. *Die Kriminalprognose auf statistischer Grundlage.* In: *Kriminalpolit. Gegenwartsfragen.* Ed. by Bundeskriminalamt Wiesbaden.

Le Gros Clark, W. E. 1954. *History of the Primates.* 4th ed., London (Brit. Mus. Nat. Hist.)

Lehmann, F. E. 1945. *Einführung in die physiologische Embryologie.* Basel (Birkhäuser).

Lehmann, O. 1906. *Flüssige Kristalle und die Theorien des Lebens.* Vers. dtsch. Naturforsch. u. Ärzte. Stuttgart.

――――. 1907. Flüssige Kristalle und ihre Analogien zu den niedrigsten Lebewesen. *Kosmos* (Stuttgart), 5-13, 36-40.

――――. 1908. *Flüssige Kristalle, ihre Entstehung, Bedeutung und Ähnlichkeit mit Lebewesen.* Frankfurt.

Lehnartz, E. 1959. *Einführung in die chemische Physiologie.* 11th ed. Berlin, Göttingen, Heidelberg (Springer).

Lehrman, D. S. 1961. Hormonal regulation of parental behavior in birds and infrahuman mammals. In: W. C. Young, *Sex and Internal Secretion*, 3d ed., No. 21, 1268-1382.

Leibniz, G. W. 1906a. Betrachtungen über die Lehre von einem einigen, allumfassenden Geiste. (Sur ce qui passe les sens et la matière.) (1702). In: *Hauptschriften zur Grundlegung d. Philos.*, vol. II. Philos. Bibl. Bd. 108, Leipzig (Dürr).

――――. 1906b. Betrachtungen über die Lebensprinzipien und über die plastischen Naturen (1705). In: *Hauptschriften zur Grundlegung der Philosophie.* Trans. by A. Buchenau, ed. by E. Cassirer, vol. II, Leipzig (Dürr).

――――. 1879. *Die Theodicee* (1710). Trans. by J. H. von Kirchmann (Philos. Bibl. Bd. 71), Leipzig (Meiner).

Lemaître, G. 1946. *Essai de Cosmologie: L'hypothèse de l'atom primitif.* Neufchâtel.

Lerner, J. M. 1954. *Genetic Homeostasis.* Edinburgh, London (Oliver and Boyd).

Liebmann, O. 1911. *Zur Analysis der Wirklichkeit.* 4th ed. Straßburg (Trübner). (1st ed. 1876).

Lindemann, E. 1922. Experimentelle Untersuchungen über das Entstehen und Vergehen von Gestalten. *Psychol. Forschung.,* **2,** 5-60.

Lipps, G. F. 1912. *Das Problem der Willensfreiheit.* Leipzig (Teubner).

Liszt, F. von. 1905. *Strafrechtliche Aufsätze und Vorträge.* 2 vols., Berlin (J. Kapp). Ges. Schriften. Leipzig (Breitkopf u. Härtel 1910).

Locke, J. 1877. *An Essay Concerning Human Understanding* (1690). In: *Philos. Works,* vol. 1, ed. by J. A. St. John. London (Bell).

Lorenz, K. 1943. Die angeborenen Formen möglicher Erfahrung. *Z. f. Tierpsychol.,* **5,** 235-409.

———. 1957. Methoden der Verhaltensforschung. In: Kükenthal, Krumbach, v. Lengerken, Helmcke, *Handb. d. Zool.* vol. 8, T. 10, Berlin (de Gruyter).

Lotze, R. H. 1852. *Medizinische Psychologie oder Physiologie der Seele.* Leipzig (Weidmann).

Lullies, H. 1959. Elektrophysiologie der Erregung und Erregungsleitung in Geweben, insbesondere in Nerven. *Elektromedizin,* **4,** 1-12, 25-34, 49-58.

Lüttgau, H. C. 1960. Die Elektrophysiologie erregbarer Membranen. *Dtsch. med. Wschr.* **85,** 2288-2292.

———. 1963. Nervenphysiologie. *Fortschritte d. Zool.,* **15,** 92-124.

Mach, E. 1906. *Erkenntnis und Irrtum.* 2d ed. Leipzig (Barth).

Machemer, H. 1966. Erschütterungsbedingte Sensibilisierung gegenüber rauhem Untergrund bei Stylonychia mytilus. *Arch. Protistenkde,* **109,** 245-256.

Mahadevan, T. M. P. 1953. *The Time and the Timeless.* Madras (Upanishad Vihar).

Malebranche, N. 1938. *De la recherche de la vérité* (6th ed. 1712). In: *Oeuvres Complètes,* T. 1, Paris (Roustan).

Mangold, O. 1935. Kombination verschiedener Geschmacksqualitäten zur Untersuchung des chemischen Sinnes des Regenwurms. *Naturwiss.,* **23,** 472-474, 1935.

Marquis, D. G. and E. R. Hilgard. 1936. Conditioned lid responses to light in dogs after removal of the visual cortex. *J. Comp. Psychol.,* **22,** 157-178.

Martin, R. u. K. Saller. 1966. *Lehrbuch der Anthropologie.* vol. IV. Stuttgart (G. Fischer).

Mayr, E. 1942. *Systematics and the Origin of Species.* New York (Columbia Univ. Press).

———. 1963. *Animal Species and Evolution.* Cambridge, Mass. (Univ. of Harvard Press).

Meesters, K. 1940. Über die Organisation des Gesichtsfeldes der Fische. *Z. f. Tierpsychol.*, **4**, 84-149.

Meisenheimer, J. 1921, 1930. *Geschlecht und Geschlechter im Tierreich.* 2 vols. Jena (G. Fischer).

Meister, H.-J. 1960. *Thermodynamik.* In: W. Gerlach, *Physik.* Fischer-Lexikon. Frankfurt a.M. (S. Fischer).

Metzger, W. 1940. *Psychologie.* Darmstadt (Steinkopff) 1940, (2d ed. 1954).

Mezger, E. 1926. *Persönlichkeit und strafrechtliche Zurechnung.* München (Bergmann).

Meyer-Abich, A. 1948. *Naturphilosophie auf neuen Wegen.* Stuttgart (Hippocrates).

Meyer-Holzapfel, M. 1956. *Das Spiel der Säugetiere.* In: Kükenthal et al., *Handb. d. Zool.*, vol. 8, Teil 10. Berlin (de Gruyter).

Meynert, Th. 1867, 1868. Der Bau der Großhirnrinde und seine örtlichen Verschiedenheiten nebst einem pathologisch-anatomischen Corollarium. *Vierteljahresschr. f. Psychiatrie, Leipzig*, 77-93, 88-113.

――――. 1872. *Zur Mechanik des Gehirnbaues.* Vortr. Naturforschervers. Wiesbaden.

Mill, J. St. 1898. *A System of Logic Ratiocinative and Inductive . . .* (1843) London (Longmans).

Miller, S. L. 1953. A production of amino acids under possible primitive earth conditions. *Science*, **117**, 528-529.

Mittmann, O. 1940. *Erbbiologische Fragen in mathematischer Behandlung.* Berlin.

Monakow, C. von. 1910. *Über Lokalisation von Hirnfunktionen.* Wiesbaden (Bergmann).

Monod, J. 1959. Biosynthese eines Enzyms. *Angew. Chemie*, **71**, 685-691.

Morris, D. 1962. *The Biology of Art.* London (Methuen).

Morus, Thomas. 1947. *Utopia, das ist Nirgendland oder von der besten Staatsform.* Trans. by H. Schiel, Köln.

Müffelmann, L. 1902. *Das Problem der Willensfreiheit in der neuesten deutschen Philosophie.* Leipzig (Barth).

Müller, J. 1840. *Handbuch der Physiologie des Menschen.* vol. 2. Coblenz (Hölscher).

Munk, H. 1890. *Über die Funktionen der Großhirnrinde.* Berlin (Hirschwald) (1881) 2d ed.

Muralt, A. von. 1945. *Die Signalübermittlung im Nerven.* Basel (Birkhäuser).

――――. 1958. *Neue Ergebnisse der Nervenphysiologie.* Berlin, Göttingen, Heidelberg (Springer).

Natorp, P. 1908. *Religion innerhalb der Grenzen der Humanität.* Tübingen (Mohr).

Newman, H. H., F. N. Freeman and K. J. Holzinger (1937). *Twins: A Study of Heredity and Environment.* Chicago (Chicago Univ. Press).

Nikhilananda, S. 1951. *The Upanishads.* London (Phoenix House).

Noer, F. A. von. 1880, 1885. *Kaiser Akbar.* 2 vols. Leiden (Brill).

Oesterreich, T. K. 1923. *Die deutsche Philosophie des XIX. Jahrhunderts und der Gegenwart.* In: F. Ueberwegs *Grundriß d. Gesch. d. Philos.* 4th Teil, 12th ed., Berlin (Mittler).

Offner, M. 1904. *Willensfreiheit, Zurechnung und Verantwortung.* Leipzig (Barth).

Oken, L. 1809. *Lehrbuch der Naturphilosophie.* I. Jena (Fromann).

Oldenburg, H. 1914. *Buddha. Sein Leben, seine Lehre, seine Gemeinde.* Stuttgart, Berlin (Cotta).

Olds, J. 1956a. Pleasure centers in the brain. *Sci. Amer.,* **193,** 105-116.

———. 1956b. Runway and maze behavior controlled by basomedial forebrain stimulation in the rat. *J. Comp. Physiol. Psychol.,* **49,** 507.

Olds, J. and P. Milner. 1954. Positive reinforcement produced by electrical stimulation of septal area and other regions of rat brain. *J. Comp. Physiol. Psychol.,* **47,** 419-427.

Oparin, A. I. 1957. *The Origin of Life on the Earth.* New York (Academic Press).

———. 1965. The pathways of the primary development of metabolism and artificial modeling of this development in coarcevate drops. In: S. W. Fox, *The Origin of Prebiological Systems.* 331-346, New York, London (Academic Press).

Parriss, J. R. and J. Z. Young. 1962. The limits of transfer of learned discrimination to figures of larger and smaller sizes. *Z. vergl. Physiol.,* **45,** 618-635.

Pearson, K. 1930, 1931. *Tables for Statisticians and Biometricians.* London (Biometric Office) 3d ed.

Penfield, W. 1951. Mechanism of memory. *Transact. Amer. Clin. Climat. Ass.,* **62,** 165-169.

———. 1954. Studies of the cerebral cortex of man. A review and an interpretation. In: J. F. Delafresnaye, *Brain Mechanisms and Consciousness.* 284-309, Oxford (Blackwell).

Penfield, W. and B. Milner. 1958. *Neur. Psychol.,* **79,** 475: quoted after J. Bures 1965.

Penfield, W. u. T. Rasmussen. 1950. *The Cerebral Cortex of Man. A Clinical Study of Localization of Function.* New York (Macmillan).

Peters, W. 1915. *Die Vererbung geistiger Eigenschaften und die psychische Konstitution.* Jena (G. Fischer).

Piaget, J. 1927. *La causalité physique chez l'enfant.* Paris.

Planck, M. 1933a. Die Kausalität in der Natur. In: *Wege zur physikalischen Erkenntnis,* 223-259, Leipzig (Hirzel).

———. 1933b. Kausalgesetz und Willensfreiheit (Vortr. Preuss. Ak. Wiss. 1923). In: *Wege zur physikalischen Erkenntnis,* 87-127, Leipzig (Hirzel).

Platon. *Theaitetos.* In: *Sämtliche Werke,* vol. II, Berlin (Lambrecht Schneider).

Poeck, K. 1959. Neurophysiologische Grundlagen des Bewußtseins. *Umschau,* 59, 299-301.

Poljak, St. 1941. *The Retina.* Chicago (Univ. of Chicago Press).

Ponnamperuma, C. 1965. A biological synthesis of some nucleid acid. In: S. W. Fox, *The Origin of Prebiological Systems,* 221-242. New York, London (Academic Press), 1965.

Ponnamperuma, C. and R. Mack. 1965. Nucleotide synthesis under possible primitive earth conditions. *Science,* 148, 1221-1223.

Prosser, C. L. and F. A. Brown, Jr. 1950. *Comparative Animal Physiology.* 2d ed. Philadelphia, London (Saunders).

Purkinje, J. 1823, 1825. *Beobachtungen und Versuche zur Physiologie der Sinne.* 2 vols., Prag.

Quarton, G. C., T. Melneschuk, F. O. Schmitt (eds.). 1967. *The Neurosciences—a Study Program.* New York.

Radhakrishnan, S. 1929, 1931. *Indian Philosophy,* 2 vols., 2d ed., New York and London (Allen and Unwin).

——. 1961. *My Search for Truth.* (Meine Suche nach Wahrheit.) Trans. by H. D. Lohnherr and A. von Bonin, Gütersloh, (Bertelsmann).

Raschke, H. 1954. *Das Christusmysterium.* Bremen (Schunemann).

——. 1966. Der ungeschichtliche Jesus. In: K. Deschner, *Jesusbilder in theologischer Sicht,* 343-444, München (P. List).

Raven, Ch. P. 1960. The formalization of finality. *Folia Biotheoret.,* 5, 1-27.

Rehfeldt, B. 1962. *Einführung in die Rechtswissenschaft.* Berlin (de Gruyter).

Rehmke, J. 1911. *Die Willensfreiheit.* Leipzig.

Reinke, J. 1905. *Philosophie der Botanik.* Leipzig (Barth).

Reiss, E. 1922. *Über erbliche Belastung von Schwerverbrechern.* Klin. Wochenschr. 1.

Remane, A. 1956. *Die Grundlagen des natürlichen Systems, der vergleichenden Anatomie und der Phylogenetik.* 2d ed. Leipzig (Akad. Verlagsges.).

Remplein, H. 1954. *Psychologie der Persönlichkeit.* München, Basel (Reinhardt).

Rensch, B. 1926. Rassenkreisstudien bei Mollusken I. Der Rassenkreis der Felsenschnecke Campylaea zonata Studer. *Zool. Anz.,* 67, 253-263.

——. 1929. *Das Prinzip geographischer Rassenkreise und das Problem der Artbildung.* Berlin (Bornträger).

——. 1954. *Neuere Probleme der Abstammungslehre.* Stuttgart (Enke) 1st ed. 1947.

——. 1952. *Psychische Komponenten der Sinnesorgane. Eine psychophysische Hypothese.* Stuttgart (Thieme).

——. 1957. Ästhetische Faktoren bei Farb- und Formbevorzugungen von Affen. *Z. f. Tierpsychol.,* 14, 71-99.

———. 1958a. Die Abhängigkeit der Struktur und der Leistungen tierischer Hirne von ihrer Größe. *Naturwiss.*, **45**, 145-154, 175-180.

———. 1958b. Die Wirksamkeit ästhetischer Faktoren bei Wirbeltieren. *Z. f. Tierpsychol.*, **15**, 447-461.

———. 1959, 1960a. *Evolution above the Species Level.* London (Methuen). New York (Columbia Univ. Press).

———. 1960b. The laws of evolution. In S. Tax, *Evolution after Darwin*, vol. 1, 95-116, Chicago (Chicago Univ. Press).

———. 1961a. Die Evolutionsgesetze der Organismen in naturphilosophischer Sicht. *Philos. Naturalis*, **6**, 288-326.

———. 1961b. Malversuche mit Affen. *Z. f. Tierpsychol.*, **18**, 347-364.

———. 1963. Probleme der Willensfreiheit in biologischer und philosophischer Sicht. *Hippokrates*, **24**, 1019-1032.

———. 1964. Die philosophische Bedeutung der Evolutionsgesetze. In: H. Kuhn u. F. Wiedmann, *Die Philosophie und die Frage nach dem Fortschritt.* 179-206. München (Pustet).

———. 1965a. *Homo sapiens. Vom Tier zum Halbgott.* (Kl. Vandenhoeck-Reihe 70 S.), 2d ed. Göttingen (Vandenhoeck u. Ruprecht) (1st ed. 1959).

———. 1965b. Die höchsten Hirnleistungen der Tiere. *Naturwiss. Rundschau*, **18**, 91-101.

———. 1965c. Über ästhetische Faktoren im Erleben höherer Tiere. *Naturwiss. u. Medizin*, **2**, 43-57 (Boehringer, Mannheim).

———. 1964, 1966. Problems of biological philosophy with regard to the philosophy of the Upanishads. *J. Univ. Bombay*, **33**, pt. 2, 89-95 (1964); *Indian J. Hist. Sci.*, **1**, 75-81.

———. 1967. The evolution of brain achievements. In: Th. Dobzhansky, M. K. Hecht, and W. C. Steere, *Evolutionary Biol.*, **1**, 26-68.

———. 1969. Die fünffache Wurzel des panpsychistischen Identismus. *Philosophia naturalis*, **11**, 129-150.

Rensch, B. u. J. Döhl. 1967. Spontanes Öffnen verschiedener Kistenverschlüsse durch einen Schimpansen. *Z. f. Tierpsychol.*, **24**, 476-489.

———. 1968. Wahlen zwischen zwei überschaubaren Labyrinthwegen durch einen Schimpansen. *Z. f. Tierpsychol.*, **25**, 216-231.

Rensch, B. u. G. Dücker. 1959a. Versuche über visuelle Generalisation bei einer Schleichkatze. *Z. f. Tierpsychol.*, **6**, 671-692.

———. 1959b. Die Spiele von Mungo und Ichneumon. *Behaviour*, **14**, 185-213.

———. 1967. Manipulierfähigkeit eines jungen Orang-Utans und eines jungen Gorillas. *Z. f. Tierpsychol.*, **23**, 874-899.

Rensch, B. u. H. Rahmann. 1966. Autoradiographische Untersuchungen über visuelle "Engramm"-Bildung bei Zahnkarpfen. I. *Pflügers Arch.*, **290**, 158-166.

Rensch, B., H. Rahmann u. K. Skrzipek. 1968. Autoradiographische Untersuchungen über visuelle "Engramm"-Bildung bei Fischen (II). *Pflügers Arch.*, **304,** 242-252.

Richter, L. C. 1943. *Jacob Böhme. Mystische Schau.* Hamburg (Hoffmann u. Kampe).

Riehl, A. 1921. Logik und Erkenntnistheorie. In: P. Hinneberg, *"Kultur der Gegenwart." Systematische Philosophie.* Teil I, Abt. VI, 68-134. 3d ed. Berlin u. Leipzig (Teubner).

———. 1908. *Der philosophische Kritizismus und seine Bedeutung für die positive Wissenschaft.* 2d ed. Leipzig (Engelmann). (1st ed. 1879).

Robson, E. A. 1965. Adaptive changes in Cnidaria. *Animal Behaviour,* Supp. 1, 54-60, London.

Rohracher, H. 1948. *Die Vorgänge im Gehirn und das geistige Leben.* 2d ed. Leipzig (Barth).

———. 1960. *Einführung in die Psychologie.* 7th ed. Wien (Urban u. Schwarzenberg) (1st ed. 1946).

Rolland, R. 1931. *Das Leben des Ramakrishna.* Erlenbach, Zürich, Leipzig (Rotapfel Verlag).

Romer, A. S. 1966. *Vergleichende Anatomie der Wirbeltiere.* Hamburg u. Berlin (Parey). 2d ed.

Rosenmöller, B. 1932. *Religionsphilosophie.* Münster (Aschendorf).

Ross, D. M. 1965. The behaviour of sessile coelenterates in relation to some conditioning experiments. *Animal Behaviour,* Supp. 1, 43-53.

Rothschuh, K. E. 1953. *Geschichte der Physiologie.* Berlin, Göttingen, Heidelberg (Springer).

———. 1959. *Theorie des Organismus.* München, Berlin (Urban u. Schwarzenberg).

Russell, B. 1957. *Why I Am Not a Christian and Other Essays on Religion and Related Subjects.* New York (Simon and Schuster).

Salisbury, F. B. 1965. Die Induktion der Blütenbildung. *Endeavour,* **24,** 74-80.

Saxena, A. 1960. Lernkapazität, Gedächtnis und Transpositions-vermögen bei Forellen. *Zool. Jahrb., Abt. allg. Zool.,* **69,** 63-94.

Schaltenbrand, G. and Cl. N. Woolsey. 1964. *Cerebral Localization and Organization.* Madison, Milwaukee (Univ. Wisconsin Press).

Scheler, M. 1921. *Vom Ewigen im Menschen.* Leipzig (D. neue Geist Verl.).

Schelling, F. W. J. 1800. *System des transzendentalen Idealismus.* Tübingen (Hamburg, Meiner 1957).

Schleiermacher, E. D. 1884. *Der christliche Glaube nach den Grundsätzen der evangelischen Kirche . . .* 6th ed. Berlin (Reimer).

Schlick, M. 1920. *Raum und Zeit in der gegenwärtigen Physik.* Berlin (Springer).

———. 1925. *Allgemeine Erkenntnislehre.* 2d ed. Berlin (Springer).

Schmalhausen, I. I. 1949. *Factors of Evolution. The Theory of Stabilizing Selection.* Philadelphia, Toronto (Blakiston).

Schmitt, F. O. 1967. Makromolekulare Datenverarbeitung im Zentralnerven-system. *Verh. Ges. Deutsch. Naturforsch. u. Ärzte,* 104. Vers. (1966), 863-868.

Schopenhauer, A. *Die Welt als Wille und Vorstellung.* 2 Teile. (1879). In: *Werke,* ed. by M. Frischeisen-Köhler, vol. 1, 2. Berlin (Weichert) without datum; Leipzig (Insel) without datum.

Schramm, G. 1965a. *Belebte Materie.* (Opuscula aus Wiss. u. Dicht. Nr. 15), Pfullingen (Neske).

———. 1965b. Synthesis of nucleosides and polynucleotides with meta-phosphate esters. In S. W. Fox, *The Origin of Prebiological Systems,* 299-315. New York, London (Academic Press).

Schrödinger, E. 1944. *What Is Life?* Cambridge (Cambridge Univ. Press).

———. 1954. Unsere Vorstellung von der Materie. *Naturwiss. Rundschau,* 7, 277-282.

Schulze-Schencking, M. 1970. Untersuchungen über visuelle Lerngeschwin-digkeit bei Bienen, Hummeln und Ameisen. *Z. f. Tierpsychol.* 27, 1970.

Schumacher, W. 1948. *Die Edikte des Kaisers Asoka.* Konstanz (Weller).

Schütz, E. 1966. *Physiologie.* 9th and 10th eds. München, Berlin (Urban u. Schwarzenberg).

Seidel, F. 1953. *Entwicklungsphysiologie der Tiere.* Samml. Göschen vol. 1162, 1163, Berlin (de Gruyter).

Seifert, H. 1966. Strukturgelenkte Grenzflächenvorgänge in der unbelebten und belebten Natur. *Naturwiss. Rundschau,* 19, 56-62.

Sewertzoff, A. N. 1931. *Morphologische Gesetzmäßigkeiten der Evolution.* Jena (G. Fischer).

Shankara. 1957. *Das Kleinod der Unterscheidung. Viveka-Chudamani.* München-Planegg (O. W. Barth).

Sheldon, W. H. and W. B. Tucker. 1940. *The Varieties of Human Physique.* New York (Harper).

Simpson, G. G. 1944. *Tempo and Mode in Evolution.* New York (Columbia Univ. Press).

———. 1949. *The Meaning of Evolution.* New Haven (Yale Univ. Press).

———. 1953. *The Major Features of Evolution.* New York (Columbia Univ. Press).

———. 1959. The nature and origin of supraspecific taxa. *Cold Spring Harbor Symposia Quant. Biol.,* 24, 255-271.

Skodak, M. and H. M. Skeels. 1949. A final follow-up study of one hundred adopted children. *J. Genet. Psychol.,* 75, 85-125.

Soederblom, N. 1964. Der Animismus. Belebung und Beseelung. In: *Das Werden des Gottesglaubens.* 10-32, Leipzig (1916). Reprinted in C. A. Schmitz, *Religions-Ethnologie,* 9-29, Frankfurt a.M. (Akad. Verlagsges.).

Spemann, H. 1924. Über Organisatoren in der tierischen Entwicklung. *Naturwiss.,* 12, 1092-1094.

————. 1936. *Experimentelle Beiträge zu einer Theorie der Entwicklung.* Berlin (Springer).

————. 1943. *Forschung und Leben.* Stuttgart (Engelhorn, Nachf.).

Spencer, H. 1898. *The Principles of Biology.* Vol. 1, London.

Spinoza, B. de. 1914. *Ethices* (1677). In: *Opera quotquot reperta sunt.* *Herausgeg. v.* J. van Vloten u. J. P. N. Land, 3d ed. vol. 1, Den Haag (Nijhoff).

————. 1871. *Die Ethik.* In: *Sämtliche Werke,* vol. II, ed. by B. Auerbach, Stuttgart (Cotta).

Spreng, M. 1964. *Die Codierung und Informationsleitung im Zentralnervensystem.* In: H. Frank, *Kybernetik,* 73-82, Frankfurt a.M. (Umschau Verl.).

Stache, W. 1958. *Erkenntnistheorie.* In: A. Diemer u. I. Frenzel, *Philosophie.* Frankfurt a.M. (S. Fischer) (Fischer-Lexikon).

Stebbins, G. L., Jr. 1950. *Variation and Evolution in Plants.* New York (Columbia Univ. Press).

Steinbuch, K. 1965. *Automat und Mensch.* 3d ed. Berlin, Heidelberg, New York (Springer).

Stern, C. 1960. *Principles of Human Genetics.* 2d ed. San Francisco (Freeman).

Stöhr, A. 1922. *Psychologie.* 2d ed. Wien u. Leipzig (Braumüller).

Störring, G. 1920. *Die Frage der Wahrheit der christlichen Religion.* Leipzig (Engelmann).

Strauss, D. F. 1864. *Das Leben Jesu.* Leipzig (Brockhaus).

————. 1873. *Der alte und der neue Glaube.* 6th ed. Bonn (Strauss).

Stumpf, C. 1907. Gefühlsempfindungen. *Z. f. Psychol.,* **44,** 1-49.

Stumpfl, F. 1935. *Erbanlage und Verbrechen.* In: E. Rüdin, *Studien über Vererbung und Entstehung geistiger Störungen.* Berlin.

Süssmann, G. 1965. 50 Jahre allgemeine, 60 Jahre spezielle Relativitätstheorie. *Umschau,* **65,** 6-11.

Szczesny, G. 1958. *Die Zukunft des Unglaubens.* München (List).

Tax, S. (ed.). 1960. *Evolution after Darwin.* 3 vols. Chicago (Chicago Univ. Press).

Taylor, J. H. (ed.). 1963, 1967. *Molecular Genetics.* 2 vols. New York, London (Academic Press).

Tellier, M. 1932. Reconnaissance par le toucher d'objets connus par la vue chez le macaque. *Bul. Soc. Roy. Liege,* **1,** 114-117.

Tembrock, G. 1964. *Verhaltensforschung.* Jena (G. Fischer).

Thompson, R. E. 1952. A validation of the Glueck social prediction scale for proneness to delinquency. *J. Crimin. Law. Criminol. and Police Sci.* (quoted after F. Bauer).

Thompson, R. and J. McConnell. 1955. Classical conditioning in the planarian Dugesia dorotocephala. *J. Comp. Physiol. Psychol.,* **48,** 65-68.

Thompson, W. d'Arcy. 1942. *On Growth and Form.* Cambridge (Univ. Press) 1917. New York (Macmillan).

Thorpe, W. H. 1965. *Science, Man and Morals.* Ithaca, New York (Cornell Univ. Press).

Tiedemann, H. 1954. Veränderung des Stoffwechsels von Tritonkeimen unter anaeroben Bedingungen. *Z. f. Naturforsch.,* **9** b, 801-802.

Tiedemann, H. u. H. Tiedemann. 1954. Einwirkungen von HCN auf die frühen Entwicklungsstadien des Alpenmolches. *Z. f. Naturforsch.,* **9** b, 371-380.

Tigges, M. 1963. Muster- und Farbbevorzugung bei Fischen und Vögeln. *Z. f. Tierpsychol.,* **20,** 129-142.

Tinbergen, N. 1951. *The Study of Instinct.* Oxford (Clarendon).

Trendelenburg, W., M. Monjé, I. Schmidt u. E. Schütz. 1961. *Der Gesichtssinn.* Berlin, Göttingen, Heidelberg (Springer).

Tugendhat, E. 1960. Tarskis semantische Definition der Wahrheit und ihre Stellung innerhalb fer Geschichte des Wahrheitsproblems . . . *Philos. Rundschau,* **8,** 131-159.

———. 1964. Zum Verhältnis von Wissenschaft und Wahrheit. In: E.-W. Böckenförde, W. Goerdt et al., *Collegium Philosophicum* 389-402, Basel, Stuttgart (Schwabe).

Vernadsky, E. 1930. *Die Biosphäre.* Leipzig.

Verschuer, O. von. 1954. *Wirksame Faktoren im Leben des Menschen. Beobachtungen an ein- und zweieiigen Zwillingen durch 25 Jahre.* Wiesbaden (F. Steiner).

———. 1959. *Genetik des Menschen.* München, Berlin (Urban u. Schwarzenberg).

Verworn, M. 1903. *Die Biogenhypothese.* Jena (G. Fischer).

———. 1912. *Kausale und konditionale Weltanschauung.* Jena (G. Fischer).

———. 1915. *Allgemeine Physiologie.* 6th ed. Jena (G. Fischer) (1st ed. 1909).

Vogell, W. 1963. Die Morphologie der Mitochondrien. In: P. Karlson, *Funktionelle u. morphologische Organisation der Zelle,* 56-68, Berlin, Göttingen, Heidelberg (Springer).

Vogt, O. 1910. Die myeloarchitektonische Felderung des menschlichen Stirnhirns. *J. Psychol. u. Neurol.,* **15.**

Volk, H. 1955. *Schöpfungsglaube und Entwicklung.* Schrift. d. Ges. z. Förd. d. Westfäl. Wilhelms-Univ. zu Münster, H. 13, Münster.

Voltaire. 1785. *Philosophie générale: Métaphysique, Morale et Théologie.* In *Oeuvres Complètes,* vol. 40, Paris (Soc. Littéraire-Typographique).

Waaser, F. 1942. *Gestalt und Wirklichkeit im Lichte Goethescher Naturanschauung.* (Die Gestalt, H. 8), Halle (S.) (Niemeyer).

Waddington, C. H. 1956. *Principles of Embryology.* London (Allen and Unwin).

————. 1957. *The Strategy of the Genes.* London (Allen and Unwin).

————. 1962. *New Patterns in Genetics and Development.* New York, London (Columbia Univ. Press).

Walter, K. 1964. Die räumlichen Grenzen des beobachtbaren Alls. *Naturwiss. Rundschau,* **17,** 49-56.

Walter, W. G. 1953. *The Living Brain.* London (Duckworth).

Warden, C. J., T. N. Jenkins and L. H. Warner. 1940. *Comparative Psychology.* Vol. II: *Plants and Invertebrates.* New York.

Warren, J. M. 1960. Oddity learning set in a cat. *J. Comp. Physiol. Psychol.,* **53,** 433-434.

Weber, A. 1955. Conflements ou stases de l'axoplasme en des zones localisées du tissu nerveux central ou périphérique. *Acta Anat. (Basel),* **23,** 33-48.

Weber, R. (ed.). 1965. *The Biochemistry of Animal Development.* New York, London (Academic Press).

Weiss, P. and A. C. Taylor. 1960. Reconstitution of complete organs from single-cell suspensions of chick embryos in advanced stages of differentiation. *Proc. Nat. Ac. Sci.,* **46,** 1177-1185.

Weiss, T. 1965. Die unspezifischen aktivierenden und hemmenden Systeme von Hirnstamm und Zwischenhirn. In W. Rüdiger, *Probleme der Physiologie des Gehirns.* Berlin (Volk u. Gesundheit).

Weizel, W. 1954. Das Problem der Kausalität in der Physik. *Arbeitsgem. f. Forsch. Nordrhein-Westfalen, H.,* **43,** 37-62, Köln, Opladen.

Weizsäcker, C. F. von. 1949. *Zum Weltbild der Physik.* 4th ed. Stuttgart (Hirzel).

Welzel, W. 1960. *Das deutsche Strafrecht.* 7th ed. Berlin (de Gruyter).

Werner, M. 1959. *Die Entstehung des christlichen Dogmas.* (Urban-Bücher Nr. 38) Stuttgart.

Wertheimer, M. 1912. Experimentelle Studien über das Sehen von Bewegungen. *Z. f. Psychol.,* **61.**

————. 1922, 1923. Untersuchungen zur Lehre von der Gestalt. *Psychol. Forsch.,* 1, 4.

————. 1925. *Drei Abhandlungen zur Gestalttheorie.* Erlangen (Philos. Ak. Verl.).

Whytt, R. 1768. *Works.* Edinburgh.

Wiener, N. 1948. *Cybernetics.* Cambridge, Mass. (Technology Press).

Wilbrandt, H. u. A. Sänger. 1904. *Die Neurologie des Auges.* Vol. 3, 1. Abt. Wiesbaden (Bergmann).

Windelband, W. 1904. *Über Willensfreiheit.* Tübingen (Mohr) (3d ed. 1918).

Wissenburgh, J. C. et P. H. C. Tibout. 1921. Choix basé sur l'aperception complexe chez les cobayes. *Arch. Néerland. Physiol.,* **6.**

Witthöft, W. 1967. Zahl und Verteilung der Zellen im Hirn der Honigbiene. *Z. f. Morph. u. Ökol.,* **61,** 160-184.

Wolf, L. u. W. Troll. 1940. *Goethes morphologischer Auftrag. Versuch einer naturwissenschaftlichen Morphologie.* Leipzig (Akad. Verlagsges.).

Wolfe, J. B. 1936. Effectiveness of token-rewards for chimpanzees. *Comp. Psychol. Monogr.* 12, No. 5.

Woltereck, R. 1940. *Ontologie des Lebendigen.* Stuttgart (Enke).

Wooldridge, C. (ed.). 1959. Genetics and Twentieth Century Darwinism. *Cold Spring Harbor Symposia Quant. Biol.,* vol. 24.

Wright, S. 1932. The roles of mutation, inbreeding, crossbreeding, and selection in evolution. *Proc. 6. Intern. Congr. Gen.,* **1,** 356-366.

———. 1940. The statistical consequences of Mendelian heredity in relation to speciation. In: J. Huxley, *The New Systematics,* 161-183, Oxford (Clarendon Press).

Wundt, W. 1886. *Ethik.* Stuttgart (Enke) 3 vols. (4th ed. 1912).

———. 1894. Über psychische Causalität und das Princip des psychophysischen Parallelismus. *Philos. Studien,* 10, 1-124.

———. 1908, 1910. *Grundzüge der physiologischen Psychologie.* 3 vols., 6th ed. Leipzig (Engelmann).

Yerkes, R. M. 1943. *Chimpanzees. A Laboratory Colony.* New Haven (Yale Univ. Press).

Yerkes, R. M. and A. W. Yerkes. 1929. *The Great Apes. A Study of Anthropoid Life.* New Haven (Yale Univ. Press).

Young, J. Z. 1961. Learning and discrimination in the *Octopus. Biol. Reviews,* **36,** 32-96.

———. 1962. Memory mechanisms of the brain. *J. of Mental Sci.,* **108,** 119-133.

Zahn, H. 1966. Über Insulin: Arbeitsgem. f. Forsch. Nordrhein-Westf., *Nat. Wiss.,* H. 161, Köln, Oplanden.

Ziehen, Th. 1924. Leitfaden der physiologischen Psychologie. 1st ed. 1890. 3d ed. 1896. Jena (G. Fischer), 12th ed.

———. 1898. *Psychophysische Erkenntnistheorie.* Jena (G. Fischer).

———. 1911. *Psychiatrie.* 4th ed. Leipzig (Hirzel).

———. 1913. *Erkenntnistheorie auf psychophysiologischer und physikalischer Grundlage.* Jena (G. Fischer).

———. 1915. *Die Grundlagen der Psychologie.* 2 vols. Leipzig u. Berlin (Teubner).

———. 1920. *Lehrbuch der Logik auf positivistischer Grundlage mit Berücksichtigung der Geschichte der Logik.* Bonn (Marcus u. Weber).

———. 1921. Die Beziehungen der Lebenserscheinungen zum Bewußtsein. *Abh. theoret. Biol. H.* 13, Berlin (Bornträger).

———. 1923, 1925. *Vorlesungen über Ästhetik.* 2 vols. Halle (S.) (Niemeyer).

———. 1927a. *Das Problem der Gesetze.* Hallische Univers. Reden Nr. 33. Halle (S.) (Niemeyer).

————. 1927b. *Sechs Vorträge zur Willenspsychologie.* Jena (G. Fischer).

————. 1928. *Die Grundlagen der Religionsphilosophie.* Leipzig (Meiner).

————. 1930. Gestalten, Strukturen und Kausalgesetze. *Arch. ges. Psychol.,* **77,** 291-306.

————. 1934, 1939. *Erkenntnistheorie.* 2d ed., 2 vols. Jena (G. Fischer).

Zimmermann, W. 1949. *Geschichte der Pflanzen.* Stuttgart (Thieme).

Author Index

Subject Index